Alternate Energy Processes in Chemical Synthesis

Microwave, Ultrasonic and Photo Activation

Alternate Energy Processes in Chemical Synthesis
Microwave, Ultrasonic and Photo Activation

V.K. Ahluwalia
R.S. Varma

Alpha Science International Ltd.
Oxford, U.K.

V.K. Ahluwalia
B.R. Ambedkar Centre for Biomedical Research
University of Delhi
Delhi, India

R.S. Varma
Clean Process Branch
National Risk Management Research Laboratory
U.S. Environmental Protection Agency
Cincinnati, Ohio 45268, USA

Copyright © 2008

ALPHA SCIENCE INTERNATIONAL LTD.

7200, The Quorum, Oxford Business Park North
Garsington Road, Oxford OX4 2JZ, U.K.

www.alphasci.com

All rights reserved. No part of this publication may be reproduced, stored in a retrieval system or transmitted in any form or by any means, electronic, mechanical, photocopying, recording or otherwise, without the prior written permission of the publisher.

ISBN 978-1-84265-434-7

Printed in India

PREFACE

Organic chemistry deals with the synthesis of molecules having diverse use in medicine, agrochemicals and biomolecules. In fact, organic chemistry brought about medical revolution till about the middle ot twentieth century in which drugs and antibiotics were discovered. The World's food supply also increased due to discovery of hybrid varieties, improved methods of farming, better seeds and the discovery of agrochemicals (insecticides, herbicides and fertilizers etc). The quality of life on earth became much better due to the discovery of dyes, plastics, cosmetics and other materials of day to day use. All these advances resulted in the average life expentancy increasing from 47 years in 1900 to 75 years in 1990's. However, the ill effects of all these became apparent when the industries manufacturing all these chemicals resulted in environmental pollution. This is because most of industrial processes use electricity for heating purposes. This electricity is mostly generated from coal, which is a non renewable source. The coal fired processes are responsible for polluting the environment with gases like oxides and nitrogen, sulphur and carbon, besides other pollutants.

It is most appropriate to use alternative sources of energy. These sources include Microwave, Ultrasound and Photochemical. Besides certain organic reactions can be performed just by mixing of the reactants without the use of energy.

This book describes alternative sources of energy i.e. microwave, ultrasound and photochemical.

Part I deals with the microwave assisted organic synthesis. These reactions could be conducted in water, in organic solvents and in solid state without the use of any solvent. Part II deals with ultrasound assisted organic synthesis. In this part, homogeneous sonochemical reactions, heterogeneous liquid, liquid reactions, heterogeneous solid liquid reactions are discussed. Part III deals with solid state solvent free organic synthesis. In this Part some reactions are discussed which could be performed by just mixing in a pestle mortar or by slightly warming the mixture. Part IV deals with photo induced organic synthesis. In this part some photochemical reactions, their industrial applications and templeted reactions in a homogeneous solution which are mediated by light are discussed.

It is hoped that this book will be extremely helpful to all students for their laboratory work. The researchers in universities, R and D Laboratories of Industries will find this book specially valuable for planning their work.

The Authors express their sincere thanks to Prof. Sukh Deo F.N.A. for various suggestions and the help rendered.

The chapter on Microwave was written by one of the authors while employed by the U.S. Government. In view of Section 105 of the Copyright Act (17 U.S.C. Section 105) this portion is not subject to U.S. Copyright protection. The views expressed in this article are those of the individual author and do not necessarily reflect the views and policies of the U.S. Environmental Protection Agency. The use of trade names does not imply endorsement by the U.S. Government.

Authors

CONTENTS

Preface　　v

PART-I MICROWAVE ASSISTED ORGANIC SYNTHESIS

1. **Introduction**　　1.3
   Notes and References　　1.6
2. **Microwave Assisted Organic Reactions in Water**　　2.1
   - 2.1　Hofmann Elimination　　2.1
   - 2.2　Hydrolysis of Benzyl Chloride　　2.1
   - 2.3　Hydrolysis of Benzamide　　2.2
   - 2.4　Hydrolysis of N-Phenyl Benzamide　　2.2
   - 2.5　Hydrolysis of Methyl Benzoate to Benzoic Acid (Saponification)　　2.2
   - 2.6　Oxidation of Toluene　　2.2
   - 2.7　Coupling of Amines with Halides　　2.3
   - 2.8　N-Heterocyclizations　　2.3
   - Notes and References　　2.3
3. **Microwave Assisted Organic Reactions in Organic Solvents**　　3.1
   - 3.1　Esterifications: Reaction of Carboxylic Acid and Alcohol　　3.1
   - 3.2　Reaction of Carboxylic Acid and Benzyl Ethers using Microwave in Presence of Ln Br$_3$ (Ln = La, Nd, Sm, Dy, Er)　　3.2
   - 3.3　Fries Rearrangement　　3.2
   - 3.4　Diels Alder Reaction　　3.2
   - 3.5　Claisen Rearrangement　　3.3
   - 3.6　Cydoaddition Reaction between Fulvenes and Some Alkenes and Alkynes, Synthesis of Polycyclic Ring Systems　　3.4
   - 3.7　Knoevenagel Condensation　　3.4
   - 3.8　Baylis-Hillman Reaction　　3.4
   - 3.9　Orthoester Claisen Rearrangement　　3.4
   - 3.10　Synthesis of 4-aryl-3, 4-dihydropyrimidine 2(1H) ones　　3.5
   - 3.11　Synthesis of β-lactams　　3.5
   - 3.12　Cycloaddition Reactions　　3.6
   - 3.13　Synthesis of Benzodiazepin-2-ones　　3.6

3.14	Aromatic Substitution Reactions	3.7
3.15	Catalytic Hydrogenation	3.7
3.16	Synthesis of Chalcones	3.8
3.17	Decarboxylations	3.8
3.18	C-Alkylation of Active Methylene Group	3.9
3.19	Methanolysis of Oligosaccharides	3.9
3.20	Preparation of Unsaturated Pyranosides	3.9
3.21	Ferrier Rearrangement	3.10
3.22	Synthesis of Jusminaldehyde	3.10
3.23	Racemization of (–)-Vincadifformine to the (+)-Isomer	3.10
3.24	Synthesis of 1, 2-dimethyl-3-hydroxy-pyrid-4-one	3.11
3.25	Synthesis of Isopropylidine Glycerol	3.11
3.26	Synthesis of Isotopically Labelled (^{11}C) Diethyl Oxalate and (^{11}C) Oxalic Acid	3.11
3.27	Stereoselective Addition of 2-Aminothiophenol to Glycidic Esters	3.12
3.28	Development and Application of a Continuous Microwave Reactor (CMR) for Organic Synthesis using Solvents	3.12
3.29	Pericyclic Reactions	3.14
3.30	Cyclisation Reactions	3.16
3.31	Oxidation	3.17
3.32	Synthesis of Alkenes	3.17
3.33	Preparation of Ferrocenyl Oxime	3.17
3.34	Synthesis of Ethers	3.17
3.35	Carbohydrates	3.18
3.36	Radical Reactions	3.18
	Notes and References	3.18
4.	**Microwave Assisted Reactions in Solid State**	**4.1**
4.1	Protection and Deprotection Reactions	4.1
	4.1.1 Formation of Acetals and Diaxolanes	4.1
	4.1.2 N-Alkylation Reactions	4.2
	4.1.3 Deacetylation	4.4
	4.1.4 Debenzylation of Carboxylic Esters	4.4
	4.1.5 Selective Cleavage of N-tert-Butoxycarbonyl Group	4.5
	4.1.6 Desilylation Reactions	4.5
	4.1.7 Dethioacetalization Reactions	4.6
	4.1.8 Deoximation Reactions	4.6
	4.1.9 Cleavage of Semicarbazones and Phenylhydrazones	4.7
	4.1.10 Dethiocarbonylation	4.8
	4.1.11 Thionotion Reactions: Synthesis of Thioketones, Thioamides, Thioesters and Thioflavonoids	4.8
	4.1.12 Saponification of Esters	4.9
4.2	Oxidations	4.9
	4.2.1 Oxidation of Alcohols	4.9
	(a) Using Clayfen	4.9

		(b) Using Activated MnO$_2$–Silica	4.10
		(c) Using Chromium Trioxide Supported on Wet Alumina	4.10
		(d) Using Iodobenzene Diacetate (IBD) 'Doped' Alumina	4.11
	4.2.2	Oxidation of Alkanes, Bromides, Carboxylic Acids, Cyanides and Amines to Carbonyl Compounds	4.11
	4.2.3	Oxidation of α-hydroxyketones to 1, 2-diketones	4.11
	4.2.4	Oxidation of Sulfides to Sulfoxides and Sulfones	4.12
	4.2.5	Oxidation of Enamines	4.13
	4.2.6	Oxidation of Arenes	4.13
	4.2.7	Aromatisation	4.13
4.3	Reductions		4.14
	4.3.1	Sodium Borohydride Reduction of Carbonyl Compounds to Alcohols	4.14
	4.3.2	Reductive Amination of Carbonyl Compounds	4.15
	4.3.3	Reduction of Carbonyl Compounds with Aluminium Alkoxides	4.15
	4.3.4	Solid State Crossed Cannizzaro Reaction	4.15
4.4	Rearrangement Reactions		4.17
	4.4.1	Pinacol-pinacolone Rearrangement	4.17
	4.4.2	Ring Expansion	4.17
	4.4.3	Beckmann Rearrangement	4.17
	4.4.4	Benzil-Benzilic Acid Rearrangement	4.18
4.5	Isomerisation Reactions		4.18
	4.5.1	Octylthiocyanate-Octylisothiocyanate Isomerisation	4.18
	4.5.2	Eugenol Isoeugenol Isomerisation	4.18
4.6	Condensation Reactions		4.19
	4.6.1	Knoevenagel Condensation	4.19
	4.6.2	Wittig Olefination Reactions	4.19
	4.6.3	Synthesis of Imines, Enamines, Nitroalkenes and N-Sulfonylimines, Hydrazones and Amides	4.20
4.7	Synthesis of Heterocyclic Compounds		4.22
	4.7.1	Aziridines	4.22
	4.7.2	Benzimidazoles	4.22
	4.7.3	Pyrazoles	4.23
	4.7.4	Pyrroles	4.24
	4.7.5	Azoles	4.25
	4.7.6	Isoxazoles	4.26
	4.7.7	Thiazoles	4.26
	4.7.8	Pyridines	4.26
	4.7.9	Quinolines	4.27
	4.7.10	Quinolones, Quinolinones and Quinazolines	4.27
	4.7.11	4-Aminoquinazolines	4.28
	4.7.12	Pyrimidines	4.28
	4.7.13	Oxadiazines	4.29
	4.7.14	Thiadiazepines	4.29
	4.7.15	β-Lactams	4.30
	4.7.16	Furans	4.31
	4.7.17	Flavones	4.32
	4.7.18	Isoflavan-3-enes	4.32
	Notes and References		4.33

5. Miscellaneous Reactions — 5.1

5.1	A Single Step Conversion of Aryl Aldehydes to Aromatic Nitriles	5.1
5.2	Synthesis of Anhydrides from Dicarboxylic Acids	5.1
5.2a	Side Chain Nitration of Styrene to β-nitrostyrene	5.2
5.3	Oxidative Coupling of β-Napthols	5.2
5.4	Methylenation of 3, 4-dihydroxybenzaldehyde	5.2
5.5	Michael Addition	5.3
5.6	Synthesis of Bridgehead Nitrogen Heterocyclic Compounds	5.3
5.7	Enzyme Catalysed Reactions	5.5
5.8	Solid-phase Organic Synthesis (SPOS)	5.5
5.9	Synthesis of Radiolabelled Compounds	5.6
5.10	Organometallic Reactions (Reactions Involving C-C Bond Formation)	5.7
5.11	Aromatic Substitution	5.8
5.12	Pericyclic Reactions	5.9
5.13	Alkylations	5.10
5.14	Condensations	5.11
5.15	Synthesis of Peptides	5.14
5.16	Reactions Involving Silicon Reagents	5.15
5.17	Synthesis of Aspirin	5.16
5.18	Microwave Assisted Combinatorial Synthesis	5.16
5.19	Synthesis of Ionic Liquids Using Microwaves	5.19
5.20	Conclusion	5.21
	Notes and References	5.21

PART-II ULTRASOUND ASSISTED ORGANIC SYNTHESIS

6. Introduction — 6.3

6.1	Instrumentation	6.3
6.2	The Physical Aspects[2]	6.4
6.3	Types of Sonochemical Reactions	6.5
	6.3.1 Homogeneous Reactions	6.6
	6.3.2 Heterogeneous Liquid-Liquid Reactions	6.6
	6.3.3 Heterogeneous Solid-Liquid Reactions	6.6
	Notes and References	6.6

7. Homogeneous Sonochemical Reactions — 7.1

7.1	Curtius Rearrangement	7.1
7.2	Sulphur Extrusion from 1, 3, 4-Thiadiazines	7.1
7.3	Isomerisation of Maleic Acid to Fumaric Acid	7.2
7.4	Organometallic Reactions	7.2
	7.4.1 Isomerisation of Alkenes	7.2

	7.4.2	Annulation	7.2
	7.4.3	Grignard Reagents	7.3
7.5	Oxidations		7.4
7.6	Solvolysis and Hydrolysis		7.4
7.7	Addition Reactions		7.5
	7.7.1	Diels-Alder Reaction	7.5
	7.7.2	1, 3-Dipolar Cycloaddition	7.6
	7.7.3	Strecker Reaction	7.6
	7.7.4	Electrophilic Addition	7.7
	7.7.5	[2 + 2] Cycloaddition Reactions	7.7
	7.7.6	Cycloaddition Reactions	7.8
	Notes and References		7.8

8. Heterogeneous Liquid-Liquid Reactions — 8.1

8.0	Introduction	8.1
8.1	Esterification	8.1
8.2	Saponification	8.2
8.3	Hydrolysis/solvolysis	8.3
8.4	Substitutions	8.4
8.5	Additions	8.5
	Notes and References	8.8

9. Heterogeneous Solid-Liquid Reactions — 9.1

9.0	Introduction		9.1
9.1	Alkylations		9.1
	9.1.1	N-Alkylation	9.1
	9.1.2	C-Alkylation	9.2
	9.1.3	O-Alkylation	9.2
	9.1.4	S-Alkylation	9.3
9.2	Oxidations		9.3
9.3	Reductions		9.5
	9.3.1	Reduction of Carbonyl Group	9.6
	9.3.2	Reduction of Aromatic Rings	9.7
9.4	Hydroboration		9.7
9.5	Hydrosilation and Hydroalkylation		9.8
9.6	Coupling Reactions		9.8
9.7	Dichlorocarbene		9.10
9.8	Some Ultrasonically Induced Organic Reactions		9.11
	9.8.1	Bouveault Reaction	9.11
	9.8.2	Cannizzaro Reaction	9.11
	9.8.3	Strecker Synthesis	9.11
	9.8.4	The Reformatsky Reaction	9.11
	9.8.5	The Barbier Reaction of Carbonyl Compounds	9.12
	9.8.6	Dieckmann Cyclisation	9.13
9.9	Cyclocondensations		9.14

9.10	Carbohydrates-Formation of Acetals and Benzylidene Derivatives of Alkylglycopyranosides	9.14
	Notes and References	9.15

10. Miscellaneous Applications 10.1
    10.1 Potassium Superoxide   10.1
    10.2 Sonolysis of $Fe(CO)_5$ 10.1
    10.3 Oxymercuration of Olefins: Synthesis of α-Terpinol   10.3
    10.4 Activation of Nickel Powder   10.3
    10.5 Ultrasonically Dispersed Potassium   10.3
    10.6 Organometallic Compounds   10.4
        10.6.1 Organolithium Compounds   10.4
        10.6.2 Organoaluminium Compounds   10.5
        10.6.3 Organo Zinc and Palladium Compounds   10.6
        10.6.4 Synthesis of Trialkyl Boranes   10.7
    10.7 Synthesis of Aldehydes from Halides   10.7
    10.8 Sonochemical Methylenation of Alkenes and Carbonyl Compounds   10.7
    10.9 Sodiumphenylselenide   10.9
    10.10 Arylamides   10.9
    10.11 Spiroketones   10.9
    10.12 β-keto-Thinoesters   10.10
    10.13 Dehalogenation   10.10
    10.14 Thioamides   10.12
    10.15 Catalysis   10.12
    10.16 Bromomethylation of Aromatics   10.12
    10.17 Enhancement of the Efficiency of Carbohydrate Synthesis   10.12
        10.17.1 Manuplication of Hydroxyl Groups   10.13
        10.17.2 Synthesis of Thioglycosides   10.14
        10.17.3 Synthesis of Azidoglycosides   10.14
        10.17.4 Dipolar Cycloaddition   10.15
        10.17.5 Reductive Cleavage of Benzylidene Protecting Group   10.16
        10.17.6 Glycosylation with Phenylthioglycosides   10.16
    10.18 Preparation of Porous Carbon Powders by Ultrasonic Spray Pyrolysis   10.18
    10.19 Future Prospects   10.18
    10.20 Conclusion   10.18
        Notes and References   10.19

11. Sonochemical Treatment of Polluted Water 11.1
    11.0 Introduction   11.1
    11.1 Treatment of Polluted Water   11.1
        11.1.1 Oxidation of Contaminants   11.1
        11.1.2 Adsorption of Trace Metals on Charcoal   11.2
        Notes and References   11.3

PART-III SOLID STATE SOLVENT FREE ORGANIC SYNTHESIS

12. Solid State Organic Synthesis at Room Temperature — 12.3
- 12.0 Introduction — 12.3
- 12.1 Synthesis of β-keto Sulfones from Ketones — 12.3
- 12.2 Synthesis of α-tosyloxy β-keto Sulfones — 12.5
- 12.3 Synthesis of 1-Aryl-4-Methyl-1, 2, 4-Trizole [4,3-a] Quinoxalines — 12.6
- Notes and References — 12.8

13. Solid State Organic Synthesis by Slight Warming of the Reactants — 13.1
- 13.1 Oxidation of Hydroxylated Aldehydes and Ketones to Hydroxylated Phenols Using Urea-Hydrogen Peroxide Adduct (UPH) — 13.1
- 13.2 Oxidation of Nitriles to Amides Using UPH — 13.2
- 13.3 Selective Oxidation of Sulfides to Sulfoxides or Sulfones Using UPH — 13.3
- 13.4 Oxidation of Nitrogen Heterocycles to N-oxides Using UPH — 13.3
- Notes and References — 13.4

14. Miscellaneous Reactions — 14.1
- 14.1 Aldol Condensation — 14.1
- 14.2 Grignard Reaction — 14.2
- 14.3 Reformatsky Reaction — 14.3
- 14.4 Post Script — 14.3
- Notes and References — 14.4

PART-IV PHOTO INDUCED ORGANIC SYNTHESIS

15. Photo Induced Organic Synthesis — 15.3
- 15.0 Introduction — 15.3
- 15.1 Photochemical Reactions — 15.4
  - 15.1.1 Photochemical Reactions at Some Carbonyl Compounds — 15.5
    - 15.1.1.1 Photolysis of Acetone (Propanone) — 15.5
    - 15.1.1.2 Photolysis of 2-Hexanone — 15.6
    - 15.1.1.3 Photolysis of Benzophenone — 15.7
  - 15.1.2 Photochemical Reactions of Olefins — 15.8
    - 15.1.2.1 Isomerization of Olefins — 15.8
    - 15.1.2.2 Photoisomerisation of cis and trans-Stilbene — 15.8
  - 15.1.3 Photochemical Cycloaddition Reactions — 15.9
  - 15.1.4 Photochemical Cycloaddition Reactions in Water — 15.13
  - 15.1.5 Photochemical Reactions in Micellar Media — 15.15
  - 15.1.6 Photochemical Reactions in Solid State — 15.19
- 15.2 Principal Industrial Applications of Photochemistry — 15.21
  - 15.2.1 Free Radical Chlorination — 15.21
  - 15.2.2 Free Radical Sulfochlorination — 15.22

	15.2.3	Photochemical Sulfoxidation	15.23

15.2.3 Photochemical Sulfoxidation — 15.23
15.2.4 Photonitrosation — 15.23
15.2.5 Photochemical Synthesis of Vitamin D and Related Compounds — 15.24
  15.2.5.1 Vitamin D_2 — 15.24
  15.2.5.2 Vitamin D_3 — 15.24
  15.2.5.3 Hydroxy Derivatives of Vitamin D_3 — 15.25
  15.2.5.4 Photoisomerisation of Vitamin A Acetate — 15.26
15.2.6 Photo-oxygenation — 15.26
  15.2.6.1 Rose Oxide — 15.27
  15.2.6.1 Ascaridole — 15.27
15.2.7 The Barton Reaction — 15.27

15.3 Miscellaneous Photochemical Reactions — 15.28
  15.3.1 Photochemical Conversion of α-pinene into Trans-Pinocarveol Using Singlit Oxygen — 15.28
  15.3.2 Photoirradiation of Dibenzoyldiazomethane in Presence to Amino Acid[54]
  15.3.3 Photochemical Aromatic Substitution — 15.29
  15.3.4 Synthesis of Dydrogesterone — 15.29
  15.3.5 Templeted Reactions in a Homogeneous Solution which are Mediated by Light — 15.30
    15.3.5.1 [2 + 2] Photocycloaddition of 4-methoxy-2-quinolones — 15.30
    15.3.5.2 [2 + 2] Photocyclisation of Protected 4-(2′-aminoethyl)-quinolones with Acrylates — 15.31
    15.3.5.3 [4 + 4] Photocyclisation of 2-Pyridone to Cyclopentadiene — 15.32
    15.3.5.4 [4π] Cyclisation of 2-Pyridones — 15.32
    15.3.5.5 Steroselective Paterno-Büchi Reaction — 15.33
  Notes and References — 15.33

Index — I.1

Part-I

MICROWAVE ASSISTED ORGANIC SYNTHESIS

1

INTRODUCTION

In the new millennium, Industrial Chemistry is adopting the concept of green chemistry to meet the scientific challenges of protecting human health and environment while maintaining commercial viability. A most important aspect is the replacement of volatile organic solvents from the reaction medium with possible substitutions by nonvolatile or recyclable alternatives. Among the emerging important tools the use of microwaves (MW) as alternative source of energy is becoming an alternative especially under the solvent free conditions[1-4].

The dramatic increase in the number of publications, books[5,6] and a growing amount of patent literature[7-16] speaks well for microwave enhanced chemical synthesis.

Microwave ovens normally have wave lengths between 1 cm and 1 m (frequencies of 30 GHz to 300 Hz). These are similar to frequencies of radar and telecommunications. In order to avoid any interference with this systems, the frequency of radiation that can be emitted by household and industrial microwave oven is regulated, most of these appliances operate at a fixed frequency at 2.45 GHz.

Microwave (MW) energy has been used[17] for heating food materials for almost half a century. Microwaves are now commonly used for heating purposes. The mechanison of how energy is given to substance which is subjected to microwave irradiation is complex. It is believed that microwave reactions involve selective absorption of electromagnetic waves by polar molecules, non-polar molecules are inert to microwaves. When molecules with a permanent dipole are subjected to an electric field, they became aligned and as the field oscillates their orientation changes, this rapid reorientation provides intense internal heating. The main difference between classical heating and microwave heating, lies in core and homogeneous heating associated with microwaves, whereas classical heating is all about heat transfer by preheated molecules.

The preferred reaction-vessel for microwave induced organic reaction, is a tall beaker (particularly for small scale preparations in the laboratory), loosely covered and the capacity of the beaker should be much greater than the volume of the reaction mixture. Alternatively, teflon and polystyrene containers can be used[18-19]. These materials are transparent to microwaves. Metallic containers should not be used as reaction vessels, as it gets heated soon due to preferential absorption and reflection of rays.

In microwave induced organic reactions, the reaction can be carried out in a solvent medium or on a solid support in which no solvent is used. For reactions in a solvent medium, the choice of the solvent is very important[2, 3]. The solvent to be used must have a dipole moment so as to absorb microwaves and a boiling point at least 20–30° higher than the desired reaction temperature. An excellent solvent in a domestic microwave oven is N, N-dimethyl formamide (DMF) (b.p. 160°, $\varepsilon = 36.7$). This solvent can retain water formed in a reaction, thus obviating the need for water separation. Some other solvents of choice are given in the table below:

Some Common Solvents for use in 17 W Ovens

Solvent	b.p.	Dielectric constant (ε)
Formamide	216	11.1
Methanol	65	32.7
Ethanol	78	24.6
Chlorobenzene	214	5.6
1, 2-Dichlorobenzene	180	1.53
1, 2, 4-Trichlorobenzene	214	1.57
1, 2-Dichloroethane	83	10.19
Ethylene glycol	196	37.7
Dioxane	101	2.20
Diglyme	162	7.0
Triglyme	216	1.42

Hydrocarbon solvents, for example, hexane ($\varepsilon = 1.9$), benzene ($\varepsilon = 2.3$), toluene ($\varepsilon = 2.4$) and xylene are unsuitable because of less dipole moment, and also because these solvents absorb microwave radiations poorly. However, addition of small amounts of alcohol or water to these solvents can lead to dramatic coupling effects. Liquids which do not have dipole moment cannot be heated by microwaves. By the addition of a small amount of a dipolar liquid to a miscible non-polar liquid, the mixture will rapidly achieve a uniform temperature under irradiation.

Microwaves may be considered as a more efficient source of heating than conventional heating (steam or oil heating), since the energy is directly imparted to the reaction medium rather than through the walls of a reaction vessel. In fact, the rapid heating capacity of the microwave leads to a considerable saving in dissolution of the reaction time. The smaller volume of the solvent required contributes to saving in cost and diminishes the waste disposal problems[20–21].

Microwave procedures are limited[22] by the presence of solvents which reach their boiling points within a very short time (~ 1 min) of exposure to microwave. Consequently, high pressures are developed, leading to damage to the vessel material or the microwave oven itself and may occasionally lead to explosion.

Well designed microwave ovens are now available. Consideration of safety aspects coupled with the limitations of the solvents imposed by microwave heating, has led to many reactions being carried out in water or more commonly under solvent free conditions. It is believed that due to high polarity and non-volatility, ionic liquids (section 5.19) might be ideal for carrying out high temperature reactions efficiently, since temperatures of over 200°C can be readily attainable.

Microwave procedures now find potential useful applications in chemical technology. These include acceleration of chemical reactions[23-35]; in this explosive growth has been witnessed, where in chemical reactions are accelerated because of selective absorption of mw energy by polar molecules, non-polar molecules being inert to MW dielectric loss[36]. As has already been started, initially the chemical experiments with microwave heating used high dielectric solvents such as dimethyl sulfoxide (DMSO) and dimethylformamide (DMF) where the rate enhancements are now believed to be due to rapid superheating of the polar solvents and pressure effects[14]. However the development of high pressures, and the use of specialized sealed vessels are some of the challenges in these solution-phase reactions, which in part, have been addressed by newly introduced commercial MW instruments with precise temperature and pressure control. The latest developments in such newer instruments had attracted the attention of several chemical companies and their heigthened interest has become obvious as exemplified recently in the application of microwaves in combinatorial chemistry that rapidly generates a library of potentially useful chemical entities[22].

According to report published[37] in science and technology (chemical and engineering news) microwaves are used in a number of industrial processes including meat tempering, beacon cooking, potato chips drying, rubber vulcanization and drying of pharmaceutical compounds. The premise of micro waves attracted more than 320 scientists, engineers and others from 23 countries. Since microwave easily penetrate materials (with the exception of metals), they can be directly and uniformly absorbed throughout the entire volume of an object, causing it to heat up evenly and rapidly. This capability has opened new opportunities in chemistry, material science, and other areas. According to Ajay K Bose, of Stevens Institute of Technology in Hoboken, N.J. a pharmaceutical company had a problem. During the course of developing a new drug, the chemists in the company had to synthesise an intermediate using a reaction that required four days of heating in an oil bath. This procedure besides consuming more time led to the loss of product in the form of degradation products. Professor Bose, a consultant to the company (Wyeth-Ayerst Research at Pearl River, N.Y.), suggested replacing the oil bath with microwave heating. With the aid of computer-controlled microwave oven that was able to maintain a constant reaction temperature, the intermediate was produced in four hours, with only a small amount of degradation product. According to Bose, microwave assisted chemistry could be a boon to the pharmaceutical, biotechnology and other industries. It is an environmentally friendly technology since it can reduce or eliminate the use of solvents. Reactants can simply be mixed together with little or no solvent. Since products often are produced in purer form with microwaves, use of solvents and chromatographic materials in purification can also be reduced.

It is interesting to note that peptide hydrolysis using conventional heating has to be run overnight in a sealed tube containing hydrochloric acid. The advant of microwave heating shortened the reaction time to 15 minutes. The hydrolysis can also be carried out in open vessels inside an ordinary microwave oven using this procedure, Bose *et al* effected complete hydrolysis of di, tri and tetrapeptide in 3 to 15 minutes using saturated barium hydroxide solution in open flask inside an ordinary microwave oven.

To prevent the volatile reagents from boiling off, the researchers devised a make shift retlux condenser which was placed on the top at the reaction flask. The condenser contained dry ice, liquid nitrogen or ordinary ice in the water socket of the condenser.

Rajinder S. Varma[38], a scientist while working at Sam Houstan State University in Hurntsville, Texas and Houstan Advanced Research Centre in the Woodlands Texas, has found ways to perform organic transformations rapidly using catalyst under solvent free 'dry' conditions. According to varma, it is the easiest possible way to run a reaction-just mix the neat reactants together with some catalyst (if possible) and heat them in a microwave for a few minutes. Varma believes that using microwave energy under solvent free conditions will lead to new procedures for conducting cleaner and efficient chemical transformations. This technology could aid pollution abatement efforts aimed at destroying hazardous wastes in the environment. Contaminated solids could be treated with catalysts (which are inexpensive) and microwaves to destroy toxic compounds.

Broadly speaking the organic synthesis using microwaves can be grouped in the following three categories:

— Microwave assisted reactions in water;
— Microwave assisted reactions in organic solvents;
— Microwave assisted reactions in solid state.

Only a limited representative description of each category is given, since, excellent review articles are available[14–16, 22].

NOTES AND REFERENCES

1. (a) R.S. Varma in Microwaves in Organic Synthesis, A. Loupy (Ed.). Chapter 6, pp 181–218, Wiley-VCH, Weinheim (2002).

   (b) R.S. Varma in *ACS Symposium* series No. 7671 Green Chemical Synthesis and Processes, P.T. Anastas, L. Heine, T. Williamson (Eds.),Chapter 23, pp 292–313, *American Chemical Society*, Washington DC (2000).

   (c) R.S. Varma in *Green. Chemistry, Challenging Perspectives*, P. Tundo, P.T. Anastas, (Eds.), Oxford University Press, Oxford, 2000, pp. 221–244.

2. R.S. Varma, *Pure Appl. Chem.*, 2001, **73**, 193.

3. R.S. Varma, *Green. Chem.*, 1999, **I**, 43.

4. (a) R.S. Varma, *Tetrahedron* 2002, **58**, 1235.

   (b) U. Pillai, E. Sahle-Demmessie, and R.S. Varma, *J. Mat. Chem.*, 2002, **12**, 3199.

5. A. Loupy. (Ed.), *Microwaves in Organic Synthesis*, Wiley-VCH, Weinheim, 2002.

6. B.L. Hayes, *Microwave Syunthesis Chemistry at the speed of light*. CEM Publishing, Mathews, NC 2002.

7. B. Herzog. (Celanese GmbH). Catalyst Based on Pd. Au and Alkali Metals for the Preparation of Vinyl Acetate. *Eur. Patent Application* EP 922, 491, (1999) (DE Application 19,754,992, 11 Dec 1997); *Chem. Abstr.*, **131**, 45225s (1999).

8. J.-R. Desmurs, J. Dubac, A. Laporterie, C. Laporte and J. Marquie (Rhodia Chimie) Method for Acylation or Sulfonylation of an Aromatic Compound. *PCT International Application* WO 40,39, (1998) (FR Application 97/2,917, 12 Mar 1997); *Chem. Abstr.*,**129**, 244928g (1998).

9. A Pöppl, S. Witt and B. Zimmermann (Henkel KgaA), Solid Perfumed Deodorant Composition Containing Alunite, *PCT International Application* WO 23, 197 (*DE Application* 19,548,067, 26 Jun 1997: *Chem. Abstr.*, **127**, 70625x (1997).

10. O. Rhode, S. Witt and I. Hardacker (Henkel K.A.). Method for Preparing of Alkyl-Glycosides Using Microwave Irradiation. *PCT International Application* WO 3,869, (1999) (*DE Application* 19,730,836, 21 Jan 1999); *Chem. Abstr.*, **130**, 110555v (1999).

11. I.T. Badejo (Bayer Corporation). Microwave Syntheses of Quinacridones, 6, 13-Dihydroquinacridones and 6, 13-Quinacridonequinones at moderate Temperatures. *Eur. Patent Application* EP 905, 199, (1999) (*U.S. Patent Application* 63,128, 20 Apr 1998); *Chem. Abstr.*, 1999, **130**, 253670q.

12. P. Coe. T. Waring and C. Mercier (Rhone-Poulene Chemicals). Preparation of Fluoro Compounds by Treatment of the Corresponding Amines with Hydrogen Flouride and a Nitrosating Agent Under Ultrasound or Microwave Irradiation. *PCT International Application* WO 41,083, (1997) (*CB Application* 96/9, 154, 1 May 1996); *Chem. Abstr.*, **128**, 13127h (1998).

13. G. Forat, J.-M., Mas and L. Saint-Jalmes (Rhone-Poulenc Chimie). Method for Grafting a Substituted Difluoromethyl Group to a Compound Containing an Electrophilic Group with Microwave Irratdiation. *PCT International Application* WO 5,609, (1998) (*FR Application* 96/9,754, 1 Aug 1996); *Chem. Abstr.*, **128**, 166999u (1998).

14. D. Semeria and M. Philippe (L'Oreal) High-Yield Preparation of Ceramides by Conducting the Aminoalcohol Amidation in the Presence of Microwave Irradiation. *Eur. Patent Application* EP 884,305, (1998) (*FR Application* 97/7,240, 11 Jun 1997): *Chem. Abstr.*, **130**, 52677y (1999).

15. G. Lindeberg. M. Larhed and A. Hallberg (Labwell AB), Method for Organic Reactions-Transition Metal Catalyzed Organic Reactions. *PCT International Application* W.O. 43,230, (1997) (*SE Application* 96/3,913, 25 Oct 1996): *Chem. Abstr.*, **128**, 34382c (1998).

16. J. Westman (Personal Chemistry), Preparation and Use of Ionic Liquids in Microwave assisted Chemical Transformations, *PCT International Application* WO 0072956, (2000).

17. C.R. Buffler, *Microwave, Cooking and Processing Van Nostrand Reinhold*, New York, pp. 1–68, 1993.

18. P. Michael, P. Mingos and D.R. Baghrust, *Chem. Soc. Rev.*, 1991, **20**, 1.

19. A.K. Bose, M.S. Manhas, B.K. Banik and E.W. Robb, *Res. Chem. Intermed.* 1994, **20(i)**, 1.

20. S.S. Bari, A.K. Bose, A.G. Chaudhary, M.S. Manhas, V.S. Raju and E.W. Robb., *J Chem.* Ed., 1992, **69(11)**, 938.

21. Shui-Terichem, Shyh-Horngchiou and Kung-Tsungwang, *J. Chem. Soc. Chem. Commun.*, 1990, 807.

22. M.N. Gedye, F.E. Smith and K.C. Westaway, Can. *J. Chem.*, 1988, **66**, 17.

23. R.S. Varma in *ACS Symposium Series No. 767/Green Chemical Synthesis and Processes*, P.T. Anastas, L. Heine and T. Williamson (Eds.), Chapter 23, pp. 292–313, *American Chemical Society*, Washington DC, (2000).

24. R.S. Varma in *Green Chemistry: Challenging Perspectives*, P. Tundo and P.T. Anastas (Eds.) pp. 221–244, Oxford University Press, Oxford, (2000).

25. R.S. Varma, *Green Chem.*, **1**, 43 (1999).

26. R.S. Varma, in *Microwaves in Organic Synthesis*, A. Loupy (Ed.) Chapter 6, pp 181–218, Viley-VCH, Weinheim (2002).

27. R.S. Varma in *Microwaves: Theory and Application in Material Processing IV*, D.E. Clark, W.H. Sutton and D.A. Lewis, (Eds.) pp. 357–365, American Ceramic Society, Westerville, Ohio, (1997).

28. R.A. Abramovich, *Org. Prep. Proc. Int.*, **23**, 683 (1991).

29. S. Caddick, *Tetrahedron*, **51**, 10403 (1995).

30. F. Langa, P. de la Cruz, A. de la Hoz, A. Díaz-Ortiz and E. Díez-Barra, *Contemp. Org. Chem.*, **4**, 373 (1997).

31. (a) R.S. Varma, *Tetrahedron*, **58**, 1235 (2002).

    (b) U. Pillai, E. Sahle-Demessie and R.S. Varma, *J. Mat. Chem.*, **12**, 3199 (2002).

32. P. Lindstrom, J. Tiernery, B. Wathey and J. Westman, *Tetrahedron*, **57**, 9225 (2001).

1.8 ■ *Alternate Energy Processes in Chemical Synthesis*

33. R.S. Varma, *Pure Appl. Chem.,* **73**, 193 (2001).
34. (a) L. Perreux, A. Loupy, *Tetrahedron,* **57**, 9199 (2001).
    (b) A. Loupy, A. Petit, J. Hamelin, F. Texier-Boullet, P. Jacquault and D. Mathé, *Synthesis*, 1213 (1998).
35. (a) C. Gabriel, S. Gabriel, E.H. Grant, B.S. J. Halstead and D.M.P. Mingos, *Chem. Soc. Rev.,* **27**, 213 (1998).
    (b) A. Stadler, S. Pichler, G. Horeis and C.O. Kappe, *Tetrahedron,* **58**, 3177 (2002).
36. G.A. Strohmeier and C.O. Kappe, *J. Comb. Chem.,* **4**, 154 (2002).
37. Ron Dagani, *C & EN News* Washington, 26 Feb. 1997, p. 26.
38. Present address: Clean Process Branch, National Risk Management Res. Lab., U.S. Environmental Protection Agency, 26, West MIK Drive, M.S. 443, Cincinnati, OH, 45268 USA.

2

MICROWAVE ASSISTED ORGANIC REACTIONS IN WATER

Following are given some microwave assisted organic reactions in water.

2.1 HOFMANN ELIMINATION

In this method, normally quarternary ammonium salts are heated at high temperature and the yield of the product is low. However, use of microwave irradiation in water-chloroform system gives high-yielding synthesis of a thermally unstable Hofmann elimination product (Scheme-1).

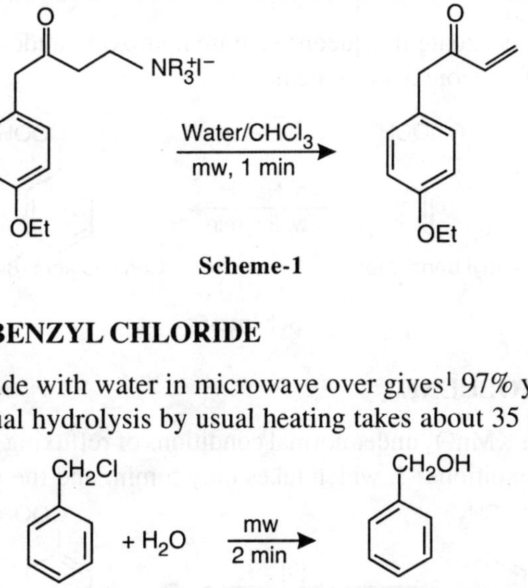

Scheme-1

2.2 HYDROLYSIS OF BENZYL CHLORIDE

Hydrolysis of benzyl chloride with water in microwave over gives[1] 97% yield of benzyl alcohol in 2 min (Scheme-2). The usual hydrolysis by usual heating takes about 35 min.

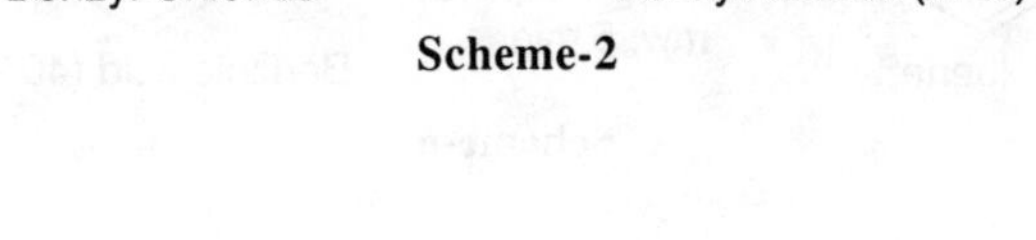

Scheme-2

2.3 HYDROLYSIS OF BENZAMIDE

The usual hydrolysis of benzamide takes 1 hr. However, under microwave, the hydrolysis is completed[2] in 7 min giving 99% yield of benzoic acid (Scheme-3).

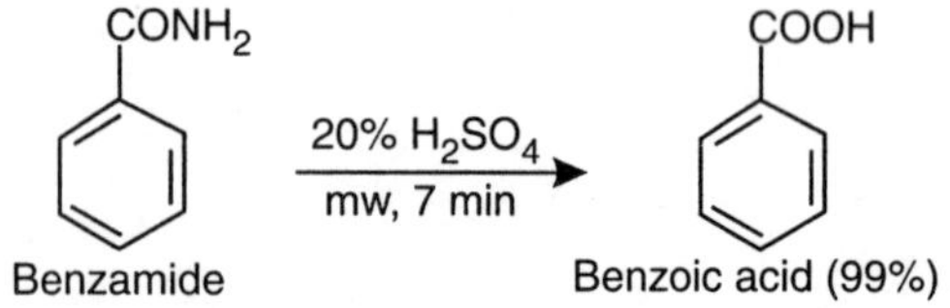

Benzamide → (20% H_2SO_4, mw, 7 min) → Benzoic acid (99%)

Scheme-3

2.4 HYDROLYSIS OF N-PHENYL BENZAMIDE

N-Phenyl benzamide on heating with 20% H_2SO_7 in an microwave oven for 12 min gives[2] 74% yield of benzoic acid (Scheme-4). The conventional heating procedure takes 18 hrs.

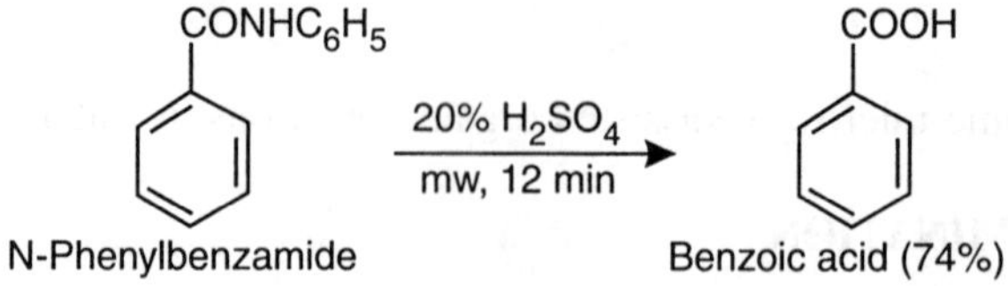

N-Phenylbenzamide → (20% H_2SO_4, mw, 12 min) → Benzoic acid (74%)

Scheme-4

2.5 HYDROLYSIS OF METHYL BENZOATE TO BENZOIC ACID (SAPONIFICATION)

Saponification of methyl benzoate in aqueons sodium hydroxide under microwave irradiation (2.4 min) gives[3] 84% yield of benzoic acid (Scheme-5).

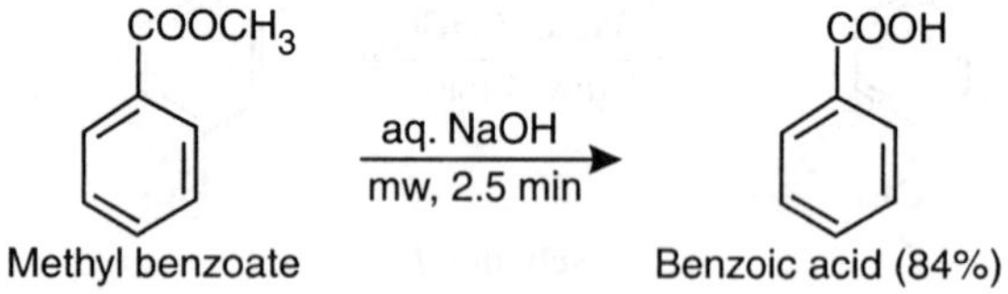

Methyl benzoate → (aq. NaOH, mw, 2.5 min) → Benzoic acid (84%)

Scheme-5

2.6 OXIDATION OF TOLUENE

Oxidation of toluene with $KMnO_4$ under normal conditions of refluxing takes 10–12 hr compared to reaction in microwave conditions[3,4], which takes only 5 min. and the yield is 40% (Scheme-6).

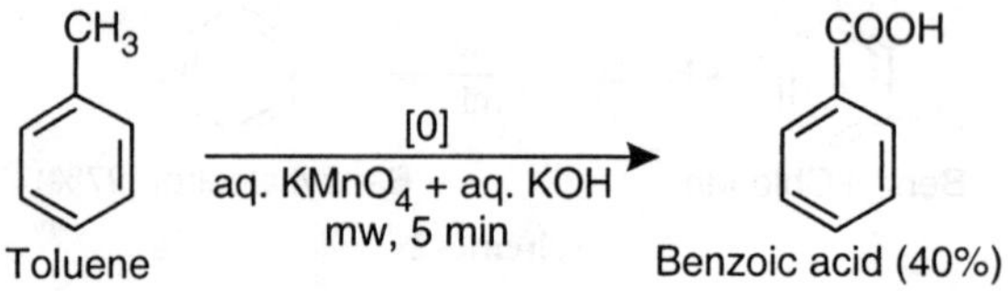

Toluene → ([O], aq. $KMnO_4$ + aq. KOH, mw, 5 min) → Benzoic acid (40%)

Scheme-6

2.7 COUPLING OF AMINES WITH HALIDES

Amines on heating in microwave with alkyl halides with $NaOH/H_2O$ gives[5] the corresponding N-alkyl derivatives (Scheme-7).

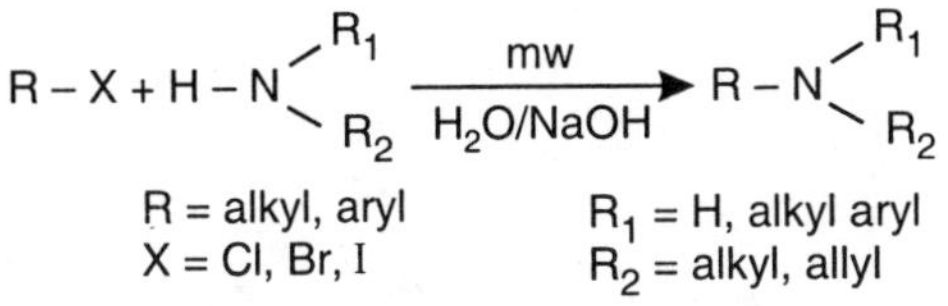

$$R-X + H-N\begin{matrix}R_1\\R_2\end{matrix} \xrightarrow[H_2O/NaOH]{mw} R-N\begin{matrix}R_1\\R_2\end{matrix}$$

R = alkyl, aryl
X = Cl, Br, I
R_1 = H, alkyl aryl
R_2 = alkyl, allyl

Scheme-7

2.8 N-HETEROCYCLIZATIONS

Microwave assisted N-heterocyclisation in aqueous media has been achieved[6] using an aromatic amine and dihalide of the type $X(CH_2)_nX$ in presence of K_2CO_3 (Scheme-8).

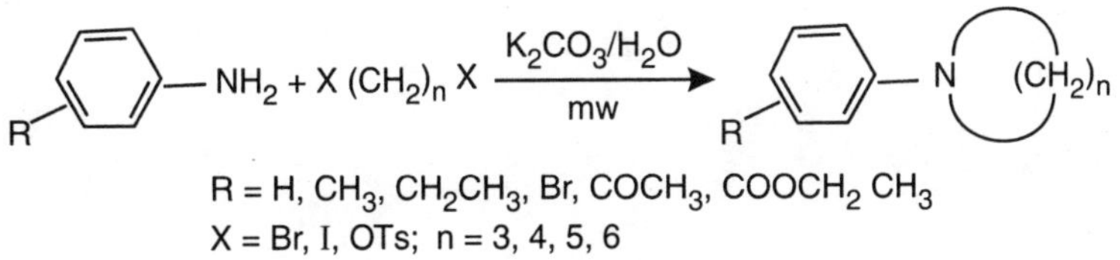

R = H, CH_3, CH_2CH_3, Br, $COCH_3$, $COOCH_2CH_3$
X = Br, I, OTs; n = 3, 4, 5, 6

Scheme-8

A typical example[6] of N-heterocyclisation is given in Scheme-9.

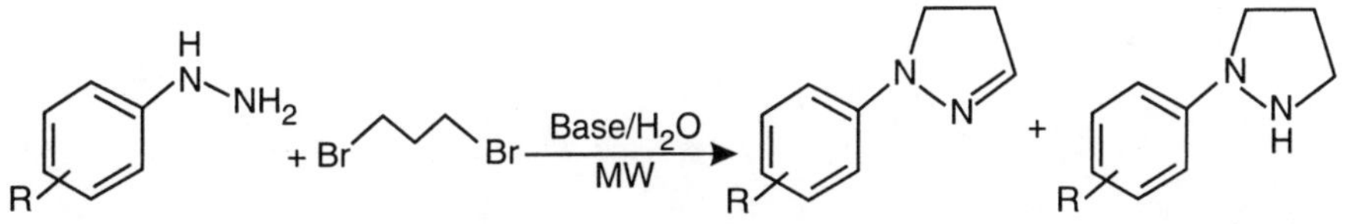

Scheme-9

NOTES AND REFERENCES

1. R.N. Gedye, W. Rank and K.C. Westaway, *Can. J. Chem.*, 1991, **69**, 706.
2. R.N. Gedye, F.E. Smith and K.C. Westaway, *Can. J. Chem.*, 1988, **66**, 17.
3. R.N. Gedye, W. Rank and K.C. Westaway, *Can. J. Chem.*, 1988, **66**, 700.
4. R.N. Gedye, F. Smith, K. Westaway, H. Ali, L. Baldisera, Laberge and Rousell, *Tetrahedron Lett.*, 1986, **26**, 279.
5. Y. Ju and R.S. Varma, *Green Chem.*, 2004, **6**, 219.
6. Y. Ju, R.S. Varma *Org. Lett.*, 2005, **7**, 2409; Y. Ju, R.S. Varma, *Tetrahedron Lett.*, 2005, **46**, 6011; Y. Ju, R.S. Varma, *J. Org. Chem.*, 2006, **71**, 135–141.

3

MICROWAVE ASSISTED ORGANIC REACTIONS IN ORGANIC SOLVENTS

This section includes those microwave induced reactions in which one or both the reactants (if liquid) act as solvent and also those reactions in which organic solvent is used to assist the reaction.

3.1 ESTERIFICATION: REACTION OF CARBOXYLIC ACID AND ALCOHOL

A mixture of a carboxylic acid (e.g. benzoic acid) and an alcohol (e.g. n-propanol) on heating in a microwave oven for 6 min in presence of catalytic amount of sulphuric acid gives[1, 2, 3] the corresponding ester (e.g. propylbenzoate) (Scheme-1).

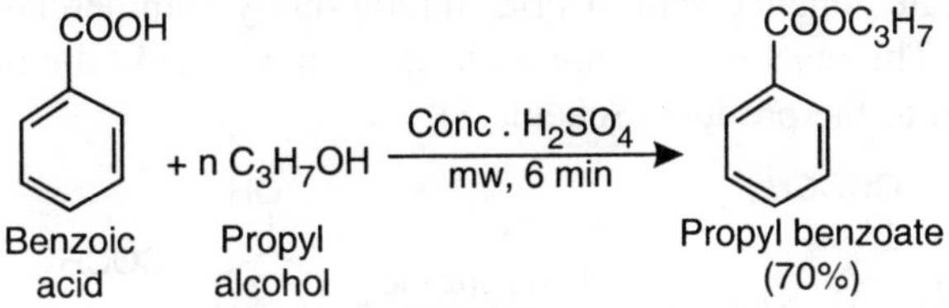

$$\text{Benzoic acid (COOH)} + n\ C_3H_7OH \xrightarrow[\text{mw, 6 min}]{\text{Conc. } H_2SO_4} \text{Propyl benzoate (COOC}_3H_7\text{) (70\%)}$$

Scheme-1

Carboxylic esters have also been prepared by a tribromolanthanoid mediated deetherification reaction (see section 3.2).

Using microwave irradiation, it has also been possible to esterify hindered acids which are difficult to esterify under usual conditions. Thus, mesiotic acid can be esterified with isopropyl alcohol using microwave irradiation in presence of catalytic amount of sulphuric acid. The product obtained is isopropyl mesitoate (Scheme-1a) in 56% yield.

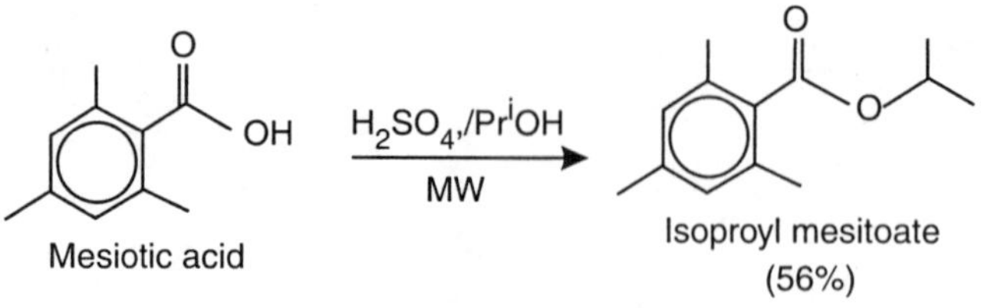

3.2 ■ *Alternate Energy Processes in Chemical Synthesis*

Mesiotic acid → Isoproyl mesitoate (56%)
(H$_2$SO$_4$/Pr^{i}OH, MW)

Scheme-1a

3.2 REACTION OF CARBOXYLIC ACID AND BENZYL ETHERS USING MICROWAVE IN PRESENCE OF Ln Br$_3$ (Ln = La, Nd, Sm, Dy, Er)

The reaction of benzyl ethers and carboxylic acids catalysed by Ln Br$_3$ (Ln = La, Nd, Sm, Ey, Ed) due to its weak lewis acid character, with microwave irradiation yeilds[4] esters (Scheme-2). The reaction is carried out without solvent and esterification in complete in 2 minutes. However, when the reaction of benzyl n-butyl ether with acetic and acid is carried out with conventional heating (118°, 10 hr) only 42% yield of benzyl acetate is formed even in presence of Nd Br$_3$ in a sealed tube. Thus, heating in microwave oven affords better results.

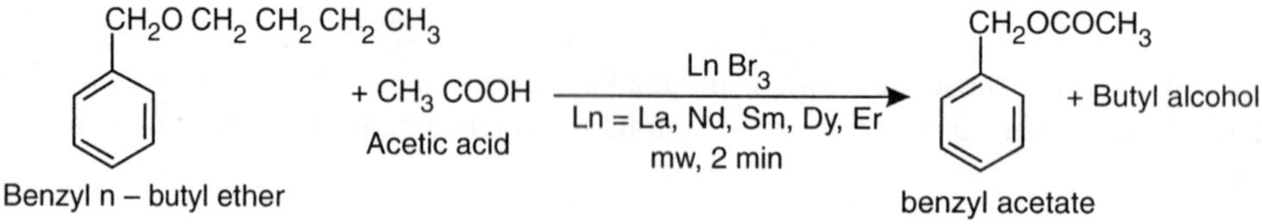

Benzyl n-butyl ether + CH$_3$COOH (Acetic acid) → benzyl acetate + Butyl alcohol
(Ln Br$_3$, Ln = La, Nd, Sm, Dy, Er, mw, 2 min)

Scheme-2

3.3 FRIES REARRANGEMENT

Fries rearrangement is used for the preparation of phenolic ketones and is usually carried out by heating a mixture of substracte (e.g. p-cresyl acetate), aluminium chloride and chlorobenzene. There is considerable rate enhancement of Fries migration by commercial microwave ovens over conventional methods. The reaction mixture on heating in a sealed tube in a microwave oven for 2 min gives[5] 85% yield to the product (Scheme-3).

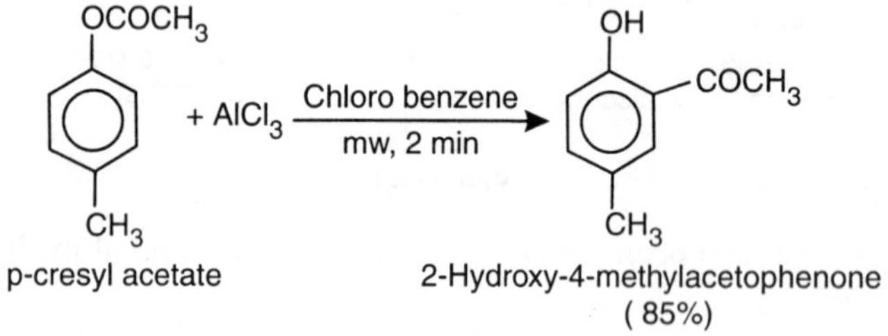

p-cresyl acetate + AlCl$_3$ → 2-Hydroxy-4-methylacetophenone (85%)
(Chloro benzene, mw, 2 min)

Scheme-3

3.4 DIELS ALDER REACTION

The reaction involves 1, 4-addition of an alkene (e.g. maleic anhydride) to a conjugated diene (e.g. anthracene) to form an adduct of six membered ring. Under usual conditions[6], the reaction requires a refluxing time of 90 min. However, under microwave irridation[7, 8] diglyme is used as a solvent and 80% yield of the adduct is obtained in 90 sec (Scheme-4).

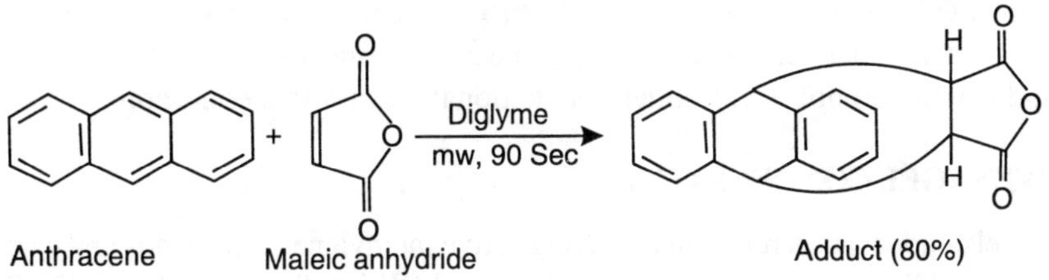

Anthracene + Maleic anhydride → (Diglyme, mw, 90 Sec) → Adduct (80%)

Scheme-4

In a similar way anthracene reacts with dimethylfumerate within 10 min in p-xylene to afford[9] 87% yield (Scheme-5) where as conventional heating conditions give only 67% yield in 5 hr.

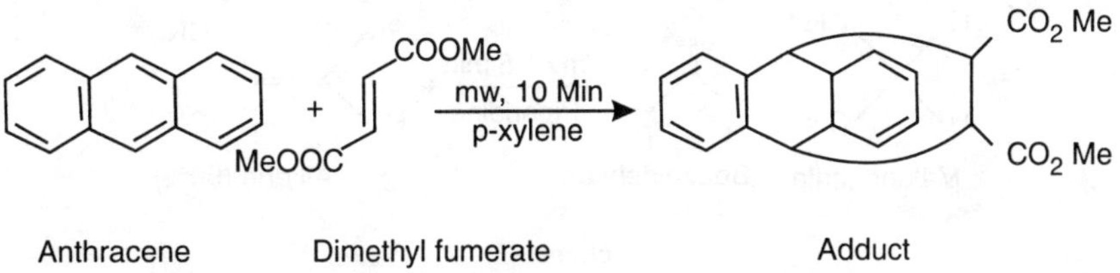

Anthracene + Dimethyl fumerate → (mw, 10 Min, p-xylene) → Adduct

Scheme-5

3.5 CLAISEN REARRANGEMENT

Claisen rearrangement of allyl ether in the absence of solvent under conventional thermolysis provides a good yield of rearranged product within a short period of time[10]. The same rearrangement using microwaves in presence of N-methylformamide affords 87% of the product (Scheme-6).

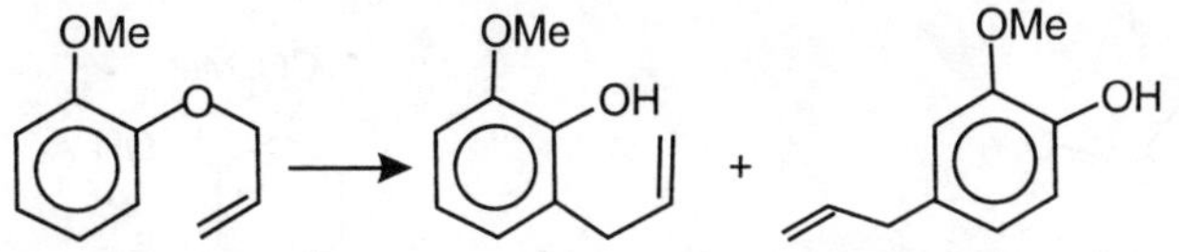

Control
At 320°, 92%, neat, 12 min
At 265°, 71%, neat, 45 min

Microwaves
At 370°C – 400°C, 71%, neat, 12 min
At 300°C – 315°C, 72%, neat, 5 min
At 276°C, 87%, NMF, 90 sec

Scheme-6

The ortho-claisen rearrangement of allyl phenyl ether has been achieved in aqueous media when 2-allylphenol is obtained[10a] exclusively after 10 min of irradiation (Scheme-6a).

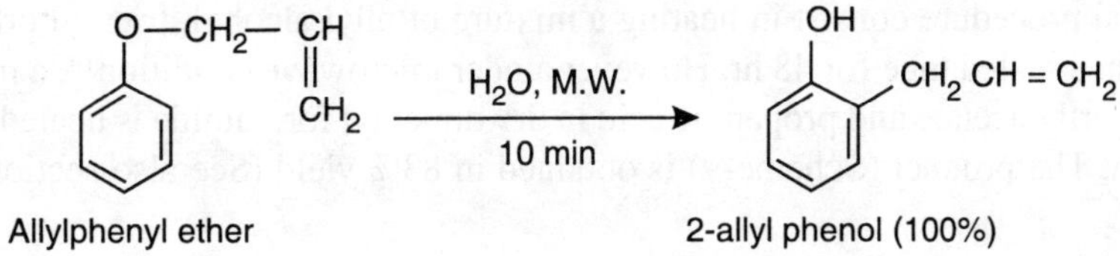

Allylphenyl ether → (H_2O, M.W., 10 min) → 2-allyl phenol (100%)

Scheme-6a

3.4 ■ *Alternate Energy Processes in Chemical Synthesis*

3.6 CYDOADDITION REACTION BETWEEN FULVENES AND SOME ALKENES AND ALKYNES SYNTHESIS OF POLYCYCLIC RING SYSTEMS

A number of polycyclic ring systems found in isobarbatene and alcyoptersin have been synthesised[11] by cydoaddition reaction between fulvenes and some alkenes and alkynes. Such reactions conducted in benzene and DMSO do not occur under conventional thermolytic conditions.

3.7 KNOEVENAGEL CONDENSATION

The Knoevenagel condensation reaction involving active methylene compounds and carbonyl groups for the synthesis of alkanes has been reported[12] using MW irradiation (Scheme-7). The reactions are conducted in open vessels that lead to the efficient removal of water, thus circumventing the use of Dean-Stark apparatus.

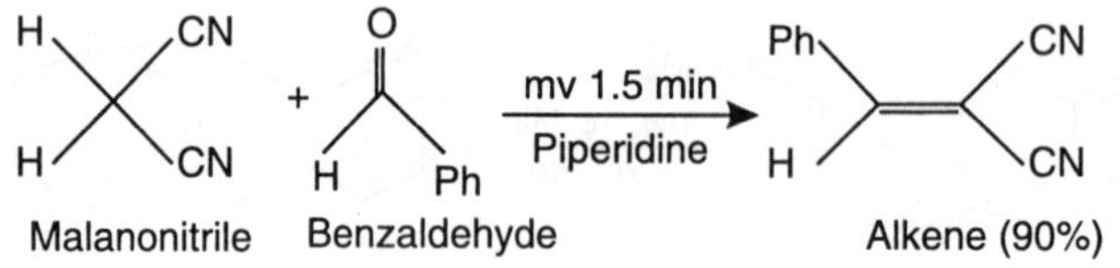

Scheme-7

3.8 BAYLIS-HILLMAN REACTION

Baylis-Hillman reaction for alkene functionalization is accelerated by altering reaction conditions such as pressure, temperature or using ultrasound. The use of MV irradiation improved[13] the yield as well as reduction in time. The reaction of benzaldehyde with methyl crotonate in presence of DABCO leads to the required product in good yield in 10 minutes (Scheme-8).

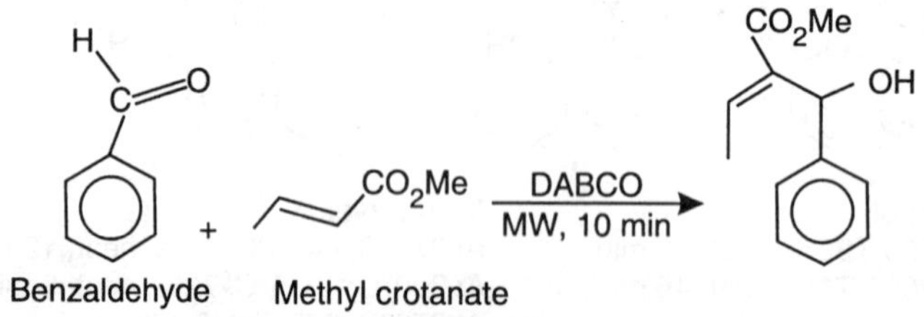

Scheme-8

3.9 ORTHOESTER CLAISEN REARRANGEMENT

The conventional procedure consist in heating a mixture of allyl alcohol, triethyl orthoacetate and propanoic acid in a sealed tube for 48 hr. However, under microwave condition[14], a mixture of allyl alcohol, triethyl orthoacetate and propanoic acid in dry dimethyl formamide is heated in microwave oven for 10 min. The product (Scheme-9) is obtained in 83% yield (See also section 5.12).

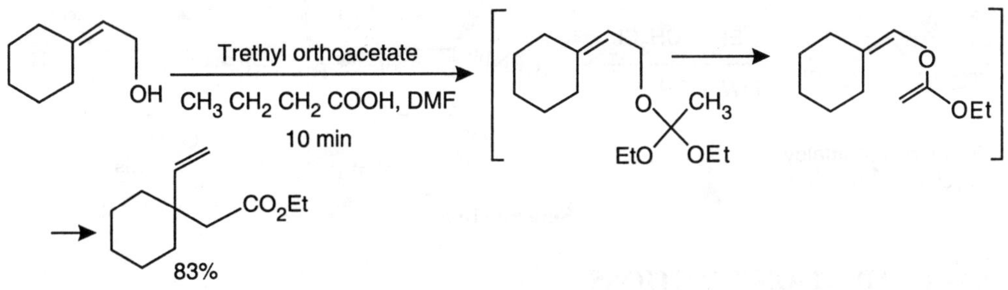

Scheme-9

3.10 SYNTHESIS OF 4-ARYL-3, 4-DIHYDROPYRIMIDINE 2(lH) ONES

A mixture of β-ketoesters, aryl halides and urea derivatives with polyphoshate esters (PPE) on heating in a microwave oven gave the corresponding 4-aryl-3, 4-dihydropyrimidine 2(1H) ones (Scheme-9a)[15, 16, 17].

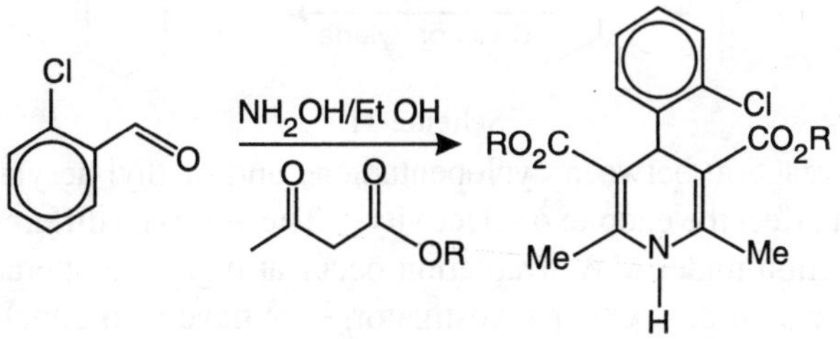

Scheme-9a

3.11 SYNTHESIS OF β-LACTAMS

Enantiomerically pure β-lactams have been prepared[18] using MW irridation (Scheme-10).

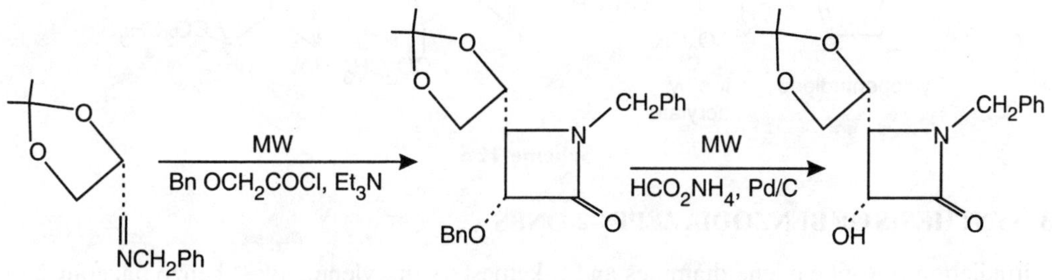

Scheme-10

Using the procedure 25g of β-lactam could be easily prepared.

Stereoselective outcome of the reaction under MV or classical conditions is different when tetraehorophthaloyl glycine chloride is reacted[18a] with an amine. There is exclusive formation of trans isomer of β-lactam under mw irradiation conditions. Variable amounts of Cis and trans isomers of β-lactams are obtained under classical conditions (Scheme-10a).

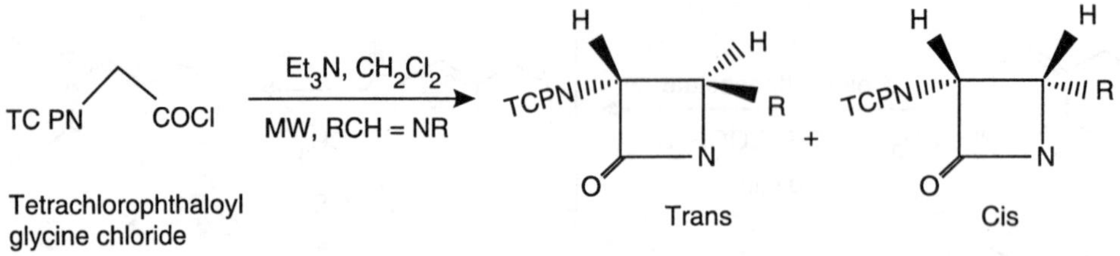

3.6 ■ Alternate Energy Processes in Chemical Synthesis

Scheme-10a

TC PN — Tetrachlorophthaloyl glycine chloride

Reagents: Et$_3$N, CH$_2$Cl$_2$, MW, RCH=NR → Trans + Cis β-lactams (TCPN, R substituents)

3.12 CYCLOADDITION REACTIONS

Some cycloaddition reactions in xylene or dibutyl ether[19] have been carried out and rate of the raction is found to be much faster under MW irradiation conditions (Scheme-11). The acceleration is more in apolar solvents that show weak dielectric losses. It is believed[19-20] that this may be due to change in entropy of the system.

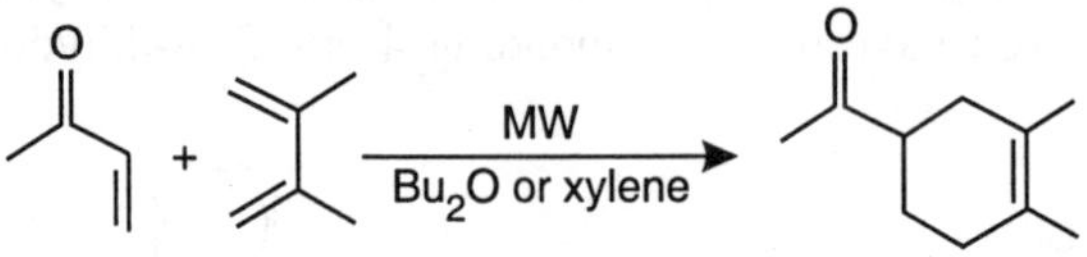

Scheme-11

MW, Bu$_2$O or xylene

The cycloaddition reaction between cyclopentadiene and methyl acrylate (Scheme-12) under MV irradiation does not affect the endo/exo selectivity[21]. The observed difference could be explained by the fact that the reaction under MW irradiation occur at higher temperatures than those under conventional refluxing conditions. Other investigators[22, 23] have also concluded that the reaction rates are identical in the presence or absence of mw irradiation, the final yield is dependent on the temperature and not on the mode of heating.

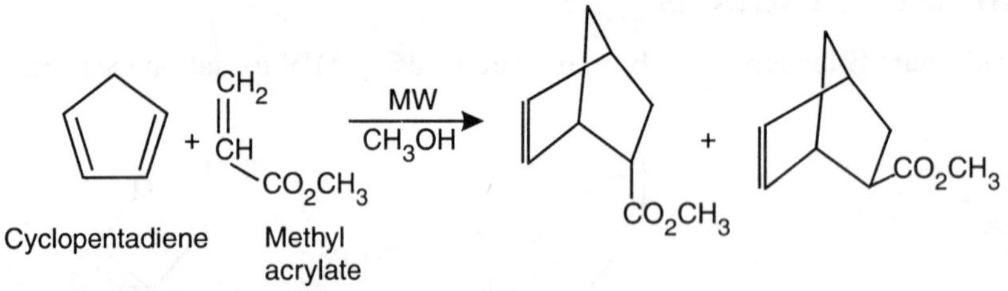

Scheme-12

Cyclopentadiene + Methyl acrylate →(MW, CH$_3$OH)→ endo + exo products (CO$_2$CH$_3$)

3.13 SYNTHESIS OF BENZODIAZEPIN-2-ONES

MW irradiation of o-phenylene diamines and β-ketoesters in xylene gave[24] benzodiazepin-2-ones (Scheme-13).

β-ketoester + o-phenylene diamine →(Xylene, MW, 10 min)→ Benzodiazepin-2-ones

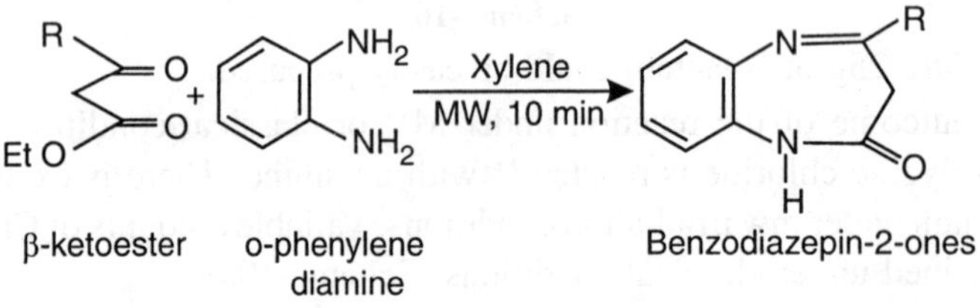

Scheme-13

The above reaction does not proceed on classical mode of heating. Other diazepines have also been prepared under the influence of microwaves[25].

3.14 AROMATIC SUBSTITUTION REACTIONS

Aromatic substitution reactions can be conveniently performed (Scheme-14) in presence of small amount of PTC under microwave irradiation; Under microwave the rate of the reaction is considerably enhanced (144–240 fold). Thus, p-chloronitrobenzene on treatment with sodium hydroxide in alcohol in presence of PTC under microwave irradiation give[26] p-ethoxynitrobenzene (Scheme-14).

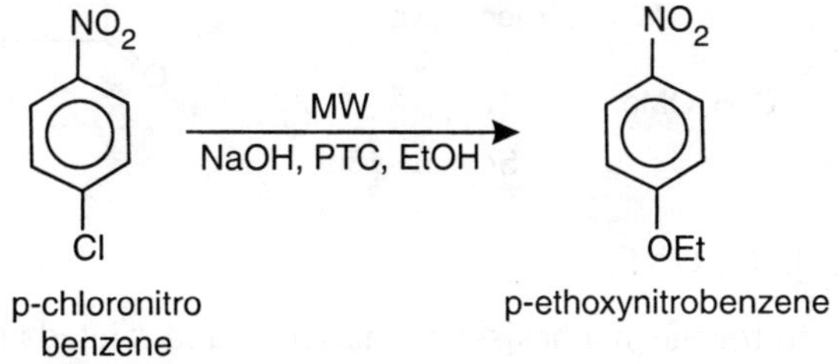

Scheme-14

The efficiency of aluminium catalysed Friedel-crafts germylation of benzene and toluene is enhanced by microwave irradiation. Thus germylation of benzene using standard reflux condition provides the substituted product (1) in 20% yield after 24 hrs. A slight improvement in the yield could be achieved using microwave irradiation. Two hours MW irradiation gave[26a] the product (1) in 25% yield (Scheme-14a).

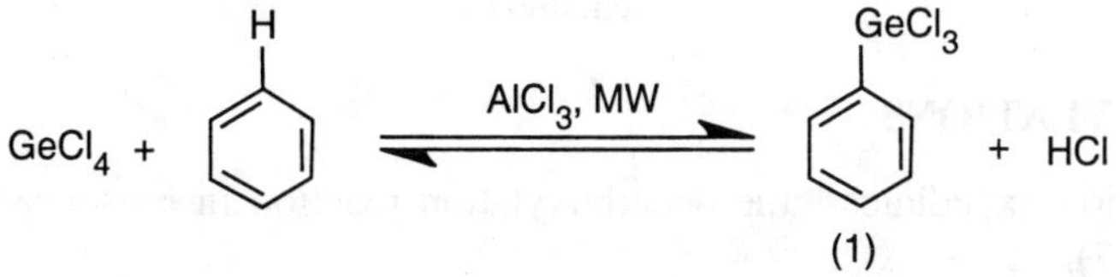

Scheme-14a

3.15 CATALYTIC HYDROGENATION

Benzaldehyde on reduction using RuHCl (CO) (PPh$_3$)$_3$ in presence of formic acid under microwave irradiation gave[27] benzyl alcohol in 7 min (conventional refluxing 3 hr) (Scheme-15).

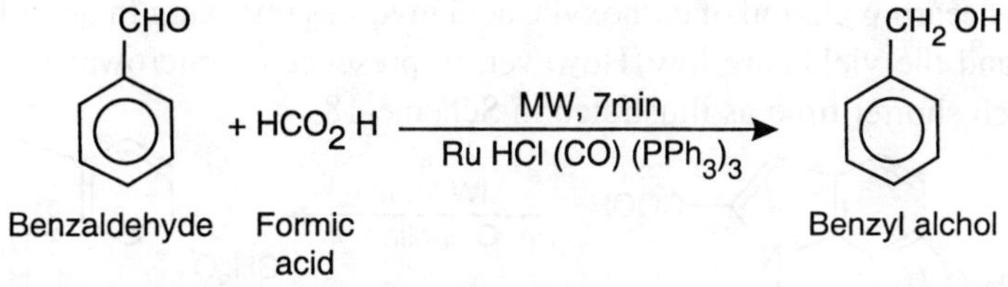

Scheme-15

Microwave-assisted hydrogenation of substituted β-lactams using Raneynickel or Pd/C have also been reported[28]. This reaction (Scheme-15a) is particularly useful on a large scale.

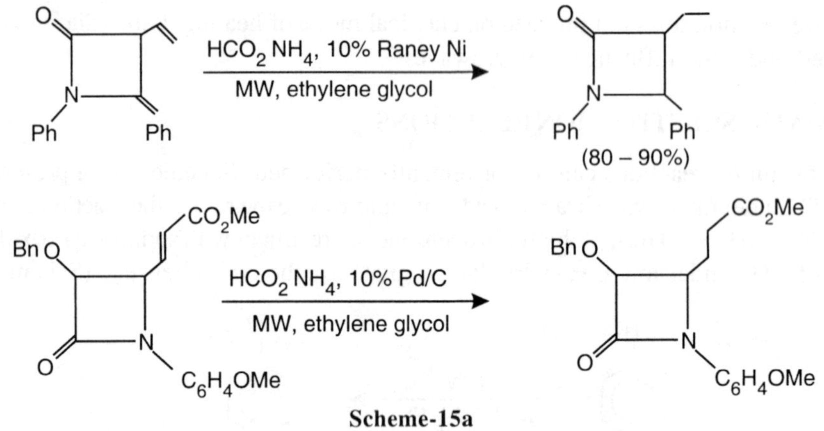

3.8 ■ *Alternate Energy Processes in Chemical Synthesis*

[Scheme showing β-lactam with vinyl and methylene groups, N-Ph, treated with HCO₂NH₄, 10% Raney Ni, MW, ethylene glycol → reduced β-lactam (80–90%)]

[Scheme showing β-lactam with BnO, CO₂Me vinyl substituent, N-C₆H₄OMe, treated with HCO₂NH₄, 10% Pd/C, MW, ethylene glycol → saturated product]

Scheme-15a

3.16 SYNTHESIS OF CHALCONES

Microwaves have been used for the synthesis[29] of chalkones and related enones. Considerable rate enhancement is observed bringing down the reaction time from hours to minutes in improved yield (Scheme-16).

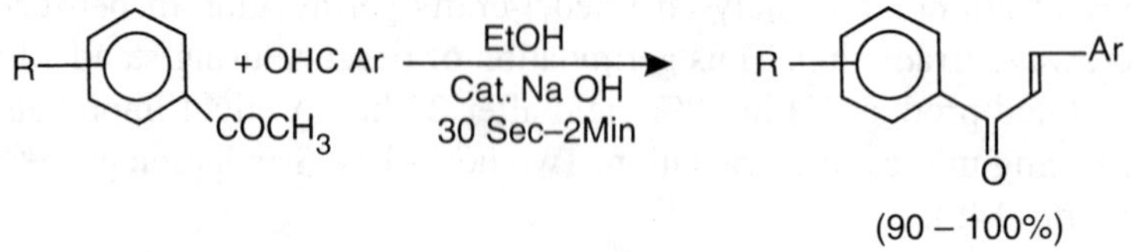

R–C₆H₄–COCH₃ + OHC–Ar $\xrightarrow[\text{30 Sec–2Min}]{\text{EtOH, Cat. NaOH}}$ R–C₆H₄–CH=CH–C(O)–Ar (90–100%)

Scheme-16

3.17 DECARBOXYLATIONS

Microwave irradiation expediates[30] the decarboxylation reaction in presence of a phase transfer catalyst (Scheme-17).

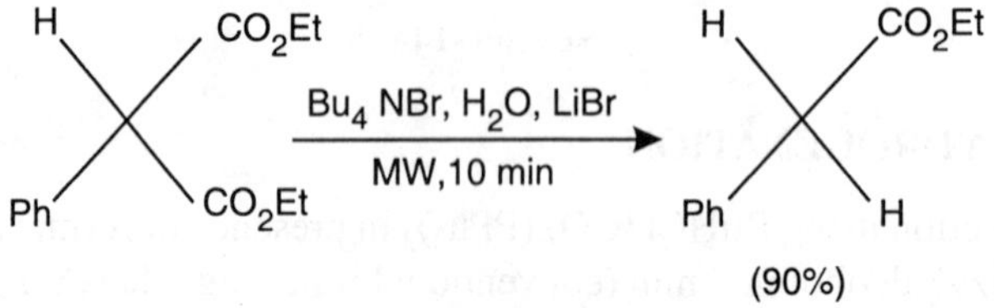

Ph(H)C(CO₂Et)₂ $\xrightarrow[\text{MW, 10 min}]{\text{Bu}_4\text{NBr, H}_2\text{O, LiBr}}$ PhCH₂(CO₂Et) (90%)

Scheme-17

Conventional decarboxylation of carboxylic acid involves refluxing in quinoline in presence of copper chromite and the yields are low. However, in presence of microwaves[31], decarboxylation takes place in much shorter time as illustrated in Scheme-18.

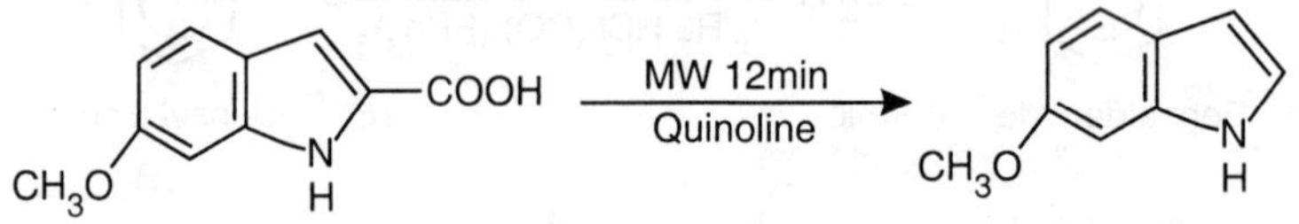

6-Methyl indole 2-Carboxylic acid (CH₃O-indole-2-COOH) $\xrightarrow[\text{Quinoline}]{\text{MW 12min}}$ 6-Methoxy indole

Scheme-18

3.18 C-ALKYLATION OF ACTIVE METHYLENE GROUP

Phase transfer catalysts promote MW assisted C-alkylation of active methylene compounds[32] (Scheme-19).

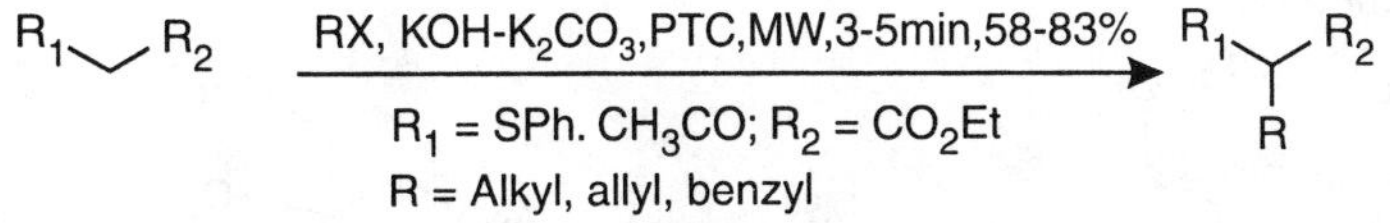

R_1 = SPh. CH_3CO; R_2 = CO_2Et
R = Alkyl, allyl, benzyl

Reagents: RX, KOH-K_2CO_3, PTC, MW, 3-5min, 58-83%

Scheme-19

There is good selectivity for monoalkylated product (Scheme-19). The reaction can be carried out in toluene to ensure efficient mixing of the reagents.

3.19 METHANOLYSIS OF OLIGOSACCHARIDES

Methanolysis of oligosaccharides by MV irradiation leads to products with anomeric inversion[33] (Scheme-20).

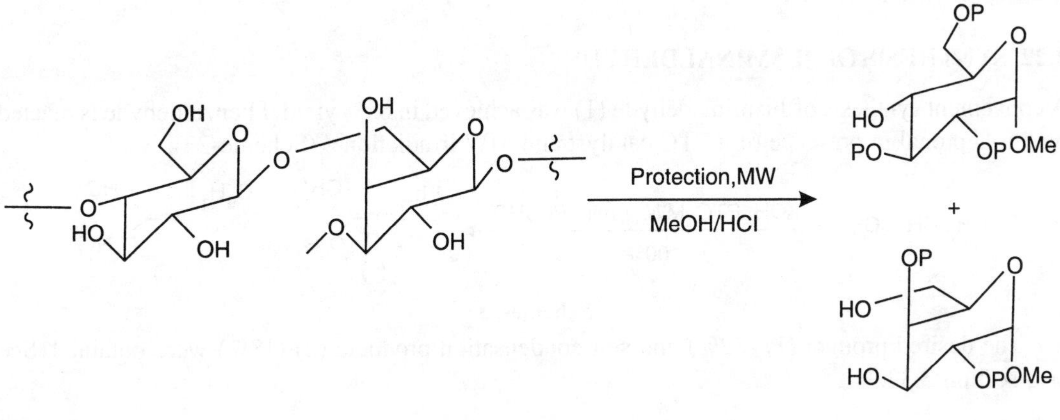

Scheme-20

Protection, MW / MeOH/HCl

3.20 PREPARATION OF UNSATURATED PYRANOSIDES

MW-irradiation of tosylate or mesylate derivatives of disaccharides leads to improved yield of unsaturated pyranosides[34] (Scheme-21) in comparison to classical heating.

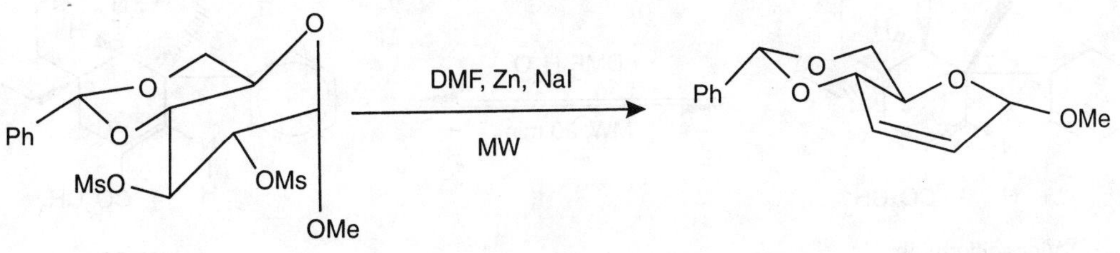

DMF, Zn, NaI / MW

Scheme-21

3.10 ■ Alternate Energy Processes in Chemical Synthesis

3.21 FERRIER REARRANGEMENT

The reaction of tri-O-acityl D-glucals with phenols in sealed vessels (with MW thermolysis) leads to the formation[35] of Ferrier rearrangement products in good yield (Scheme-22), compared to classical heating methods.

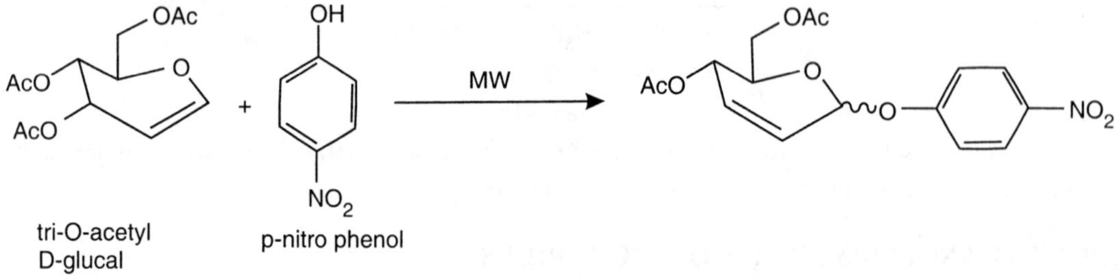

Scheme-22

Montmorillonite K10 is known to catalyse Ferrier rearrangement of tri-O-acetyl-D-glucal with different alcohols and phenols in open vessels with high α-selectivity and without the formation of d-deoxy-D-lyxohexapyranoside[36].

3.22 SYNTHESIS OF JUSMINALDEHYDE

A convenient synthesis of Jusminaldehyde (1) was achieved in 82% yield if benzaldehyde is reacted with n-heptonal in presence of a PTC catalyst and MW irradiation[37] (Scheme-23).

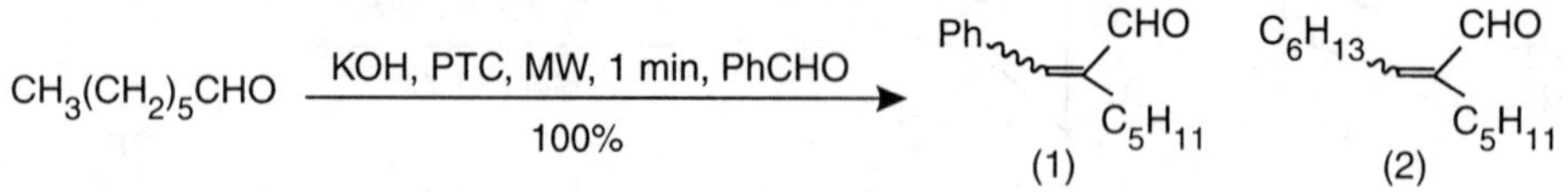

Scheme-23

The desired product (1) (82%) and selt condensation producte (2) (18%) were obtained. See also section 3.32.

3.23 RECEMIZATION OF (–)-VINCADIFFORMINE TO THE (+)-ISOMER

A complete recemization of (–)-vincadifformine has been achieved[38] under MW irradiation in DMF (Scheme-24). The reaction involves two consecutive Diels-Alder cycloreversion and cycloaddition steps. The formed (+) isomer is useful in the preparation of pharmacologically important alkaloid vicamine. Under classical condition, there is significant amount of decomposition.

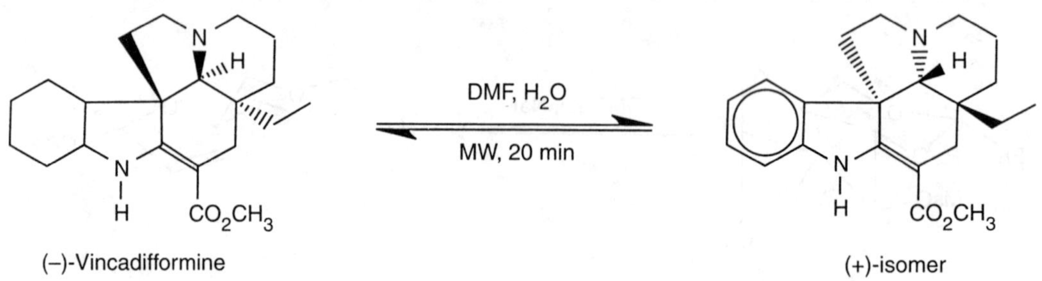

Scheme-24

3.24 SYNTHESIS OF 1, 2-DIMETHYL-3-HYDROXY-PYRID-4-ONE

The reaction of 3-hydroxy-2-methyl-4-pyrone with aquius methyl amine under MW irradiation condition gives[39] 1, 2-dimethyl-3-hydroxy-pyrid-4-one (65%) (Scheme-25). In contrast, under classical heating, the yield was only 50% after 6 hr.

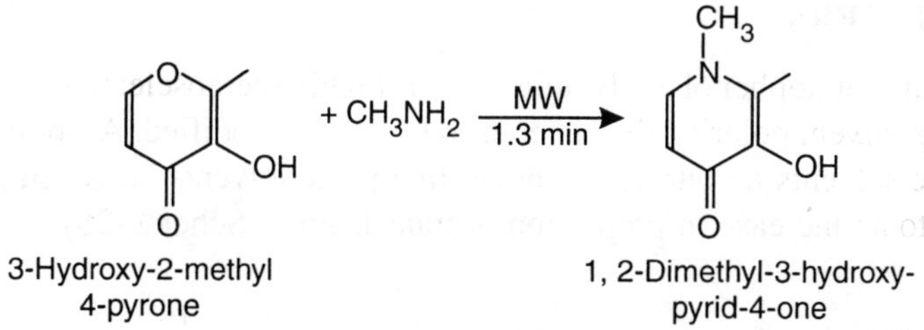

Scheme-25

3.25 SYNTHESIS OF ISOPROPYLIDENE GLYCEROL

A solution of glycerol in acetone in presence of p-toluenesulfonic acid catalyst on mw irradiation give[39] isopropylidene glycerol (Scheme-26).

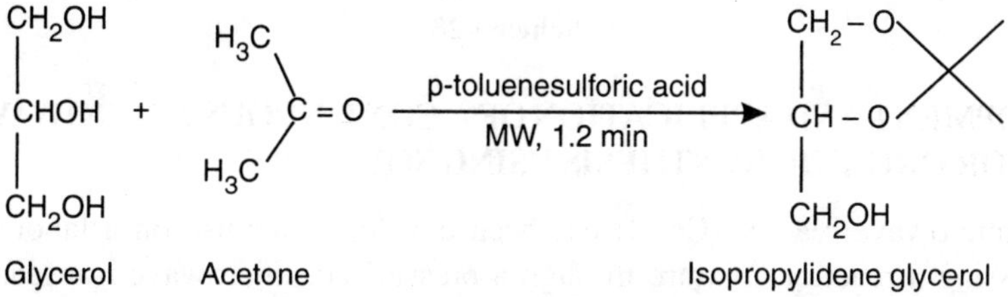

Scheme-26

3.26 SYNTHESIS OF ISOTOPICALLY LABELLED (^{11}C) DIETHYL OXALATE AND (^{11}C) OXALIC ACID

Diethyl oxalate and oxalic acid, both isotopically labelled (^{11}C) have been prepared[40] as given in Scheme-27.

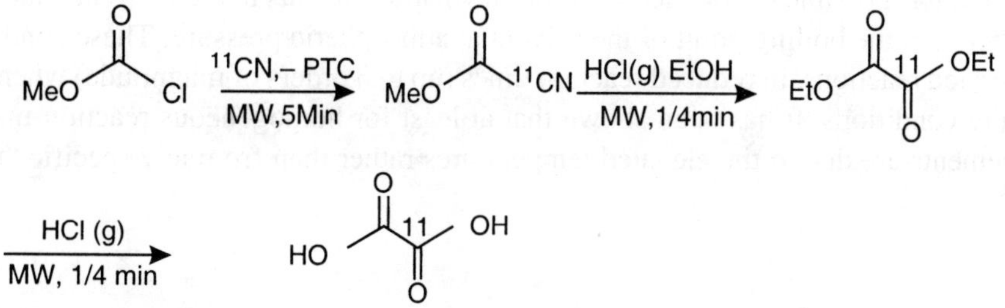

Scheme-27

An interesting application is the preparation of isotopically labelled drugs of short half life (^{11}C, t = 20 min, ^{122}I = 3.6 min and ^{18}F, $t_{1/2}$ = 100 min). The procedure is successful in terms of reduction in reaction time by a factor of 20 and in doubling the radioactivity to the final product[41].

3.27 STEREOSELECTIVE ADDITION OF 2-AMINOTHIOPHENOL TO GLYCIDIC ESTERS

The addition of 2-aminothiophenol to glycidic esters is highly seereoselective under MW irradiation conditions[42]. Using solvent polarity, the cis: trans ratio can be modified. Aprotic solvents favour cis isomer while protic solvents favour trans isomer. In a polar solvents however, gradual increase in power level leads to an increase in proportion of trans isomer (Scheme-28).

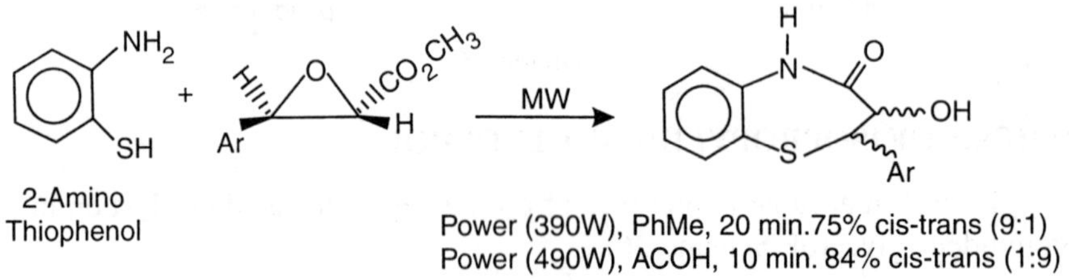

Power (390W), PhMe, 20 min. 75% cis-trans (9:1)
Power (490W), ACOH, 10 min. 84% cis-trans (1:9)

Scheme-28

3.28 DEVELOPMENT AND APPLICATION OF A CONTINUOUS MICROWAVE REACTOR (CMR) FOR ORGANIC SYNTHESIS USING SOLVENTS

A continuous microwave reactor (CMR) has been developed for use on a laboratory scale[43]. It operates by passing a reaction mixture through a pressurised, microwave-transparent coil that is held in microwave cavity[44]. It has been used to conduct organic synthesis routinely, rapidly and safely in a range of solvents, under pressure (up to 1400 Kpa) and at temperatures up to 200°C. A number of reactions have been carried out using CMR. These reactions included nucleophilic substitution, addition, esterification, transesterification, acetalisation, amidation, decarboxylation, and elimination. Some other reactions which could be performed by CMR include Michael addition, Hofmann degradation, Williamsons ether synthesis and the Mannich, Frinkestein, Baylis-Hillman and Knoevenagel reactions.

Table 1 gives a number of reactions which have been performed in the solvents and with the reactants listed. It is possible to conduct reactions in volatile solvents in the CMR at temperatures up to 100° higher than the boiling point of the solvent at atmospheric pressure. These conditions have given accelerated reactions, in reduced reaction times (up to 3 orders of magnitude) when compared with literature conditions. It has been shown that at least for homogeneous reaction-mixtures, the rate enhancements are due to the elevated temperatures rather than from any specific "microwave effect"[45, 46].

Some Examples in Solvents

Reactants (Nature of the reaction)	Temp. (°C)	Mean Press. (K pa)	Time min	Product	Yield %	Reference
1. HOAc + i – PrOH/H+ (esterification)	152-5	1200	1.3	i-Pr OAc	98	47
2. 2, 4, 6-Trimethylbenzoic acid + Me OH/H+ (esterification)	148-59	1250	4[a] × 1.3	methyl 2, 4, 6 trimethyl benzoate	83	48
3. 2, 4, 6-Trimethylbenzoic acid + i –PrOH/H+ (Esterification)	155-64	1300	4 × 1.6	isopropyl 2, 4, 6-trimetyl benzoate	81	45, 49
4. PhCOOMe + 5% Aq. NaOH (hydrolysis)	166-8	700	1.0	Ph COOH	100	50
5. Glycerol in Me_2CO/H+	132-5	1175	1.2	2, 2-dimethyl-1, 3-dioxalane-4-methanol	84	51
6. Ph COOEt + MeOH/H+ (transesterification)	145-50	1050	4 × 1.4	Ph COOMe	46	–
7. Paraformaldehyde in 3% aqueous HCl (De Polymerisation)	160-70	1100	1.2	aqueous HCHO	100	52
8. Benzophenone + $NH_2OH \cdot HCl$ in pyridine/EtOH (Oxime formation)	164	500	1.5	benzophenone oxime	93	50
9. Ph Me in aqueous $KMnO_4$ + K OH	180	1050	1.3	Ph COOH	41	50, 53
10. Bu Cl + Na OPh in Me OH (Williamson ether synthesis)	144-7	1000	1.5	Bu OPh	67	54
11. p-Chlorobenzaldehyde in EtOH/H+ (acetalization)	142-4	950	1.4	p-chlorobenzaldehyde diethyl acetal	59	55
12. Methyl acrylate in aqueous HCHO/DABCO (Michael addition)	158-64	950	1.5	Methyl 2-(hydroxymethyl) arcylate	30	56
13. Bu Cl in NaI/Me_2 CO (Finkelstein reaction)	150	1000	1.5	Bu I	36	57
14. s-BuCl in NaI/Me_2CO (Finkelstein reaction)	145	950	5 × 1.5	s BuI	5	47, 57
15. 3-Hydroxy-2-methoxy-4-pyrone is 25% aqueous $MeNH_2$	160-2	200	1,3	1, 2-dimethyl-3-hydroxy pyrid-4 one	65	58
16. Indole in Me NH_2/aqueous HCHO	160-70	700	1.2	Gramine	97	47

3.14 ■ Alternate Energy Processes in Chemical Synthesis

Reactants (Nature of the reaction)	Temp. (°C)	Mean Press. (K pa)	Time min	Product	Yield %	Reference
17. 2-Methylfuran in Me_2NH/ aqueous HCHO	164-5	350	1.4	5-methylfurfuryldi-methylamine	48	52
18. Citronellal + NH_2OH in pyridine + EtOH	164	500	1.5	Citronellaldoxime	82	50, 59
19. Citronellaldoxime + chloramine T in EtOH	128-33	800	1.5	3, 3, 6-Trimethyl--3, 3a, 4, 5, 6, 7 hexahydro-2, 1-benzisoxazole	78	60
20. Carvone + IM aqueous H_2SO_4 with emulsisier	165-75	740	4 × 1.5	Carvacrol	32	61
21. Furfural in diethyl malonate/pyridene (Knoevengel reaction)	165	1200	1.6	2-Furanacrylic acid	18	62
22. PhCOMe in Me_2NH/HCHO	180-90	400	1.0	Ph CO CH_2 CH_2 NMe_2	29	47
23. [Ph $COCH_2CH_2NMe_3$]+I- in H_2O (Hofmann degradation)	90-95	100	1.6	Ph COCH = CH_2	96	47
24. 2-Formyl phenoxyacetic acid in Ac_2O/HOAc with NaOAc	180	550	1.0	benzofuran	24	63
25. Ph CH_2 CH_2 Br in Na OMe/Me OH	95	730	1.7	Ph CH = CH_2 (75%) + Ph CH_2 CH_2 OMe (9%)	84	64, 65
26. Ph CH_2 CH_2 Br in Me OH	138-45	1100	60 × 1.8	Ph CH_2 CH_2 OMe	38	65

a The integer refers to the number of passes of the reaction mixture through the microwave zone.

3.29 PERICYCLIC REACTIONS

Some pericyclic reactions like Diels-Alder and Claisen rearrangements have already been discussed (see section 3.4 and 3.5). Some more pericyclic reactions are discussed in this section.

1, 4-Hexadiene condenses in presence of MW with demethyl acetylene dicarboxylate to give[9] the adduct 82% yield (Scheme-29).

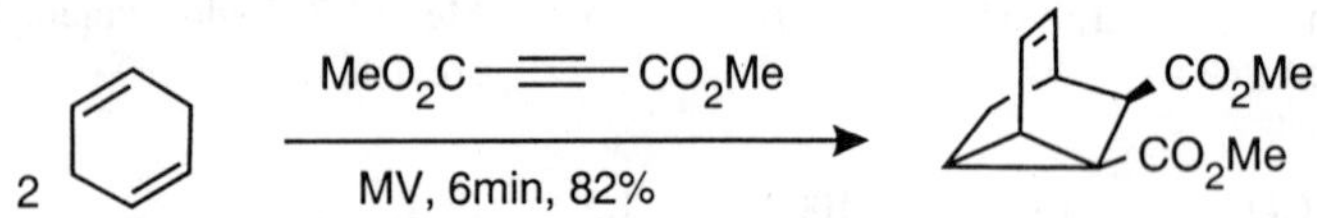

Scheme-29

Claisen rearrangement of the Ketene acetal derived from the alcohol A gives[66] 83% yield in 10 min as compared to 48 hr in a sealed tube (Scheme-30).

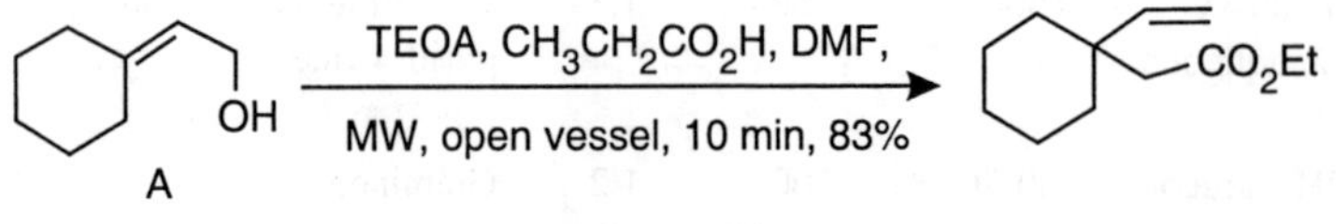

Scheme-30

Hetro-Diels-Alder reaction[67] involving cydoaddition of diene (1) with dienophile (2) and MW irradiation gave two products in good yield, whereas under conventional heating no product could be isolated after 4 hr at 140° (Scheme-31).

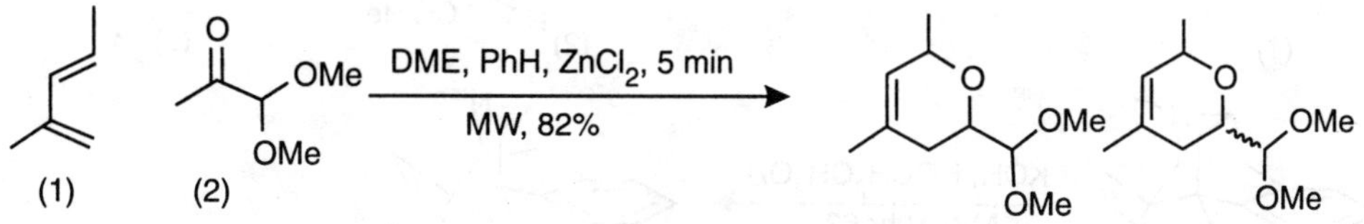

Scheme-31

Fisher Cyclisation[68] involving [3, 3] sigmatropic rearrangement of cyclohexanone phenyl hydrazone gave 100% yield of the cyclised product (Scheme-32).

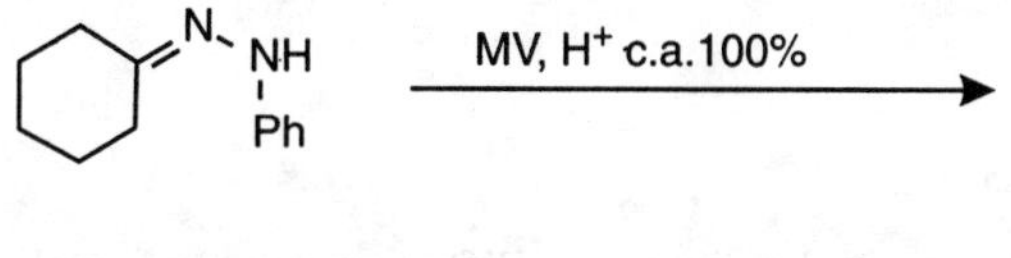

Scheme-32

The above reaction was carried out by irradiation of the hydrazone in formic acid in a Parr bomb.

Cydoreversion of (1) to secodine intermediate (2) furnishes[69] (3) via (4 + 2) cydoaddition. The reaction takes place less effeciently under conventional heating. The optimum conditions requires microwave irradiation in DMF for 20 minutes (Scheme-33).

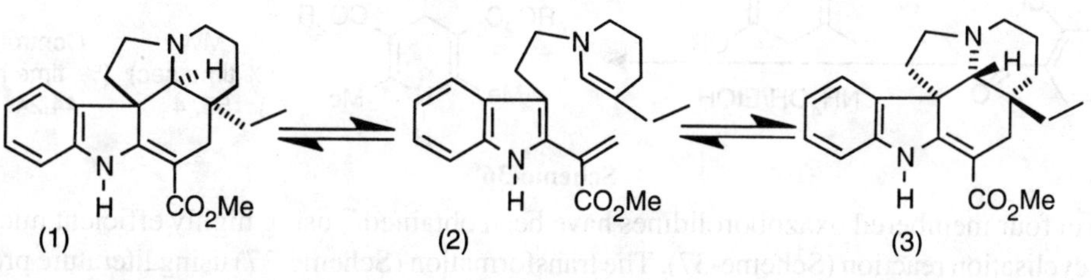

Scheme-33

Microwave assisted intramolecular Diels Alder Reaction[70] is a Key step in the synthesis of (+) Longifolene (Scheme-34).

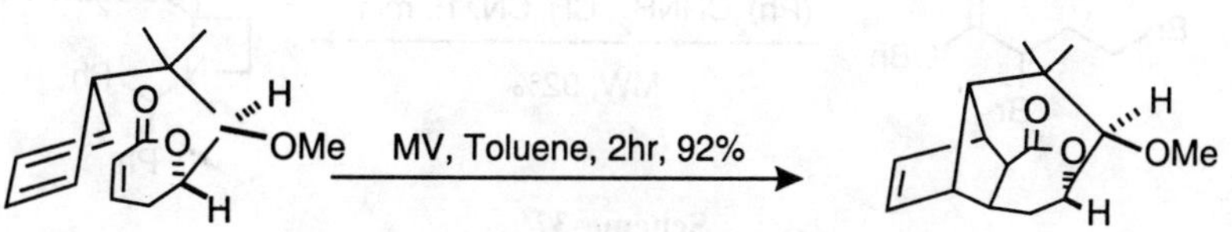

Scheme-34

Morphinan analogues have been prepared[71] by utilising microwave assisted reactions. Thus, the reaction of thebanine derivative (1) with methyl Vinyl Ketone is simplified and requires reduced reaction times, the products obtained are the cycloadducts (2) and (3). Subsequently, microwave assisted demethylation procedure yielded the target molecule (4) in reasonable yield (Scheme-35).

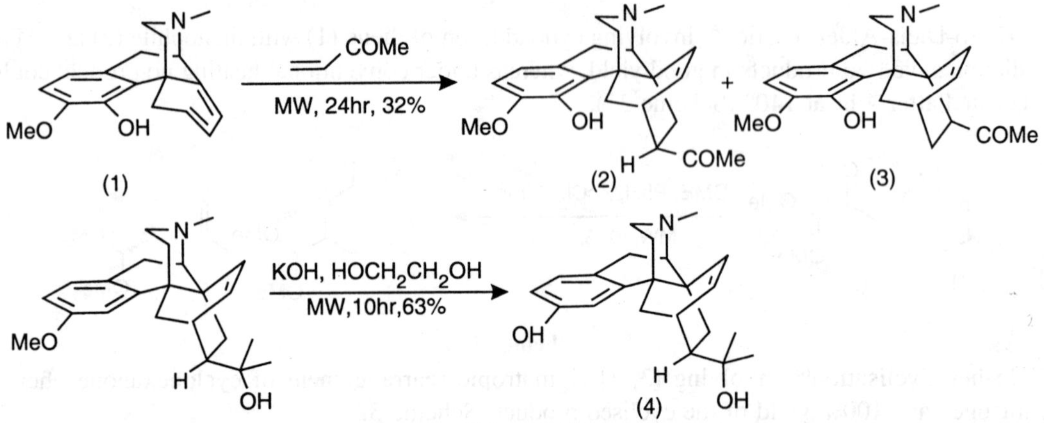

3.16 ■ Alternate Energy Processes in Chemical Synthesis

Scheme-35

See also section. 5.12.

3.30 CYCLISATION REACTIONS

Microwave assisted reactions have been used to improve different types of cyclisation reactions. Thus, in Hantzch-1, 4-dihydropyridine synthesis[72], reduced reaction times and improved yields are associated with this procedure as shown in Scheme-36 by the reaction of o-chlorobenzaldehyde and β-ketoesters.

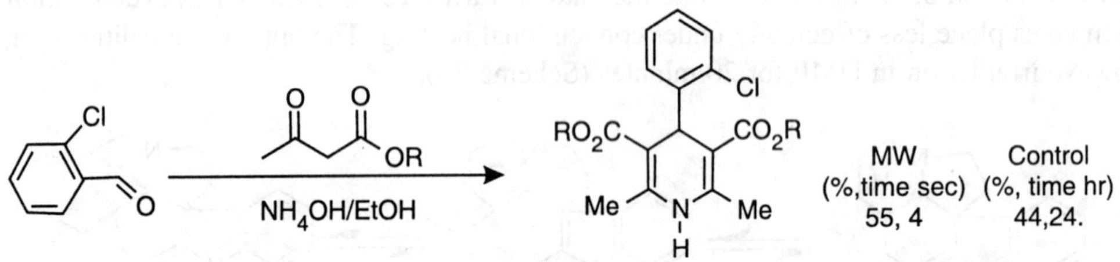

Scheme-36

Novel four membered oxazoborolidines have been obtained[73] using highly efficient microwave assisted cyclisation reaction (Scheme-37). The transformation (Scheme-37) using literature procedure required 24 hours refluxing (82%). However, microwave irradiation reduced the reaction time to 15 minutes with an improvement in yield.

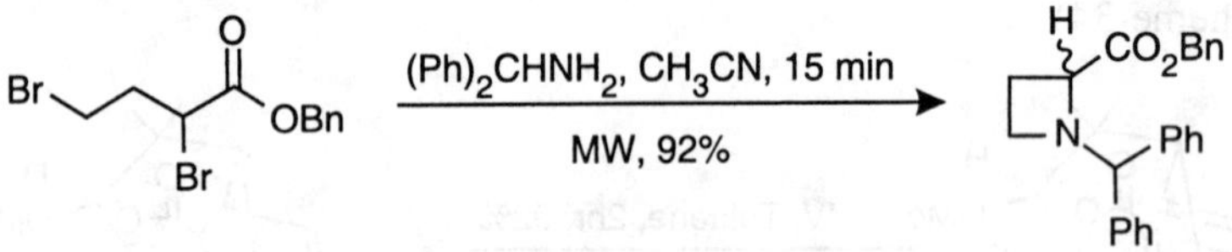

Scheme-37

Microwaves have been utilised in the synthesis of (Fe) substituted heteroaromatic systems[74]. Thus, the reaction of ferrocenyl substituted acrylaldehyde (1) with the appropriate ester (2) gave the heterocyclic compound (3) in good yield (Scheme-38).

[Scheme-38 reaction diagram]

Scheme-38

3.31 OXIDATION

Oxidation of toluene with KMnO$_4$ has also been achieved under microwave conditions[75] (Scheme-39). The reaction gives 40% yield after 5 minutes of microwave irradiation.

[Scheme-39 reaction: toluene → benzoic acid with [MnO4]⁻ MW, 40%]

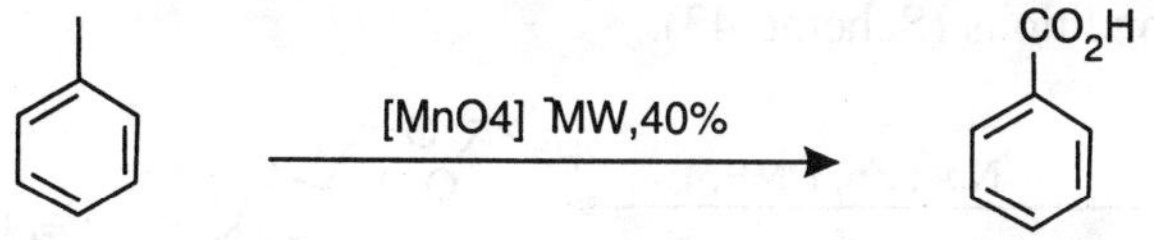

Scheme-39

3.32 SYNTHESIS OF ALKENES

Alkenes are obtained in a simple way by knoevenagel condensation of active methylenes with carbonyl groups as given below (Scheme-40).

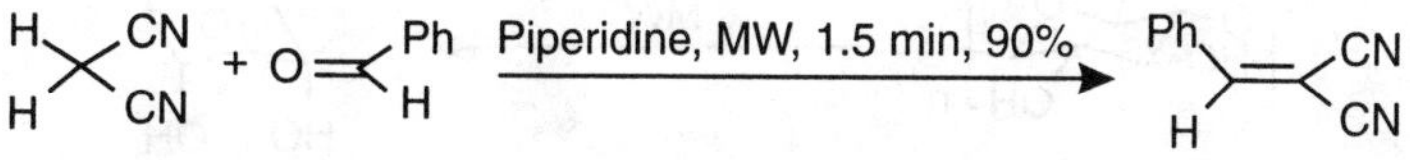

[Scheme-40 reaction: CH$_2$(CN)$_2$ + PhCHO → Ph-CH=C(CN)$_2$, Piperidine, MW, 1.5 min, 90%]

Scheme-40

Oxidation reactions are carried out in open vessels from which water is vapourised, thus, avoiding the requirement of a Dean-stark apparatus. The formed alkene is isolated and purified by washing with solvent or by short path distillation[76].

The synthesis of Jusminaldehyde, an alkene has already been reported (See Section 3.22).

3.33 PREPARATION OF FERROCENYL OXIME

In contrast to conventionally heated reactions, the microwave assisted reaction give[77] only the thermodynamically stable isomer (Scheme-41).

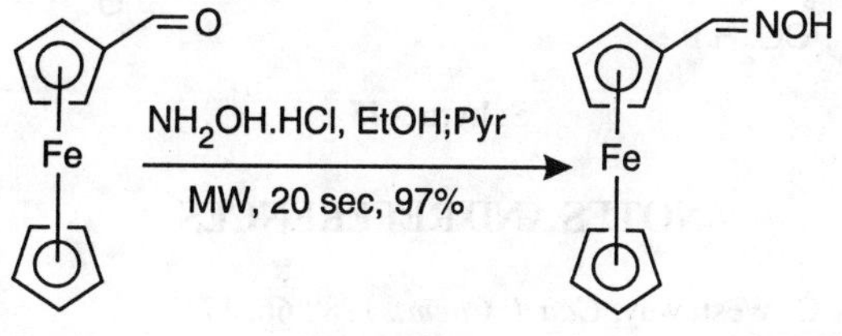

[Scheme-41 reaction: ferrocene carboxaldehyde + NH$_2$OH.HCl, EtOH; Pyr, MW, 20 sec, 97% → ferrocenyl oxime]

Scheme-41

3.34 SYNTHESIS OF ETHERS

The effect of microwave irradiation on phase transfer assisted ether synthesis was explored[78]. In the examples investigated the effect was useful, leading to much shorter reaction time and better yields. Thus, the reaction of benzyl chloride with ethanol signicificantly accelerated using microwaves (Scheme-42).

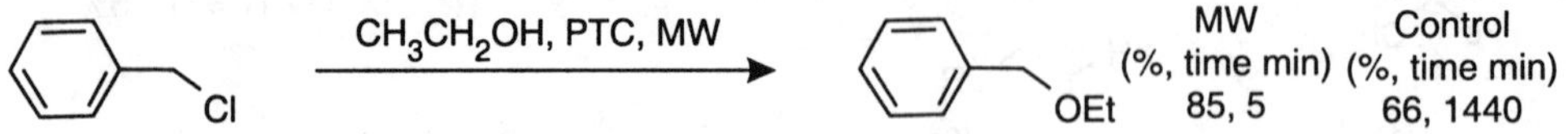
CH_3CH_2OH, PTC, MW
Cl
OEt
MW
(%, time min)
85, 5
Control
(%, time min)
66, 1440

Scheme-42

3.35 CARBOHYDRATES

Unsaturated pyranosides have been prepared by microwave assisted **Tipson-Cohen reaction**. Thus, irradiation of a mixture of mesylates or tosylates with sodium iodide and zinc dust in DMF with microwaves led in general to improved yields[79] and reduced reaction times as compared to more conventional heating conditions (Scheme-43).

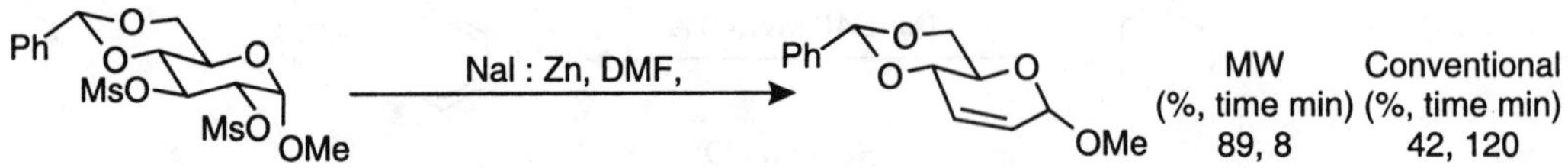
NaI : Zn, DMF,
MW
(%, time min)
89, 8
Conventional
(%, time min)
42, 120

Scheme-43

1, 6-Anhydroglucose has been prepared[80] from (1–4)-D-glucans by the microwave irradiation of starch (Scheme-44).

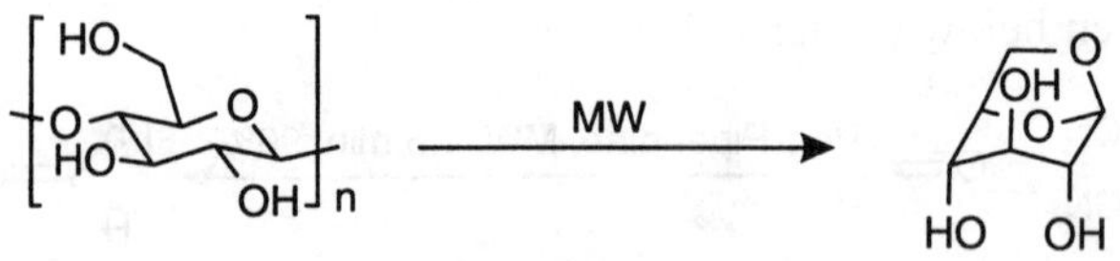
MW

Scheme-44

The simplicity of the procedure (Scheme-44) makes it an alternative method for the production of the desired product in small amounts, though the yields are low (C. 0.5 to 2%).

3.36 RADICAL REACTIONS

TBTH mediated reduction of β-lactams have been developed[81] as shown in Scheme-45.

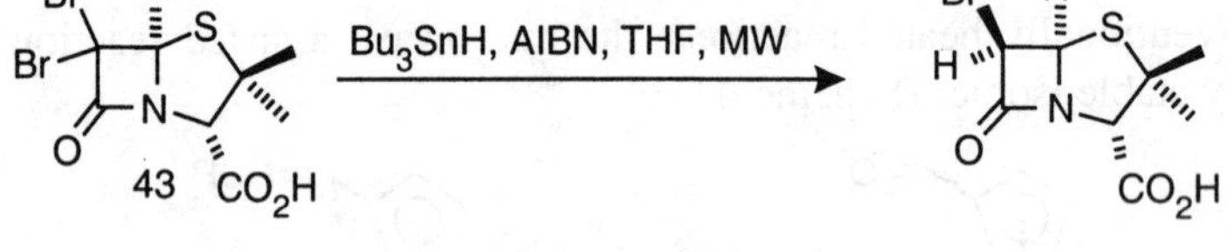
Bu_3SnH, AIBN, THF, MW
43

Scheme-45

NOTES AND REFERENCES

1. R.N. Gedye, F.E. Smith, K.C. Westaway, *Can J. Chem.*, 1988 **66**, 17.
2. R.N. Gedye, W. Rank, K.C. Westaway, *Can. J. Chem.*, 1991, **69**, 706.
3. R.N. Gedye, F. Smith, K. Westaway, L. Ali, L. Baldisera, J. Laberge, J. Rousell, *Tetrahedron Lett.*, 1986, **27**, 279.

(a) K.D. Raner and C.R. Strauss, *J. Org. Chem.*, 1992, **57,** 6231;
J.A. Vega, S. Cueto, A. Ramos, J.J. Vaquero, J.L. Garcia-Navio, J. Alvarez-Builla and J. Ezquerra, *Tetrahedon Lett.*, 1996, **37,** 6413.

4. J. Yulin, Y. Yuncheng, *Synthetic Commun.*, 1994, **24**(7), 105.
5. V. Sridar and V.S. Sundara Rao, *Indian J. Chem.*, 1994, **33B**, 184.
6. O.C. Dermer and J. King, *J. Am. Chem. Soc.*, 1941, **63**, 3232.
7. S.S. Bari, A.K. Bose, A.G. Chaudhary, M.S., Manhas, V.S. Raju and E.W. Robb, *J. Chem.*, Ed., 1992, **69**(11), 938.
8. R.J. Giguere, T.L. Bray and S.M. Duncan, *Tetrahedron Lett.*, 1986, **27**(41), 4945.
9. R.J. Giguere, A.M. Namen, B.D. Klopez, A. Arepally, D.E. Ramos, G. Majetich and I. Defauw, *Tetrahedron Lett.* 1987, **28**, 6553.
10. R.J. Giguere, A.M. Namen, O. Lopez, A. Arepally, D.E. Ramos, G. Majetich and J. Defauw, *Tetrahedron Lett.* 1987, **28**, 6553.

    (a) K.D. Raner, C.R. Strauss, R.W. Trainer and J.S. Thern, *J. Org. Chem.*, 1995, **60**, 2456.

11. B,-C. Hong, Y. -J. Shr, J.-H. Liao, *Org. Lett.*, 2002, **4**, 663.
12. S.A. Ayoubi, F. Texier-Boullet and J. Hamelin, *Synthesis*, 1994, 258.
13. M.K. Kundu, S.B. Mukherjee, N. Balu, R. Padmakumar and S.V. Bhal, *Synlett.*, 1994, 444.
14. A Srikrishna and S. Nagaraju, *J. Chem. Soc.* Perkin Trans. I, 1992, 311.
15. A Alajarin, J.J. Vaquero, J.L. Garcia Navio and J. Alverezbuilla, *Synlett*, 1992, 297.
16. C.O. Kappe, D. Kumar and R.S. Varma, *Synthesis*, 1999, 1799.
17. A Stadler and C.O. Kappe, *J. Comb. Chem.*, 2001, **3**, 624.
18. B.K. Banik, M.S. Manhas, Z. Kalruza, K.J. Barakat, and A.K. Bose, *Tetrahedron Lett.*, 1992, **33**, 3603.

    (a) A.K. Bose, B.K. Banik and M.S. Manhas, *Tetrahedron Lett.*, 1995, **36**, 213.

19. J. Borlon, P. Giboreau, S. Lateuvre and C. Merchand, *Tetrahedron Lett.*, 1991, **32**, 2363.
20. T.J. Mason and J.P. Lorimer, Sonochemistry, Theory, Applications and Uses of Ultrasound in Chemistry, Ellis Horwood, Chichester, 1988.
21. R.N. Gedye, W. Rank and K.C. Westaway, *Can J. Chem.*, 1991, **69**, 706.
22. S.D. Pollington, Y. Bond, R.B. Moyes, D.A. Whan, J.P. Candlin and J.R. Jennings, *J. Org. Chem.*, 1991, **56**, 1313.
23. (a) K.D. Raner and C.R. Strauss, *J. Org. Chem.*, 1992, **57**, 6231.

    (b) D. Constable, K. Raner, P. Somlo and C. Strauss, J. Microwave Power Electromagnetic Energy, 1992, **27**, 195.

24. K. Bougrin, A.K. Bennani, S.F. Tetouani and M. Soufiaoui, *Tetrahedron Lett.*, 1994, **35**, 8373.
25. A.C.S. Reddy, P.S. Rao and R.V. Venkataraman, *Tetrahedron Lett.*, 1996, **37**, 2843.
26. Y. Yuncheng, G. Dabin and J. Yulin, *Synth. Commun.*, 1992, **22**, 2117.

    (a) Y. Yuncheng, G. Dabin and J. Yulin, *Synth. Commun.*, 1992, **22**, 2117.

27. E.M. Gordan, D.C. Gaba, K.A. Jebber and D.M. Zackarias *Organometallics*, 1993, **12**, 5020.
28. A.K. Bose, B.K. Banik, K.J. Barakat and M.S. Manhas, *Synlett.*, 1993, 575.
29. G. Opitz and E. Tempel, *Liebigs Ann. Chem.*, 1966, 699, 68.
30. A. Loupy, P. Pigeon, M. Ramdani and P. Jacquault, *J. Chem., Res.*(S), 1993, 36.

31. G.B. Jones and B.J. Chapman, *J. Org. Chem.*, 1993, **58**, 5558.
32. D. Runhua, W. Yuliong and J. Yaozhong. *Synth., Commun.*, 1994, **24**, 111 and 1917.
33. M. Chang, H.V. Meyers, K. Nakanishi, M. Ojika, J.H. Park, M.H. Park, R. Takeda, J.T. Vazquez and W.T. Wiesler, *Pune Appl. Chem.*, 1989, **61**, 1193.
34. L.H.B. Baptistella, A.Z. Onaga and E.A.M. Godoj, *Tetrahedron Lett.*, 1993, **34**, 8407.
35. S. Sowmya and K.K. Balasubramanian, *Synth. Commun.*, 1994, **24**, 2097.
36. B. Shanmugasundaram, A.K. Bose and K.K. Balasubramanian, *Tetrahedron Lett.*, 2002, **43**, 6797.
37. D. Abenhaim, C.P. Nagoc Son, A. Loupy and N. Ba Hiep, *Synth. Commun.*, 1994, **24**, 1199.
38. S. Takano, T. Kijima, S. Satoh and K. Ogasawara, *Chem. Lett.*, 1989, 87.
39. T. Cablewski, A.F. Faux and R. Strauss, *J. Org. Chem.*, 1994, **59**, 3408.
40. J.O. Thorell, S.Stone-Elander and N.Elander, *J. Labelled Compd. Radiopharm*, 1993, **33**, 995.
41. D.R. Huang, S.M. Moerlein, L. Lang and M.J. Welch, *J. Chem. Soc. Chem. Commun.*, 1987, 1799.
42. J.A. Vega,. S. Cueto, A. Ramos, J.J. Vaquero, J.L. Garcia-Navio, J. Alvarez-Builla and J. J. Ezquerra, *Tetrahedron Lett.*, 1996, **37**, 6413.
43. T. Cablewski, A.F. Faux and C.R. Strauss, *J. Org. Chem.*, 1994, **59**, 3408-3412.
44. C.R. Strauss, A.F. Faux, International Patent Application PCT/AU 89/00437, 1989.; CMR was presented in part at the RACI Organic Division, National Conference, Tounsville, 1989, and at ACS Pacifichem '89 Conference, Honolulu, 1980.
45. K.D. Raner and C.R. Strauss, *J. Org. Chem.*, 1992, **57**, 6231.
46. K.D. Raner, C.R. Strauss, F. Vyskoc and I. Mokbel, *J. Org. Chem.*, 1993, **58**, 950.
47. B.S. Furnis, A.J. Hannaford, V. Rogers, F.W. G. Smith and A.R. Tatchell, *Vogel's Textbook of Practical Organic Chemistry*, 4th Ed., Longman, New York 1978.
48. M.S. Newman, *J. Am. Chem. Soc.*, 1941, **63**, 2431.
49. D. Constable, K. Raner, P. Somlo, and C. Strauss, Journal of Microwave Power and Eletrcomagnetic Energy, 1992, **26**, 195.
50. R.N. Gedye, F.E. Smith and K.C. Westaway, *Can. J. Chem.*, 1988, **66**, 17.
51. M. Ranoll, M.S. Newman, Organic Synthesis; E.C. Horning (Ed.); John Wiley and Sons, New York, Collect Vol. III, p. 502.
52. Reagents Chemicals American Chemical Society Specification 5th ed; American Chemical Society, Washington DC, 1974, p. 274.
53. R. Gedge, F. Smith, K. Westaway, H. Ali, L. Baldisera, L. Laberge and J. Rousell, *Tetrahedron Lett.*, 1986, **27**, 279.
54. S.T. Chen, S.-H. Chiou and K.T. Wang, *J. Chem. Commun.*, 1990, 807.
55. J.M. Sayer, and W.P. Jencks, *J. Am Chem. Soc.*, 1977, **99**, 465.
56. S.H. Kusefoglu, A.O. Kress and L.J. Mathias, Macromolecules, 1987, **20**, 2326.
57. W.K.R. Musgrave, Rodds Chemistry of Carbon Compounds, S. Coffeg, (Ed.); Elsevier Amsterdam, 1964, Vol 1A, p. 482.
58. G.J. Kontoghiorges and I. Sheppard, Inorg. *Chim. Acta*, 1987, 136, L11.
59. D. Agigoni and O. Jegar, *Helv. Chim. Add*, 1954, **37**, 881.
60. A. Hasaner and K.L.M. Rai, *Synthesis*, 1989, 57.
61. A. Sattar, R. Ahmad and S.A. Khan, *Pak. J. Sci. Res.*, 1980, **23**, 177.

62. S. Rajgopalan, and P.V. Raman, *Organic Synthesis, E. Horning* (Ed.), John Willy, New York, 1955; Collect Vol. III, p. 425.
63. A.W. Burgatahler and L.R. Worden, Organic synthesis, H.E. Baumgarian (Ed.), Organic synthesis, H.E. Baumgarian, (Ed.), John Wiley, New York, 1973; collect. Vol. V, p. 251.
64. Sh Mamedov, D.N. Zh, Obsheh Khim, 1962, **32**, 1427; *Chem. Abstr.*, 1963, **58**, 4453e.
65. V.K. Bhalerao, B.S. Nanjundiah, H.R. Sonawana and P.M. Nair, *Tetrahedron*, 1986, **42**, 1987.
66. A. Srikrishna and S. Nagaraju, *J. Chem. Soc. Perkin Trans I*, 1992, 311.
67. A. Stamboule, M. Chastrette and M. Soufiaoui, *Tetrahedron, Lett.*, 1991, 32, 1723.
68. R.A. Abramovitch and A. Bulman, *Synlett*, 1992, 795.
69. S. Takano, A. Kijima, T. Sugihara, S. Sathoh and K. Ogasawara, *Chem. Lett.,* 1989, 87.
70. B. Lei, A.G. Fallis, *J. Am. Chem. Soc.*, 1990, **112**, 4609; B. Lei, A.G. Fallis, *J. Org. Chem.*, 1993, **58,** 2186.
71. J.T.M. Linders, J.P. Kokje, M. Overand, T.S. Lie, and L. Maat, Rec., T*rav. Chim. Pays-Bas*, 1988, **107**, 449.
72. R. Alajarin, J.J. Vaquero, J.L. Garefa Novfo and J. Alvarez-Builla, *Synlett*, 1992, 297.
73. A.V. Rama Rao, M.K. Gurjar and V. Kaiwar, *Tetrahedron Assmm.*, 1992, **3**, 859.
74. M. Puciova, P. Ertl and S. Toma, Collect. *Czech. Chem. Commun.*, 1994, **59**, 175.
75. R. Gedye, F. Smith, K. Westaway, A. Humera, L. Baldisera and L.R. Laberge, *Tetrahedran Let*t., 1986, **26**, 279.
76. S.A. Ayoubi, F. Texier-Boullet and J. Hamelin, *Synthesis*, 1994, 258.
77. M. Puciova and S. Toma, Collect. *Czech. Chem. Commun.*, 1992, **57**, 2407.
78. G.B. Jones, and B.J. Chapman, *J. Org. Chem.*, 1993, **58**, 5558.
79. L.H.B. Baptistella, A.Z. Neto, H. Onaga and E.A.M. Godoi, *Tetrahedron Lett.*, 1993, **34**, 8407.
80. A.J.J. Sraahof, H. Van Bekkum, A.P.G. Kieboom *Rect. Trav. Chem. Pays-Ba*s, 1988, **107**, 647; M. Ubukuta, R. Kimura, H. Isono, C.C. Nelson, J.M. Gregson and J.A. MeCloskey, *J. Org. Chem.*, 1992, **57**, 6392.
81. A.K. Bose, M.S. Manhas, M. Ghosh, M. Shah, V.S. Raju, S.S. Bari, S.N.Newaz, B.K. Banik, A.G. Chaudhary and K.J. Karakat, *J. Org. Chem.*, 1991, **56**, 6968.

4

MICROWAVE ASSISTED REACTIONS IN SOLID STATE

Application of microwave irradiation in organic reactions has added a new dimension to solid phase synthesis. Using this technique, it is now possible to carry out reactions without the use of toxic or other solvents, which is one of the main problems associated with green synthesis and the pollution of the environment. In these reactions, the reactants are dissolved in a suitable solvent like water, alcohol, methylene chloride etc., and the solution stirred with a suitable adsorbent or solid support like silica gel, alumina or phyllosilicate (M^{n+} – montmorillonite). After stirring the solvent is removed in vacuo and the dried solid support on which the reactants have been adsorbed are used for carrying out the reaction under microwave irradiation.

Most of the reactions in solid state are performed in open glass containers (test tubes, beakers and round bottomed flasks) using neat reactants under solvent free conditions in an unmodified household MW oven. Some of the supporting reagents, vig. clay supported iron (III) nitrate (Clayfen) and copper (II) nitrate (Claytop) are prepared by literature procedures[1].

Some of the important applications of solid support synthesis are given below:

4.1 PROTECTION AND DEPROTECTION REACTIONS

Protection and deprotection are important steps for the preparation of monomer building blocks, fine chemicals and synthesis for pharmaceuticals. These reactions often involve the use of acidic, basic or hazardous reagents and toxic metal salts[2]. The solid support technique (MW-accelerated) provides an alternative to the conventional reactions.

4.1.1 Formation of Acetals and Diaxolanes

The acetals of 1-galacto-1, 4-lactone could be prepared[3] in excellent yields by absorbing the lactone and the aldehyde on montmorillonite K-10 or KSF clay followed by heating the reaction mixture in a MW oven. (Scheme-1).

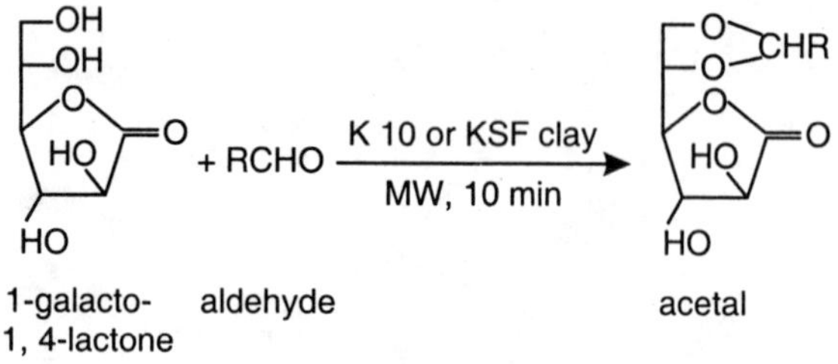

4.2 ■ Alternate Energy Processes in Chemical Synthesis

Scheme-1

1-galacto-1,4-lactone + RCHO → acetal (K 10 or KSF clay, MW, 10 min)

In a similar way, thioacetals are prepared from active methylene compounds that are adsorbed an KF-alumina, admixed with methanesulphonothioate[4] (Scheme-2).

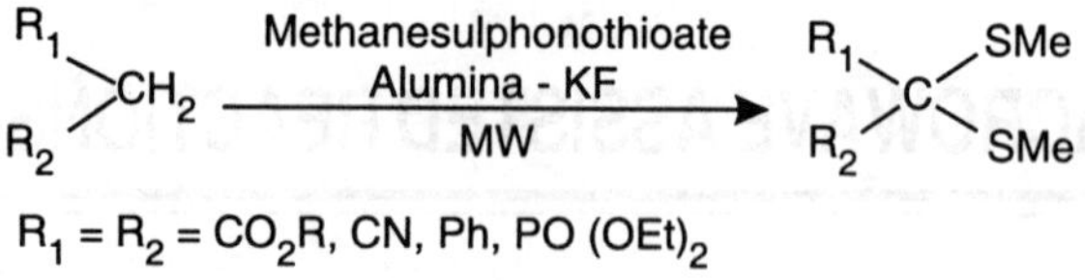

$R_1R_2CH_2$ → $R_1R_2C(SMe)_2$ (Methanesulphonothioate, Alumina - KF, MW)

$R_1 = R_2 = CO_2R, CN, Ph, PO(OEt)_2$

Scheme-2

Alkylation of reactive methylene group can be conveniently achieved in a microwave oven[5,6] using tetrabutyl ammonium Chloride (TBAC) as PTC without solvent (Scheme-3).

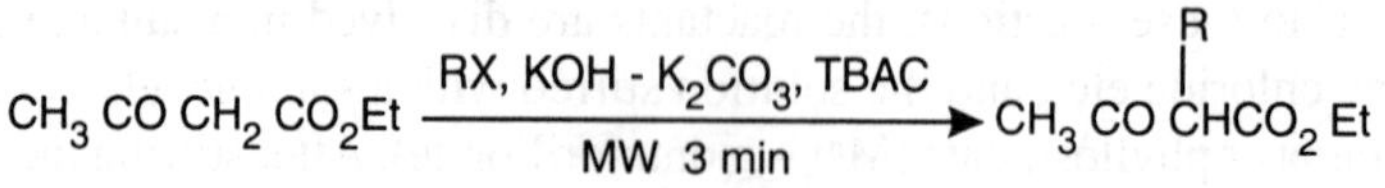

$CH_3COCH_2CO_2Et$ →(RX, KOH - K_2CO_3, TBAC, MW 3 min)→ $CH_3COCHRCO_2Et$

Scheme-3

In a similar way[7] alkylation of ethyl mercapto acetate can be achieved (Scheme-4).

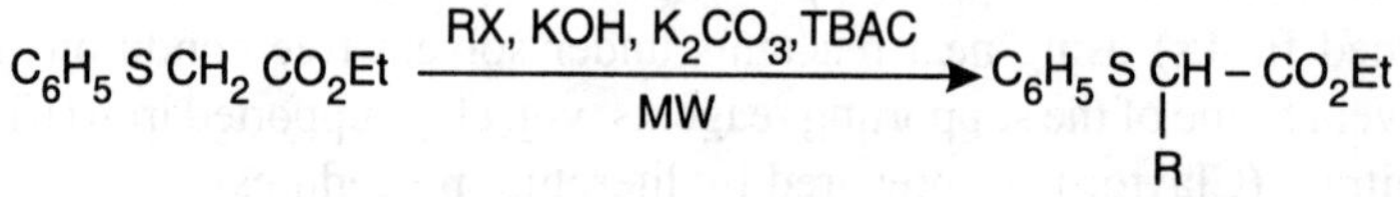

$C_6H_5SCH_2CO_2Et$ →(RX, KOH, K_2CO_3, TBAC, MW)→ $C_6H_5SCHRCO_2Et$

Scheme-4

Aldehydes and ketones can be protected as acetals and dioxolanes using orthoformates, 1,2,-ethylene dithiol or 2,2-dimethyl-1,3-dioxolane[7]. This acid catalysed reaction proceeds in presence of p-toluenesulfonic acid (PTSA) or KSF clay under solvent-free conditions (Scheme-5). The yields obtained are better than that obtained using conventional heating mode (oil bath).

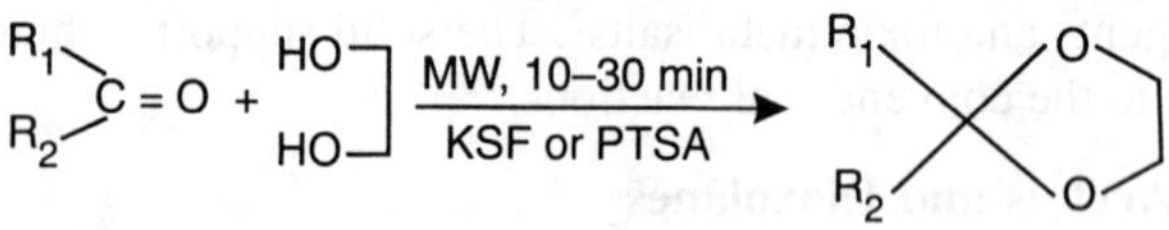

$R_1R_2C=O$ + HO-/-OH →(MW, 10–30 min, KSF or PTSA)→ dioxolane

Scheme-5

4.1.2 N-Alkylation Reactions

A number of N-alkylation reactions have been reported in solvent free conditions using phase transfer catalysts such as tetrabutylammonium bromide (TBAB) under microwave irradiation conditions. Using this procedures, N-alkylation of phthalimides[8] (Scheme-6) or its polatassium salt[9] (Scheme-7), carbazole[10] (Scheme-8) azaheterocycles[11] (Scheme-9) including pyriolidino[60] fullerenes[12] (Scheme-10) have been carried out using K_2CO_3/KOH and TBAB.

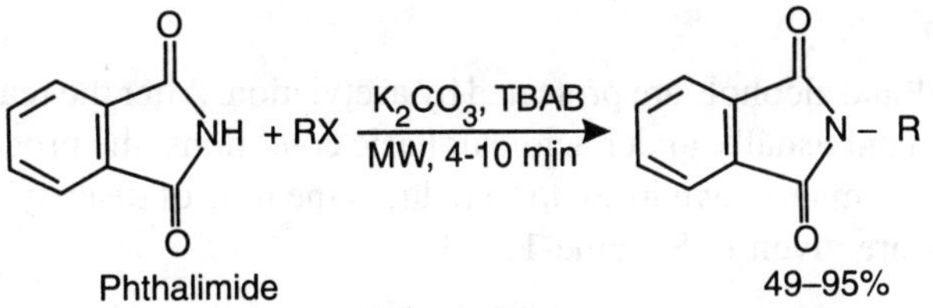

Scheme-6

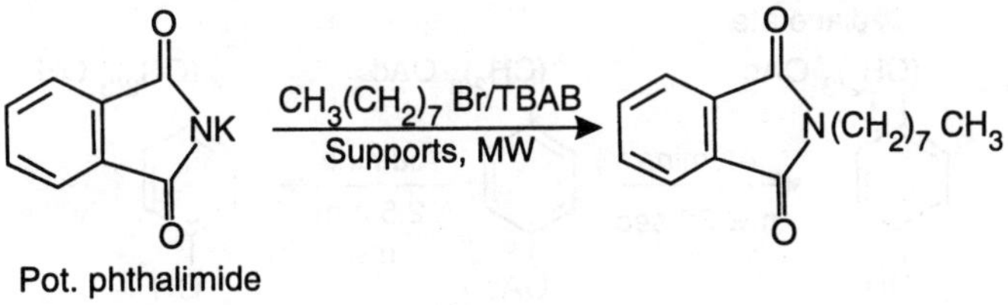

Phthalimide + RX → (N-R phthalimide), K₂CO₃, TBAB, MW, 4-10 min, 49–95%

Scheme-7

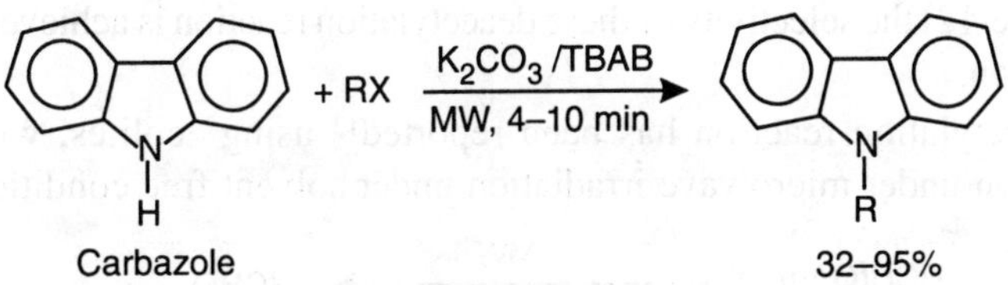

Pot. phthalimide + CH$_3$(CH$_2$)$_7$Br/TBAB, Supports, MW → N-(CH$_2$)$_7$CH$_3$ phthalimide

Scheme-8

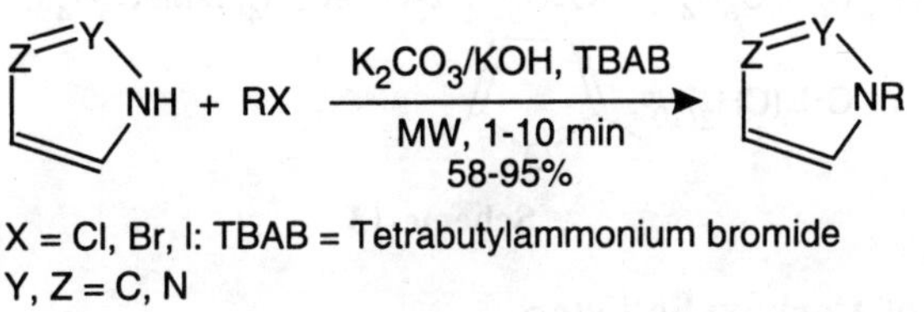

Carbazole + RX → N-R carbazole, K$_2$CO$_3$/TBAB, MW, 4–10 min, 32–95%

Scheme-9

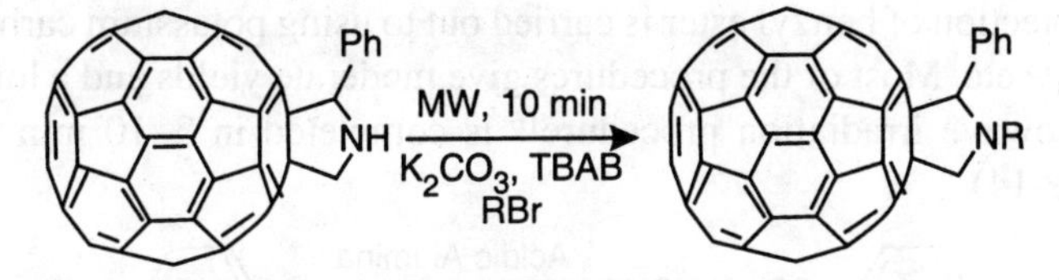

Five-membered heterocycle-NH + RX → heterocycle-NR, K$_2$CO$_3$/KOH, TBAB, MW, 1-10 min, 58-95%

X = Cl, Br, I; TBAB = Tetrabutylammonium bromide
Y, Z = C, N

Scheme-10

C$_{60}$-fullerene adduct (Ph, NH) + RBr, MW, 10 min, K$_2$CO$_3$, TBAB → N-R adduct

N-Alkylation also proceed smoothly under microwave condition. Thus, irradiation of a mixture of saccharin and alkyl halide give the N-alkylated product[12a] in good yield (Scheme-11).

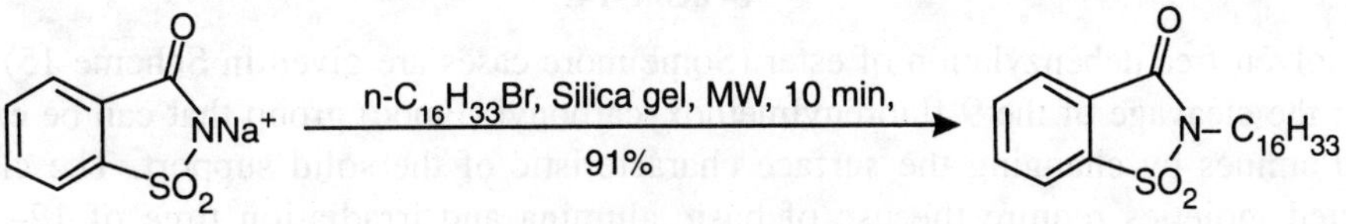

Scheme-11

Saccharin-NNa⁺ + n-C$_{16}$H$_{33}$Br, Silica gel, MW, 10 min → N-C$_{16}$H$_{33}$ saccharin, 91%

4.1.3 Deacetylation

Aldehydes[11], phenols[10] and alcohols are protected by acetylation. After the reaction, the deacetylation of the acetate is carried out usually under acidic or basic conditions; the process takes long time and the yields are low. Use of microwave irradiation reduces the time of deacetylation and the yields are good. Some examples are given in Scheme-12.

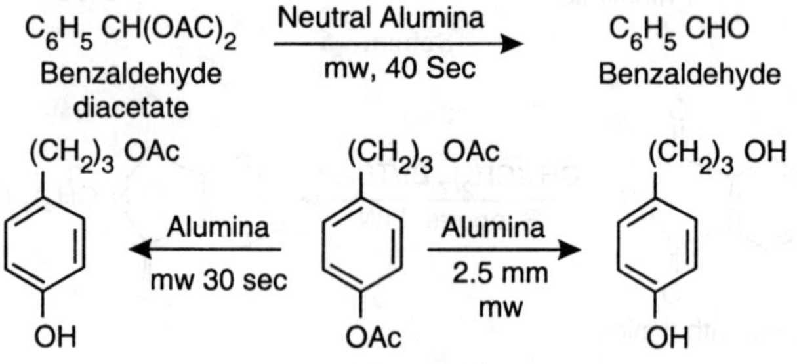

$$C_6H_5 CH(OAc)_2 \xrightarrow[\text{mw, 40 Sec}]{\text{Neutral Alumina}} C_6H_5 CHO$$
Benzaldehyde diacetate → Benzaldehyde

$$\text{p-HOC}_6H_4(CH_2)_3 OAc \xleftarrow[\text{mw 30 sec}]{\text{Alumina}} \text{p-AcOC}_6H_4(CH_2)_3 OAc \xrightarrow[\text{2.5 mm mw}]{\text{Alumina}} \text{p-HOC}_6H_4(CH_2)_3 OH$$

Scheme-12

As seen (scheme-12) the selectivity of these deacetylation reaction is achieved by simply adjusting the time of irradiation.

A similar deacetylation reaction has been reported[13] using zeolites, wherein 1, 1-diacetate undergo deprotection under microwave irradiation under solvent free conditions (Scheme-13).

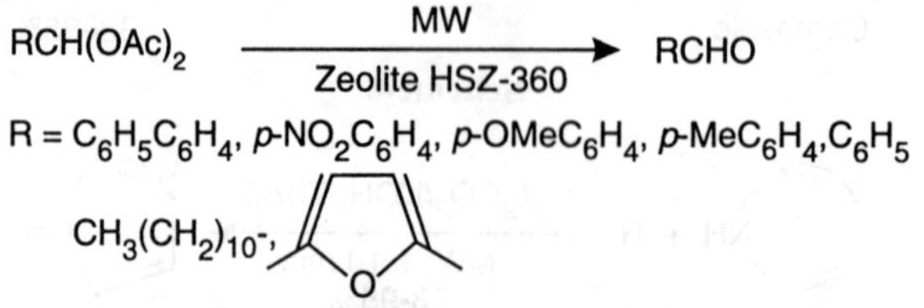

$$RCH(OAc)_2 \xrightarrow[\text{Zeolite HSZ-360}]{\text{MW}} RCHO$$

$R = C_6H_5C_6H_4, p\text{-}NO_2C_6H_4, p\text{-}OMeC_6H_4, p\text{-}MeC_6H_4, C_6H_5$

$CH_3(CH_2)_{10}$-furyl

Scheme-13

4.1.4 Debenzylation of Carboxylic Esters

The carboxylic function is generally protected by the benzyl protecting group. After the reaction sequence, the deprotection of benzyl ester is carried out to using potassium carbonate[13], Aluminium chloride[14], Na-NH$_3$[15] etc. Most of the procedures give moderate yields and a longer reaction time is required. The microwave irradiation procedure[16] is completed in 3–10 min and yields are high (89–92%) (Scheme-14).

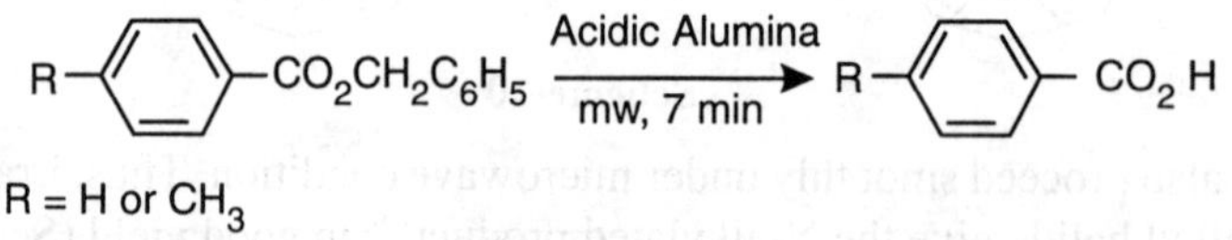

$$R\text{-}C_6H_4\text{-}CO_2CH_2C_6H_5 \xrightarrow[\text{mw, 7 min}]{\text{Acidic Alumina}} R\text{-}C_6H_4\text{-}CO_2H$$

R = H or CH$_3$

Scheme-14

The solven-free debenzylation of ester (Some more cases are given in Scheme-15) paves the way[17] for the cleavage of the 9-fluorenylmethoxycarbonyl (Fmoc) group that can be extended to protected amines by changing the surface characteristic of the solid support. The cleavage of N-protected moieties require the use of basic alumina and irradiation time of 12–13 min at ≈130-140°C.

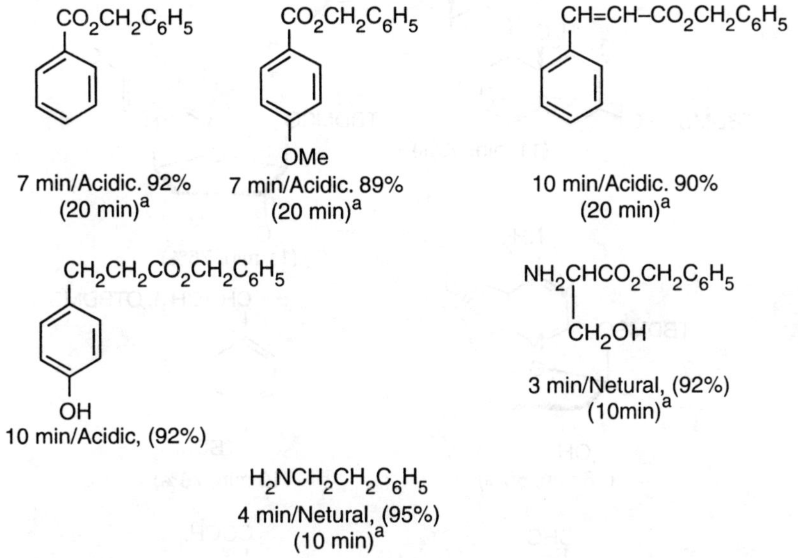

Scheme-15

(Time in parentheses refer to deprotection is an oilbath at the same temperature).

This approach finds application in peptide bond formation that may eliminate the use of the irritating and corrosive chemicals with Trifluoroacetic acid (TFA) and piperidine, as has been demonstrated for the deprotection of N-boc group (Scheme-16).

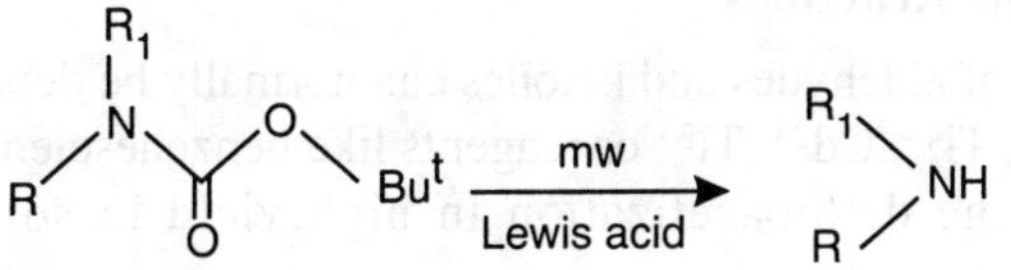

Scheme-16

4.1.5 Selective Cleavage of N-tert-Butoxycarbonyl Group

The cleavage of the N-tert-butoxycarbonyl (Nboc) group can be readily achieved in presence of aluminium chloride 'doped' natural alumina upon exposure to microwave irradiation[18] (Scheme-16).

4.1.6 Desilylation Reactions

Tert-Butyldimethylsilyl (TBDMS) ether derivatives of different type of alcohols can be deprotected to regenerate the corresponding hydroxy compounds on alumina surface under MW irradiation conditions[19] (Scheme-17). This approach circumvents the use of corrosive fluoride ions that are normally employed for cleaving of such silyl protecting groups.

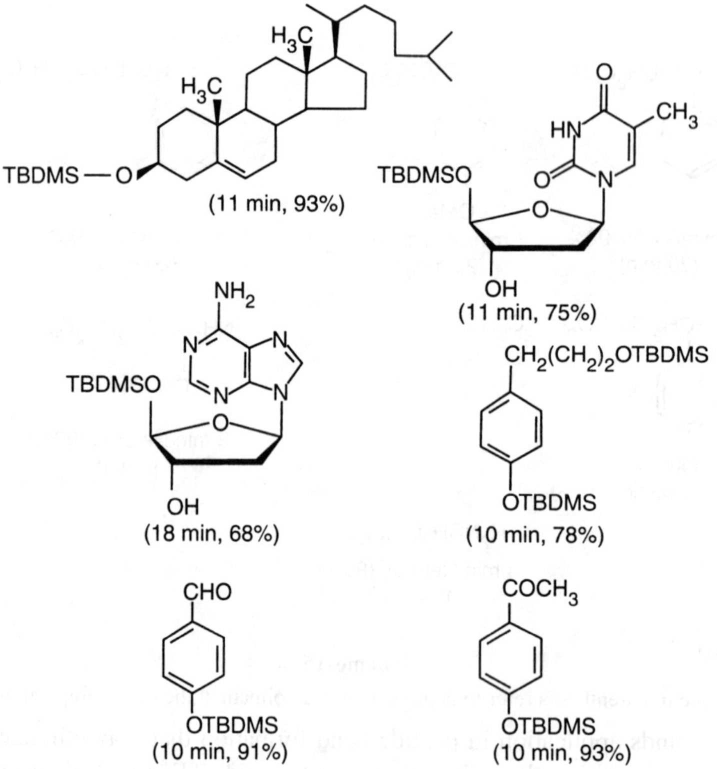

4.6 ■ Alternate Energy Processes in Chemical Synthesis

Scheme-17

(Structures with conditions:)
- TBDMS-O-cholesterol derivative (11 min, 93%)
- TBDMSO-thymidine derivative (11 min, 75%)
- TBDMSO-adenosine derivative (18 min, 68%)
- p-CH$_2$(CH$_2$)$_2$OTBDMS phenyl-OTBDMS (10 min, 78%)
- p-CHO phenyl-OTBDMS (10 min, 91%)
- p-COCH$_3$ phenyl-OTBDMS (10 min, 93%)

4.1.7 Dethioacetalization Reactions

Thioacetals and thioketals of aldehydes and ketones can normally be deprotected using toxic heavy metals such as Hg^{2+}, Ag^{2+}, Ti^{4+}, Cd^{2+}, Tl^{3+} or reagents like benzeneseleninic anhydride[20]. It is now possible to accomplish the dethioacetization in high yield in solid state using clayfen[10] (Scheme-18).

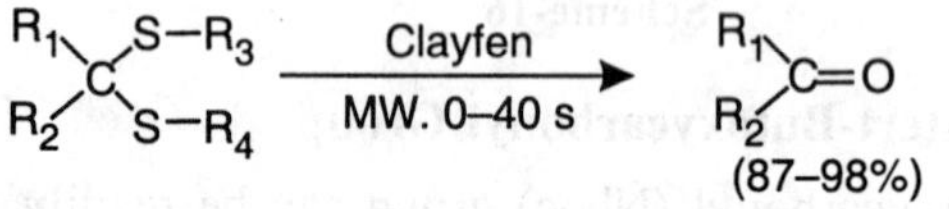

$$\begin{array}{c} R_1 \\ R_2 \end{array}\!\!C\!\!\begin{array}{c} S-R_3 \\ S-R_4 \end{array} \xrightarrow[\text{MW. 0-40 s}]{\text{Clayfen}} \begin{array}{c} R_1 \\ R_2 \end{array}\!\!C\!=\!O$$

(87–98%)

R_1 = Ph, p-anisyl. p-NO$_2$C$_6$H$_4$; R_2 = H ; R_3-R_4 =–(CH$_2$)$_2$–
R_1 = R_2 = Et ; R_3-R_4 = –(CH$_2$)$_2$– ; R_1 = R_2 =Ph; R_3 =R_4 = Et
R_1 = Ph; R_2=Me ; R_3-R_4 = –(CH$_2$)$_2$–
R_1-R_2 = 2-Methylcyclohexyl, isoflavanolyl ; R_3-R_4 =–(CH$_2$)$_2$

Scheme-18

4.1.8 Deoximation Reactions

Oximes have been used as protecting groups for carbonyl compounds owing to their hydrolytic stability. The oximes were earlier deoximated by reagents like Raney nickel, pyridinium chlorochromate, pyridinium chlorochromate-H_2O_2,, triethylammonium cholochromate, dinitrogen tetroxide, H_2O_2 over titanium silicalite-1, zirconium sulfophenyl phosphonate and bismuth chloride[20].

The solvent-free deprotection of protected carbonyl compounds has been effected using relatively benign ammonium persulfate on silca[20] (Scheme-19).

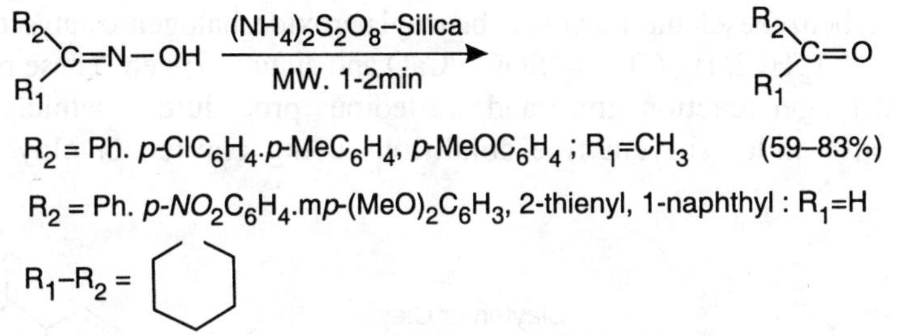

$$\begin{array}{c} R_2 \\ R_1 \end{array}\!\!C=N-OH \xrightarrow[\text{MW. 1-2min}]{(NH_4)_2S_2O_8-\text{Silica}} \begin{array}{c} R_2 \\ R_1 \end{array}\!\!C=O$$

R_2 = Ph. p-ClC_6H_4.p-MeC_6H_4, p-$MeOC_6H_4$; R_1=CH_3 (59–83%)

R_2 = Ph. p-$NO_2C_6H_4$.mp-$(MeO)_2C_6H_3$, 2-thienyl, 1-naphthyl : R_1=H

R_1-R_2 = (cyclohexyl)

Scheme-19

In the above procedure, neat oximes are mixed with a solid supported reagent and the contents are irradiated in a MW oven at full power to regenerate free adehydes and ketones. The surface of the support plays a critical role since the same reagent on a clay surface yields predominantly the Bechmann rearrangement products (i.e. the amides)[21].

Ketoximes could be deoximated[22] with sodium periodate imprignated with moist silca gel. Aldoximes are not affected by this procedure (Scheme-20).

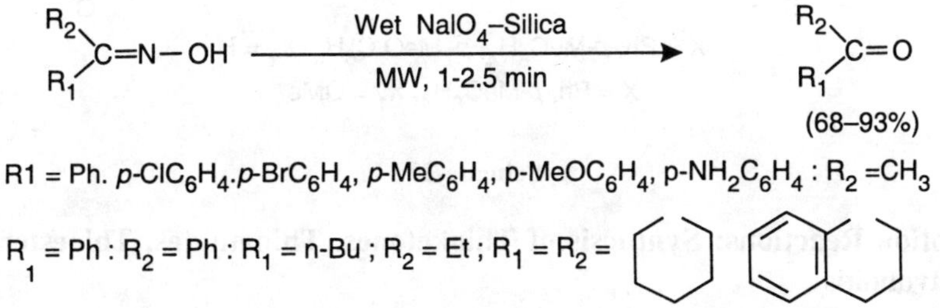

$$\begin{array}{c} R_2 \\ R_1 \end{array}\!\!C=N-OH \xrightarrow[\text{MW, 1-2.5 min}]{\text{Wet NaIO}_4-\text{Silica}} \begin{array}{c} R_2 \\ R_1 \end{array}\!\!C=O$$

(68–93%)

R1 = Ph. p-ClC_6H_4.p-BrC_6H_4, p-MeC_6H_4, p-$MeOC_6H_4$, p-$NH_2C_6H_4$: R_2 =CH_3

R_1 = Ph : R_2 = Ph : R_1 = n-Bu ; R_2 = Et ; R_1 = R_2 = (cyclohexyl) (tetralinyl)

Scheme-20

4.1.9 Cleavage of Semicarbazones and Phenylhydrazones

The semicarbazones and phenylhydrazones could be cleaved to regenerate the corresponding carbonyl compounds by using ammonium persulphate impregnated on montmorillonite K10 clay (Scheme-21) under microwave or ultrasound irradiation conditions[23].

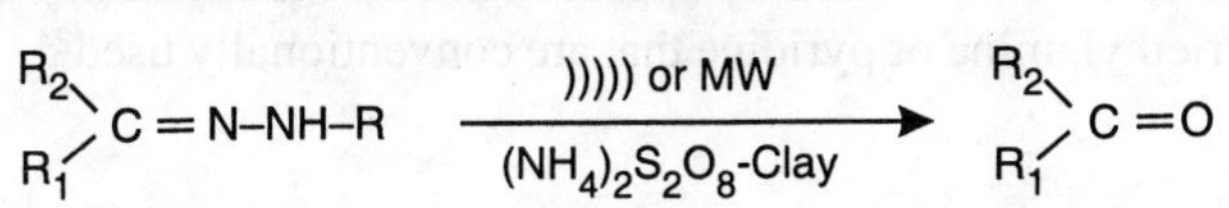

$$\begin{array}{c} R_2 \\ R_1 \end{array}\!\!C=N-NH-R \xrightarrow[{(NH_4)_2S_2O_8-\text{Clay}}]{\text{))))) or MW}} \begin{array}{c} R_2 \\ R_1 \end{array}\!\!C=O$$

Scheme-21

The above cleavage could be acheived in minutes by the microwave exposure compared to 1–3 hr for completion in the Ultrasound promoted reaction.

4.8 ■ Alternate Energy Processes in Chemical Synthesis

4.1.10 Dethiocarbonylation

Dethiocarbonylations have normally been effected[24] with a number of reagents like trifluoroacetic anhydride, CuCl/MeOH/Na OH, tetrabutylammonium hydrogen sulfate/NaOH, clay/ferric nitrate, $NOBF_4$, bromate and iodide solutions, alkaline, H_2O_2, sodium peroxide, bases like KOBu, thiophosgene, DMSO, trimethyloxonium fluoroborate, tellurium based oxidants, photochemical transformations, dimethyl selenoxide, benzeneselenic anhydride, benzoyl peroxide, halogen-catalysed alkoxides under PTC conditions, $NaNO_2/HCl$, $Hg(OAc)_2$, $SOCl_2/CaO$ and singet oxygen. These reagents are often toxic and required longer reaction times and are tedious procedures. Dethiocarbonylation has been accomplished[23] under solvent free conditions using clayfen or clayon (Schemes 22 and 23).

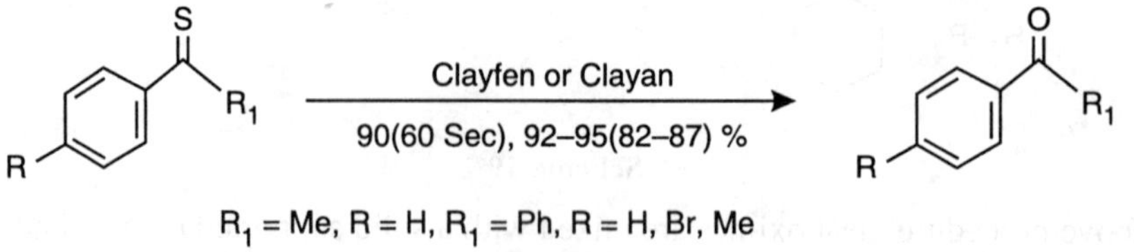

Clayfen or Clayan
90(60 Sec), 92–95(82–87) %

R_1 = Me, R = H, R_1 = Ph, R = H, Br, Me

Scheme-22

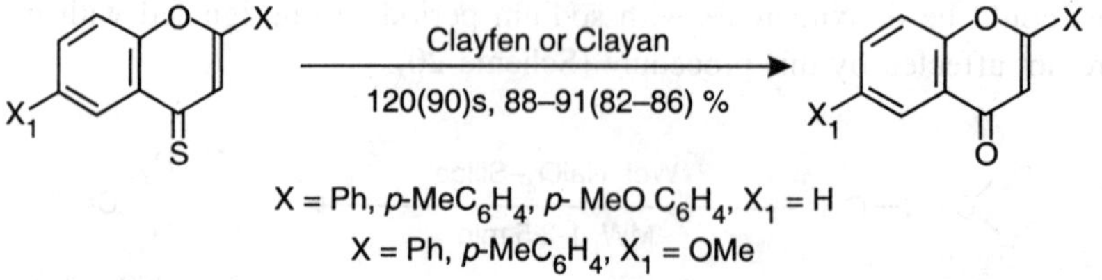

Clayfen or Clayan
120(90)s, 88–91(82–86) %

X = Ph, p-MeC_6H_4, p- MeO C_6H_4, X_1 = H
X = Ph, p-MeC_6H_4, X_1 = OMe

Scheme-23

4.1.11 Thionotion Reactions: Synthesis of Thioketones, Thioamides, Thioesters and Thioflavonoids

Conventionally the conversion of carbonyl compounds to the corresponding thioanalogues involve the reaction of the substractes with phosphorous pentasulphide under basic conditions, hydrogen sulphide in presence of acid or Lawesson's reagent. In MW approach, no acidic or basic media is used. The carbonyl compounds are mixed with neat Lawesson's reagent (0.5 equiv.) and irradiated with MW under solvent free condition. Using this approach, the conversion of ketones flavones, isoflavones, lactones, amides and esters to the corresponding thioanalogues could be achieved (Scheme-24) in high yields. This eco friendly approach avoids the use of dry hydrocarbon solvents like benzene, xylene, triethylamine or pyridine that are conventionally used[24].

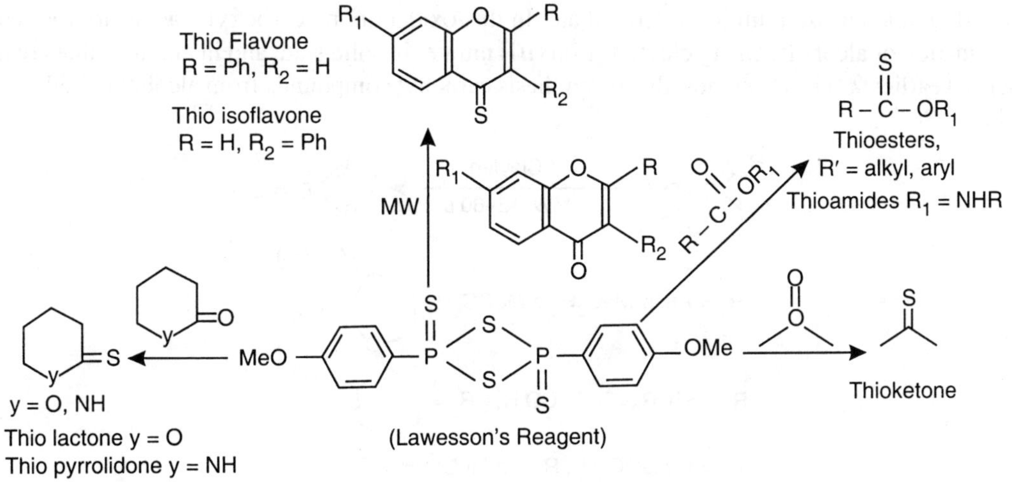

Scheme-24

[Scheme-24: Reactions using Lawesson's Reagent under MW conditions converting carbonyls to thiocarbonyls — Thio Flavone (R = Ph, R₂ = H), Thio isoflavone (R = H, R₂ = Ph), Thio lactone (y = O), Thio pyrrolidone (y = NH), Thioesters (R' = alkyl, aryl), Thioamides (R₁ = NHR), Thioketone]

4.1.12 Saponification of Esters

Hindered esters which take 5 hr under classical heating with alkali can be easily saponified under microwave conditions[27] using KOH-Aliquat (Scheme-25).

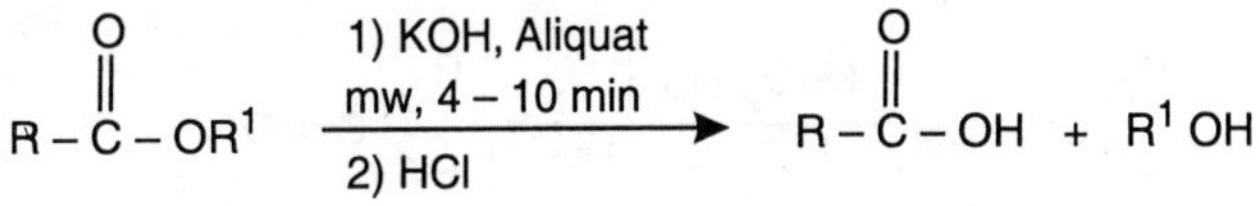

$$R-\overset{O}{\underset{\|}{C}}-OR^1 \xrightarrow[\text{2) HCl}]{\text{1) KOH, Aliquat, mw, 4-10 min}} R-\overset{O}{\underset{\|}{C}}-OH + R^1OH$$

Scheme-25

4.2 OXIDATIONS

A number of conventional oxidizing agents have been used[28] for various types of oxidations. These include peracids, peroxides, manganese dioxide (MnO_2), potassium permanganate ($KMnO_4$) and potassium dichromate. These reagents have their own limitations in terms of toxicity, workup and associated waste disposal problems.

A number of metal-based reagents have also been used in organic synthesis. Such reagents are toxic, have cumbersome properties and have potential danger (ignition or explosion) in handling of their complexes and difficulties encountered in the isolation of required products and waste disposal. However, the introduction of metallic reagents on solid support have overcome some of these problems and provided an attractive alternative in organic synthesis due to selectivity and ease of manuplation. Also, the immobilation of metals on the surface avoids their leaching into the environment.

4.2.1 Oxidation of Alcohols

(a) Using Clayfen

A facile method has been developed for the oxidation of alcohols to carbonyl compound using montmorillonite K10 clay supported iron (III) nitrate (Clayfen) in solvent free conditions. The process is accelerated many folds by using MW irradiation[29]. The above oxidation presumably

4.10 ■ Alternate Energy Processes in Chemical Synthesis

proceed via the nitrosonium ion intermediate. In this oxidation, no carboxylic acids are formed in the oxidation of alcohols. Using clayfen [Iron (III) nitrate] in solid state and in amounts that are half that used earlier[30], it has been possible to synthesis carbonyl compounds from alcohol in high yields (Scheme-26)[29].

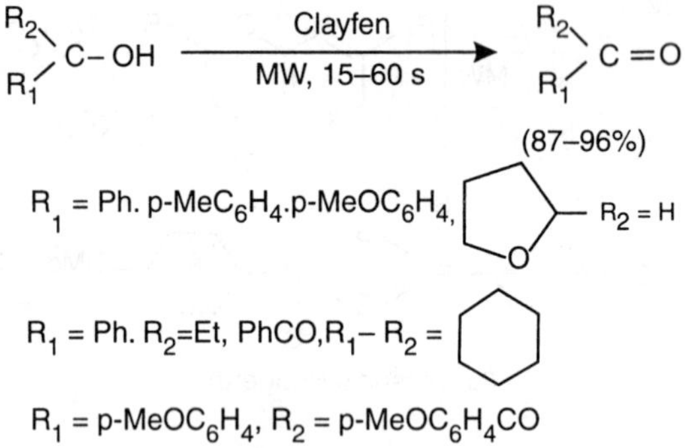

$$\underset{R_1}{\overset{R_2}{\diagdown}}C-OH \xrightarrow[\text{MW, 15–60 s}]{\text{Clayfen}} \underset{R_1}{\overset{R_2}{\diagdown}}C=O$$

(87–96%)

R_1 = Ph. p-MeC$_6$H$_4$.p-MeOC$_6$H$_4$, ⟨furan⟩ – R_2 = H

R_1 = Ph. R_2=Et, PhCO,R_1– R_2 = ⟨cyclohexyl⟩

R_1 = p-MeOC$_6$H$_4$, R_2 = p-MeOC$_6$H$_4$CO

Scheme-26

(b) Using Activated MnO$_2$-Silica

Carbonyl compounds have been obtained[30] in good yield using 35% MnO$_2$ 'doped' silica gel under MW irradiation conditions. By this procedures, benzyl alcohols are selectively oxidized to carbonyl compounds (Scheme-27).

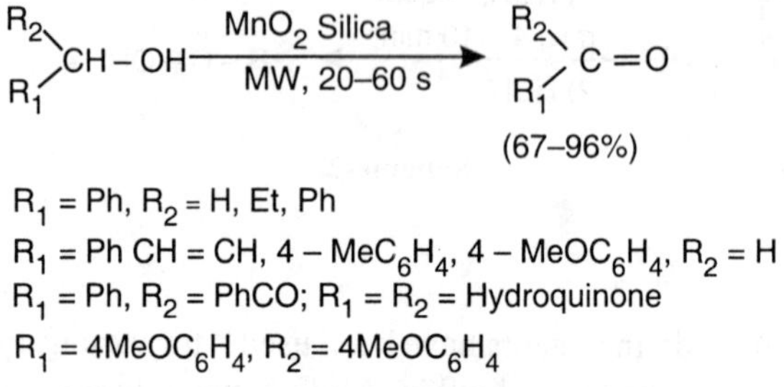

$$\underset{R_1}{\overset{R_2}{\diagdown}}CH-OH \xrightarrow[\text{MW, 20–60 s}]{\text{MnO}_2 \text{ Silica}} \underset{R_1}{\overset{R_2}{\diagdown}}C=O$$

(67–96%)

R_1 = Ph, R_2 = H, Et, Ph
R_1 = Ph CH = CH, 4 – MeC$_6$H$_4$, 4 – MeOC$_6$H$_4$, R_2 = H
R_1 = Ph, R_2 = PhCO; R_1 = R_2 = Hydroquinone
R_1 = 4MeOC$_6$H$_4$, R_2 = 4MeOC$_6$H$_4$

Scheme-27

(c) Using Chromium Trioxide Supported on Wet Alumina

The use of chromium reagents in the oxidations is limited to toxicity of chromium, preparation of its various complexes (with acetic anhydride or pyridine) and crumbersome workup procedures. Chromium Oxide (CrO$_3$) immobilized in pre-moistened alumina is very efficient for the oxidation of alcohols to carbonyl compounds by simple mixing[31] (Scheme-28). This oxidation is clean and does not involve any over oxidations to carboxylic acids.

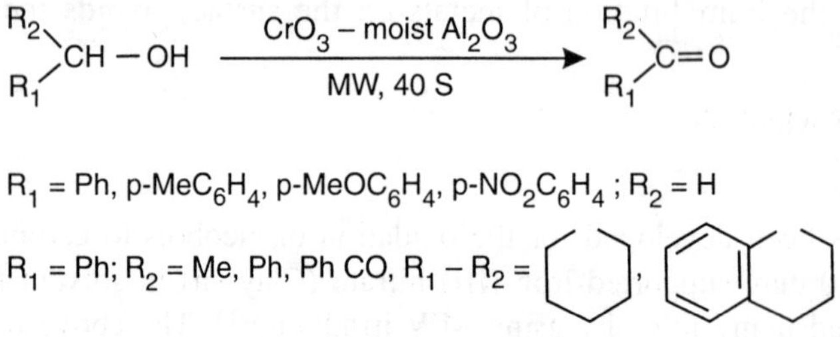

$$\underset{R_1}{\overset{R_2}{\diagdown}}CH-OH \xrightarrow[\text{MW, 40 S}]{\text{CrO}_3 - \text{moist Al}_2\text{O}_3} \underset{R_1}{\overset{R_2}{\diagdown}}C=O$$

R_1 = Ph, p-MeC$_6$H$_4$, p-MeOC$_6$H$_4$, p-NO$_2$C$_6$H$_4$; R_2 = H

R_1 = Ph; R_2 = Me, Ph, Ph CO, R_1 – R_2 = ⟨cyclohexyl⟩, ⟨decalin⟩

Scheme-28

Using the above procedure, acyclic α-nitro kitones are obtained in one-pot operation utilising *in situ* oxidation of the nitroalkanols with premoistened alumina supported chromium trioxide[32].

(d) Using Iodobenzene Diacetate (IBD) 'Doped" Alumina

Alcohols and phenols have earlier been odidised with iodoxy benzene, o-Iodoxybenzoic acid(1BX), bis (trifuoroacetoxy) iodobenzene (BTI) and Dess-Martin periodiamate. Most of these reactions are conducted in high boiling DMSO and toxic acetonitrile media. Further, IBX has been reported to be explosive on heavy impact and heating over 200°C.

It has been found that oxidation of alcohols to carbonyl compound occurs rapidly with alumina-supported IBD under solvent free conditions and MW irradiations in quantitative yields[33] (Scheme-29).

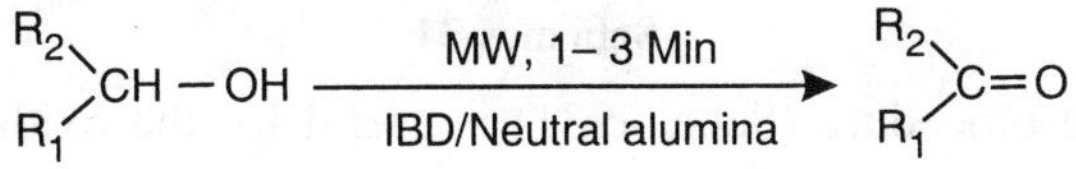

$$\underset{R_1}{\overset{R_2}{>}}CH-OH \xrightarrow[\text{IBD/Neutral alumina}]{\text{MW, 1-3 Min}} \underset{R_1}{\overset{R_2}{>}}C=O$$

Scheme-29

It is interesting to note that 1, 2-benzenedimethanol undergoes cyclisation to afford 1(3H)-isobenzofuranone.

4.2.2 Oxidation of Alkanes, Bromides, Carboxylic Acids, Cyanides and Amines to Carbonyl Compounds

Copper (II) nitrate impregnated on K10 clay (Claycop)-hydrogen peroxide is an effective reagent for the oxidations of a variety of substractes like alkanes, bromides, carboxylic acids, cyanides and amines to carbonyl compounds (Scheme-30) in excellent yields[34].

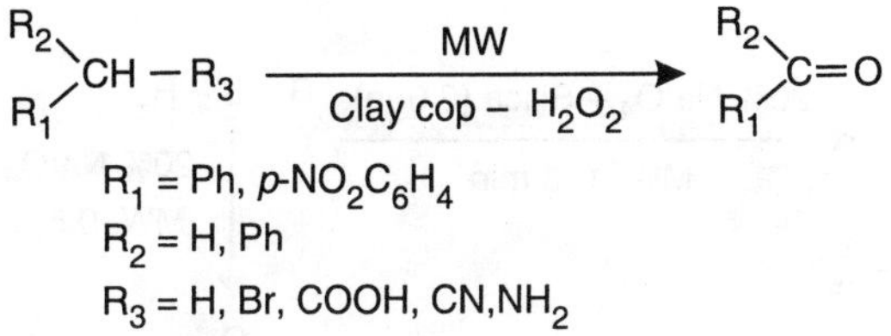

$$\underset{R_1}{\overset{R_2}{>}}CH-R_3 \xrightarrow[\text{Clay cop}-H_2O_2]{\text{MW}} \underset{R_1}{\overset{R_2}{>}}C=O$$

$R_1 = Ph, p\text{-}NO_2C_6H_4$
$R_2 = H, Ph$
$R_3 = H, Br, COOH, CN, NH_2$

Scheme-30

It may be appropriate to mention that earlier similar oxidations, were carried out by using copper (II) nitrate and hydrogen per oxide (eq.1).

$$2Cu(NO_2)_2 + H_2O_2 + H_2O \rightarrow 2CuO_2H + 4HNO_3 \qquad \ldots(1)$$

However, in this case neutralisation of the reagent was necessary by $KHCO_3$ to maintain a PH ≈ 5. In the procedure mentioned above in Scheme-30 in which claycop-H_2O_2 was used (mw irradiation) it was not ncessary to maintain the pH of the reaction mixture.

4.2.3 Oxidation of α-hydroxyketones to 1, 2-diketones

For the oxidation of α-hydroxyketones to 1, 2-diketones, a number of reagents have been used[35]. They include nitric acid, Fehling's solution, Thallium (III) nitrate (TTN), ytterbium (III) nitrate, clayfen and ammonium chlorochromate-alumina. Most of these reagents suffer from drawbacks, such as use of corrosive acids and toxic metallic compounds that generate undesirable waste materials. Recently solid support systems, copper (II) sulfate-alumina[35] or oxone®-wet alumina[36] under the

4.12 ■ *Alternate Energy Processes in Chemical Synthesis*

influence of microwaves (Scheme-31) have been used for the oxidation of α-hydroxyketones to 1, 2-diketones.

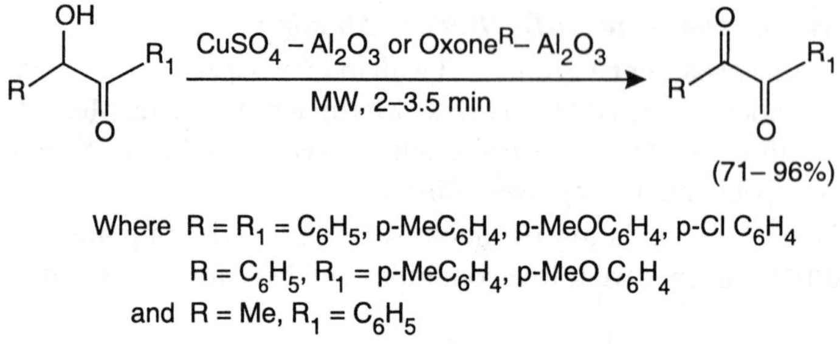

$$R-\underset{O}{\underset{\|}{C}}-\underset{OH}{\underset{|}{CH}}-R_1 \xrightarrow[\text{MW, 2-3.5 min}]{\text{CuSO}_4-\text{Al}_2\text{O}_3 \text{ or Oxone}^R-\text{Al}_2\text{O}_3} R-\underset{O}{\underset{\|}{C}}-\underset{O}{\underset{\|}{C}}-R_1$$

(71– 96%)

Where R = R$_1$ = C$_6$H$_5$, p-MeC$_6$H$_4$, p-MeOC$_6$H$_4$, p-Cl C$_6$H$_4$
R = C$_6$H$_5$, R$_1$ = p-MeC$_6$H$_4$, p-MeO C$_6$H$_4$
and R = Me, R$_1$ = C$_6$H$_5$

Schemes-31

The above oxidative procedure (Scheme-31) is useful for the oxidation of only α-hydroxy Ketones. Primary alcohols used as benzyl alcohol and secondary alcohols e.g. 1-phenylpropan-1-ol undergo only limited oxidative conversion which is of little practical utility. Mixed benzylic/aliphatic α-hydroxyketone, e.g. 2-hydroxypropiophenone gives the corresponding vicinal diketone[35].

4.2.4 Oxidation of Sulfides to Sulfoxides and Sulfones

Generally sulfides are oxidized to sulfoxides under strenuous conditions using strong oxidants like nitric and, hydrogen peroxide, chromic acid, paracids and periodate[36]. This oxidation can be conveniently carried out with desdesired selectivity to either sulfoxides or sulfones, using silica 'doped' with 10% sodium periodate under reduced power and reaction time (pulsed techniques)[36]. So a much reduced amount of the active oxidizing agent is required which is safer and easier to handle (Scheme-32).

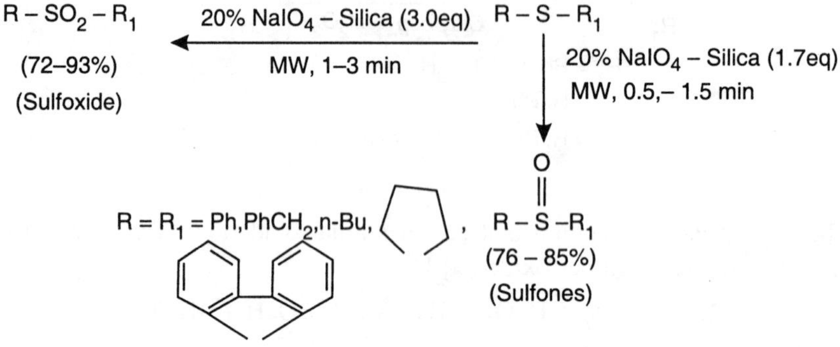

$$R-SO_2-R_1 \xleftarrow[\text{MW, 1-3 min}]{20\% \text{ NaIO}_4-\text{Silica (3.0eq)}} R-S-R_1 \xrightarrow[\text{MW, 0.5,- 1.5 min}]{20\% \text{ NaIO}_4-\text{Silica (1.7eq)}} R-\underset{\|}{\underset{O}{S}}-R_1$$

(72–93%) (Sulfoxide)

R = R$_1$ = Ph, PhCH$_2$, n-Bu, cyclopentyl, biphenylmethyl

(76 – 85%) (Sulfones)

Scheme-32

An industrial application of the above procedure (Scheme-32) is that various refractory thiophenes that are often not reductively removed by conventional refining processes can be oxidised under the above conditions e.g., benzothiophenes are oxidized to the corresponding sulfoxides and sulfones using ultrasonic and microwave irradiation respectively, in the presence of NaIO$_4$-Silica[36]. A special feature of the above procedure is its applicability to long chain fatty sulphides which are insoluble in most solvents and are consequently difficult to oxidize.

Selective oxidation of alkyl, aryl and cyclic sulphides to the corresponding sulfoxides can be achieved by the solid reagent system iodozenzene diacetate (IBD)-alumina upon microwave activation (Scheme-32a)[37].

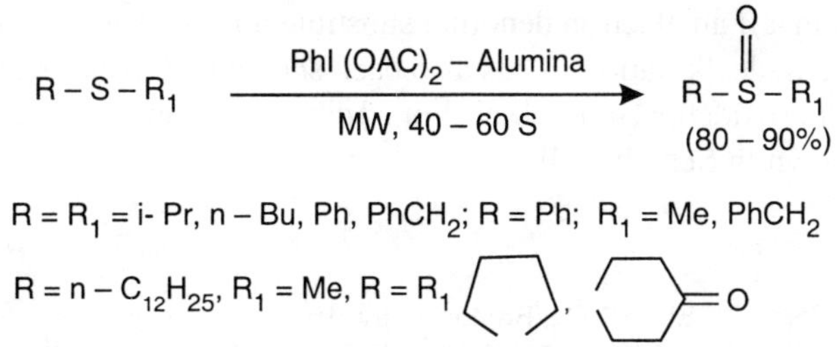

$$R-S-R_1 \xrightarrow[\text{MW, 40 – 60 S}]{\text{PhI (OAC)}_2 - \text{Alumina}} \underset{(80-90\%)}{R-\overset{\overset{O}{\|}}{S}-R_1}$$

R = R$_1$ = i- Pr, n – Bu, Ph, PhCH$_2$; R = Ph; R$_1$ = Me, PhCH$_2$

R = n – C$_{12}$H$_{25}$, R$_1$ = Me, R = R$_1$ [cyclopentyl], [cyclohexanone]

Scheme-32a

In a solid state reaction with clayfen, a variety of alkyl, aryl and cyclic sulfoxides are easily oxidised to the corresponding sulfoxides in high yield upon microwave thermolysis[38].

4.2.5 Oxidation of Enamines

β,β-Disubstituted enamines on oxidation with KMn O$_4$-Al$_2$O$_3$ in domestic (255 W, 82°C) as well as in focussed (330W, 140°C) microwave ovens under solvent free conditions give[39] the carbonyl compounds. Use of focussed (330W, 140°C) microwave oven gave better yields (Scheme-33).

In the reactions conducted in an oil bath at 140°C, no ketone formation was observed.

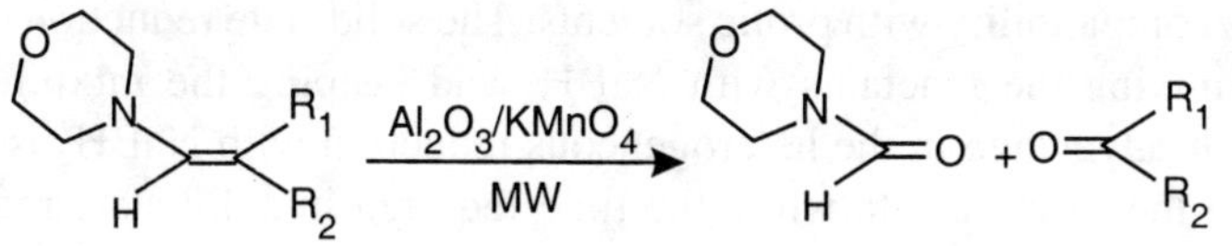

Scheme-33

4.2.6 Oxidation of Arenes

KMnO, impregnated alumina oxidises arenes to ketones within 10–30 min in solvent free conditions using focussed microwaves (Scheme-34).

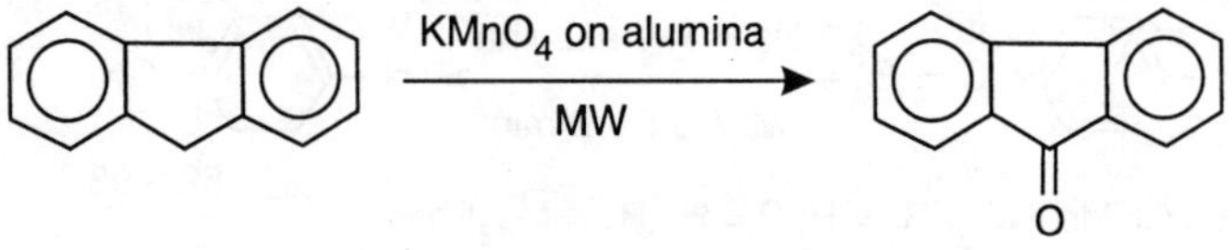

Scheme-34

4.2.7 Aromatisation

A highly efficient aromatisation process has been achieved[39a] using microwave irradiation as shown below in Scheme-34a.

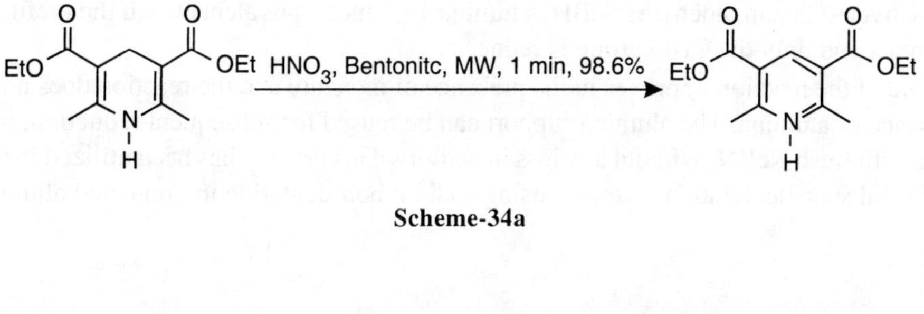

Scheme-34a

4.14 ■ Alternate Energy Processes in Chemical Synthesis

It has been shown that if an electron denoting substituent is present at position 4, oxidation can proceed with concomitant dealkylation. It has also been shown that in contrast with conventionally promoted dihydropyridine oxidation, 4-alkyl substituted dihydropyridines give a mixture of products[39b] on the oxidation as shown in Scheme-34b.

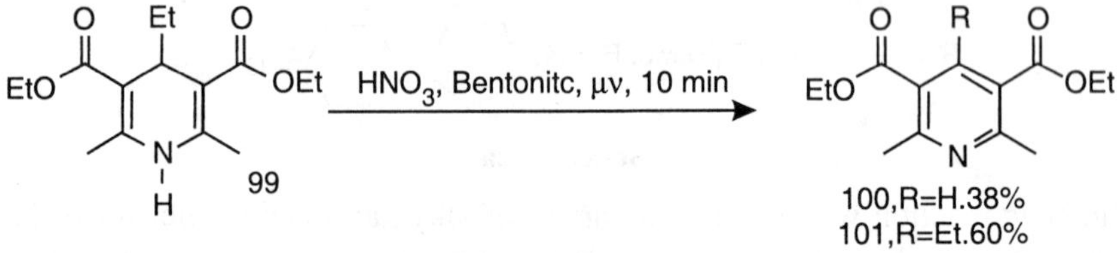

Scheme-34b

4.3 REDUCTIONS

4.3.1 Sodium Borohydride Reduction of Carbonyl Compounds to Alcohols

Sodium borohydride ($NaBH_4$), a relatively inexpensive material has been widely used as a reducing agent because of its compatibility with protic solvents. The solid state reduction of Ketones has also been achieved by mixing the reactants with $NaBH_4$ and keeping the mixture in a dry box for 5 days. The major disadvantage in the heterogeneous reduction with $NaBH_4$ is that the use of the solvent slows down the reaction rate while the time required in solid state reduction is too long (5 days) for it to be of any practical utility[40].

The reduction of carbonyl compounds with alumina supported $NaBH_4$ using microwaves proceeds satisfactorily. The process involves[40] mixing of carbonyl compound with (10%) $NaBH_4$-alumina in the solid state and irradiating the mixture in a MW oven for 0.5–2 min (Scheme-34c).

R = Cl; Me, NO_2, R_1 = H; R = H; R_1 = Me_3Ph
R = Ph; R_1 = Ph CH(OH); R_1 = R_2 = Me,
R = p-$MeOC_6H_4$, R_1 = p- $MeOC_6H_4CH(OH)$

Scheme-34c

The chemoselectivity of reduction with $NaBH_4$ is evident from the reduction of trans-cinnamaldehyde (Cinnamaldehyde/$NaBH_4$-alumina 1 : 1 mol equivalent) when the olefinic moiety remains intact and only the CHO group is reduced.

The rate of the reaction improves in the presence of moisture and the reaction does not proceed in the absence of alumina. The alumina support can be reused for subsequent reduction, repeatedly by mixing with fresh $NaBH_4$ without any loss in activity. This process has been utilized for the MW-enhanced solid state deuteration reactions using sodium borodeuteride impregnated alumina[40].

4.3.2 Reductive Amination of Carbonyl Compounds

Reductive amination of carbonyl compounds has been effected by sodium cyanoborohydride[41], sodium triacetoxyborohydride[42] or Na BH_4 coupled with sulfuric acid[43]. These reagents involve the use of corrosive acids and result in waste generation. The environmentally benign methods developed earlier[44, 45] have been extended to a Solvent-free reductive amination of carbonyl compounds using wet montmorillonite K10 clay supported sodium borohydride that is facilated by MW irradiation (Scheme-35)[46].

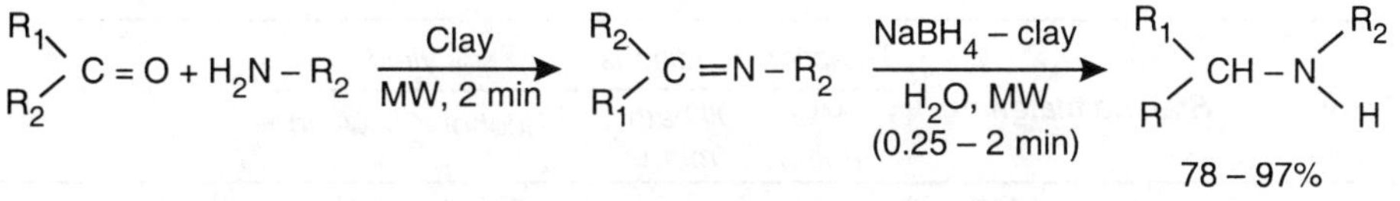

$$\begin{array}{c}R_1\\R_2\end{array}C=O + H_2N-R_2 \xrightarrow[\text{MW, 2 min}]{\text{Clay}} \begin{array}{c}R_2\\R_1\end{array}C=N-R_2 \xrightarrow[\substack{H_2O, MW\\(0.25-2\text{ min})}]{NaBH_4-\text{clay}} \begin{array}{c}R_1\\R\end{array}CH-N\begin{array}{c}R_2\\H\end{array}$$

78 – 97%

R = i - Pr, Ph, o- HOC_6H_4, p- $MeOC_6H_4$; P-$NO_2C_6H_4$, R_1 = H; R_2 = Ph
R and R_1 = – $(CH_2)_5$; R_2 = Ph ; R and R_1 = - $(CH_2)_6$ - ; R_2 = n-Pr
R = n - C_5H_{11}; R_1 = Me I R_2 = Marpholine, piperidine
R = i Pr ; R_1 = H ; R_2 = n-$C_{10}H_{21}$

Scheme-35

The solid-state reductive amination of carbonyl compounds on various inorganic solid supports such as alumina, clay, silica etc. and especially a K10 clay surface rapidly affords secondary and tertiary amines[46]. Clay behaves as a Lewis acid and also provides water from its interlayers thus enhancing the reducing ability of $NaBH_4$.

4.3.3 Reduction of Carbonyl Compounds with Aluminium Alkoxides

The reduction of carbonyl compounds with isopropyl alcohol and alumina is very well known[47]. It has now been possible to carry out the above reduction without any solvent under microwave irradiation conditions[48] (Scheme-36).

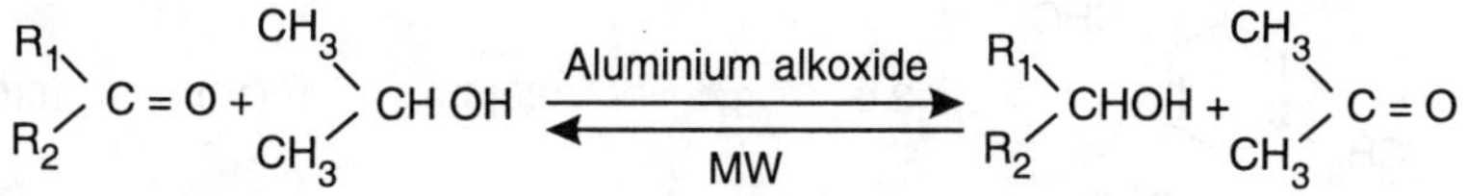

$$\begin{array}{c}R_1\\R_2\end{array}C=O + \begin{array}{c}CH_3\\CH_3\end{array}CHOH \underset{MW}{\overset{\text{Aluminium alkoxide}}{\rightleftarrows}} \begin{array}{c}R_1\\R_2\end{array}CHOH + \begin{array}{c}CH_3\\CH_3\end{array}C=O$$

Scheme-36

4.3.4 Solid State Crossed Cannizzaro Reaction

The well known cannizzaro reaction is the disproportionation of an aldehyde to an equimolar mixture of primary alcohol and carboxylic acid[49, 50] and is restricted to aldehydes which lack the α-hydrogens. This oxidation-reduction reaction is generally conducted under strongly basic conditions and has the inherent disadvantage of the lower yields of the desired products[51, 52]. After the discovery of Li AlH_4 in 1946, the popularity of cannizzaro reaction in synthetic organic chemistry was lost. The crossed cannizzaro reaction[51] using a scavenger and cheap paraformaldehyde provided improved yields of alcohol prior to the introduction of hydride reducing agent. This reaction is normally conduct in solution phase. It does not proceed under solventless MW-irradiation conditions using alumina surface with calcium hydroxide or in presence of strong base like NaOH.

The solid state crossed cannizzaro reaction has been found to proceed very well[53] on barium hydroxide [Ba $(OH)_2.8H_2O$] surface upon mv irradiation (Scheme-37) using an aldehyde (which is free of α hydrogen) and formaldehyde.

4.16 ■ Alternate Energy Processes in Chemical Synthesis

$$R\,CHO + (CH_2O)_n \xrightarrow[MW]{Ba(OH)_2,\,8H_2O} R\,CH2OH + R\,CO\text{-}OH$$
$$\phantom{R\,CHO + (CH_2O)_n \xrightarrow[MW]{Ba(OH)_2,\,8H_2O}} 80-99\% \qquad 1-20\%$$

Scheme-37

The table below gives the results obtained in case of a number of aldehydes.

Product Distribution in Solvent-free Crossed Cannizzaro Reaction using Barium Hydroxide *via* Microwave as well as Classical Conditions

Entry	Starting material	Reaction conditions		% yield		Others
		MW (min.)	Oil bath (min.)	Alcohol	Acid	
1	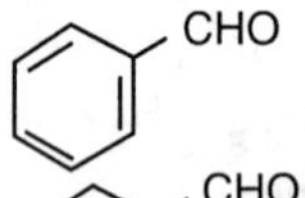 PhCHO	0.5	10	99(98)	01(02)	00(00)
2	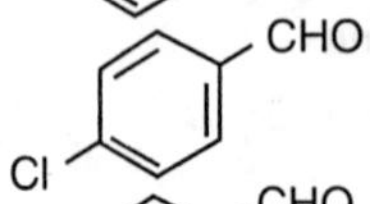 4-Cl-C6H4-CHO	0.5	10	94(95)	05(04)	01(01)
3	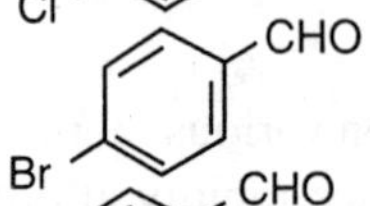 4-Br-C6H4-CHO	0.5	15	83(85)	16(14)	01(01)
4	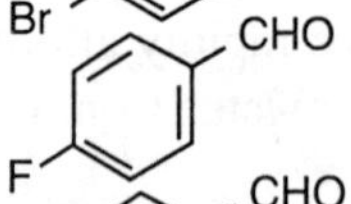 4-F-C6H4-CHO	0.5	12	85(88)	14(11)	01(01)
5	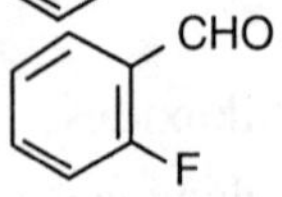 2-F-C6H4-CHO	0.25	10	80(90)	20(01)	00(00)
6	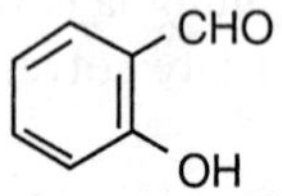 2-OH-C6H4-CHO	2.0	60	83(83)	13(11)	04(06)
7	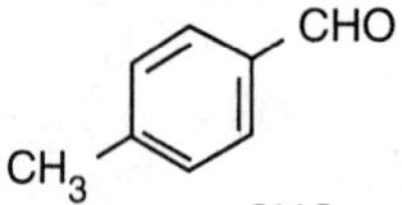 4-CH3-C6H4-CHO	2.0	37	80(85)	10(10)	10(05)
8	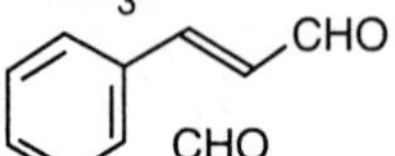 PhCH=CH-CHO	1.5	15	80(85)	19(10)	01(05)
9	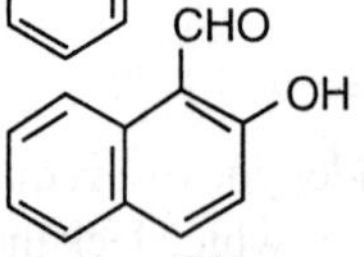 2-OH-naphthyl-1-CHO	1.0	180	85(86)	10(10)	05(04)
10	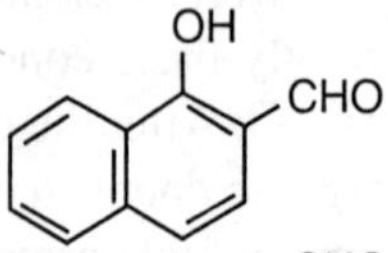 1-OH-naphthyl-2-CHO	2.0	240	83(85)	09(06)	08(06)
11	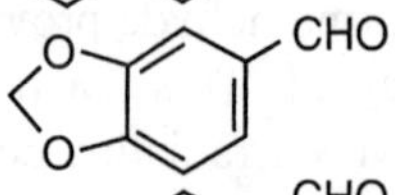 piperonal	1.5	40	82(80)	15(15)	03(05)
12	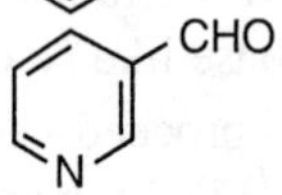 pyridine-3-CHO	2.0	35	97(97)	03(03)	00(00)

Entry	Starting material	Reaction conditions		% yield		Others
		MW (min.)	Oil bath (min.)	Alcohol	Acid	
13	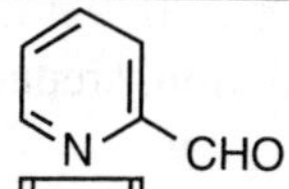 pyridine-2-CHO	1.0	20	97(98)	03(02)	00(00)
14	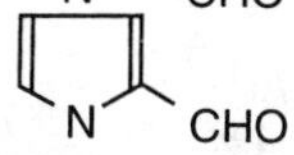 pyrrole-2-CHO	0.5	06	97(99)	03(01)	00(00)

a. The relative amounts of product formation are determined by GC-MS analysis and the results in the parentheses refer to the corresponding yields obtained using oil bath; the products exhibited physical and spectral properties (NMR and IR spectra) in accord with the assigned structures. Data taken from reference 53.

4.4 REARRANGEMENT REACTIONS

4.4.1 Pinacol-pinacolone Rearrangement

Solventless pinacol-pinacolone rearrangement using microwave irradiation[54] has been reported. The process involves the irradiation of the gemdiols with Al^{3+} montmorillonite K 10 clag for 15 min to give the rearrangement products in good yields (Scheme-38). These results are comparable to conventional heating is an oil bath wherein the reaction takes too long (15 hr).

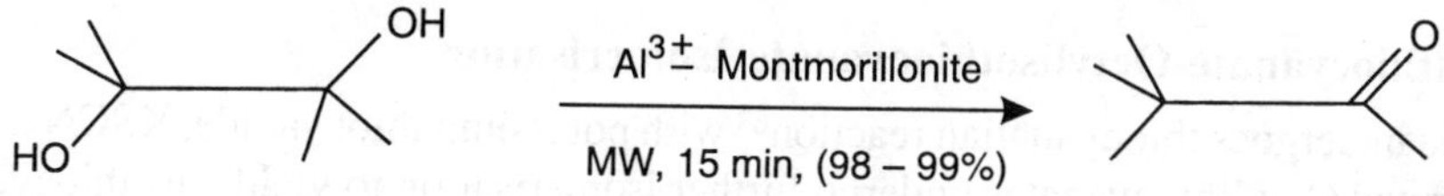

Scheme-38

4.4.2 Ring Expansion

An efficient ring expansion transformation is described[55] under solventless conditions (Scheme-38a). This procedure is superior than the reaction conducted in conventional methanolic solution.

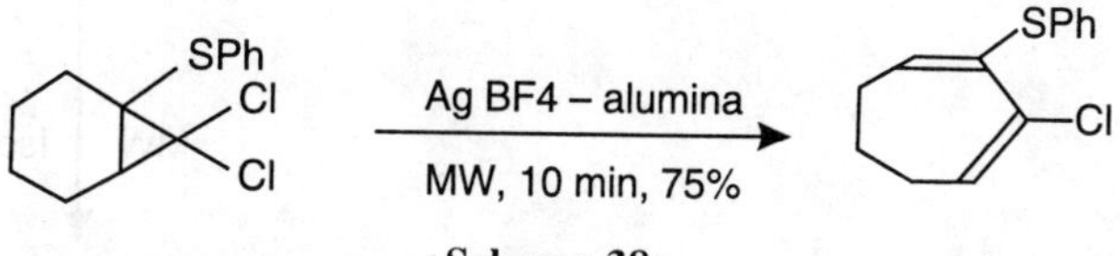

Scheme-38a

4.4.3 Beckmann Rearrangement

Usually, Beckmann rearrangement of oximes of ketones are converted into anilides by heating with acidic reagents like PCl_5, HCOOH, $SOCl_2$ etc. However, a solid-state microwave assisted Beckmann Rearrangement has been reported[56, 57]. In this method, oxime of a kelone is mixed with montmorillonite K10 clay in dry media and mixture irradiated for 7 min. in a microwave oven to give the corresponding anilide in 91% yield (Scheme-39).

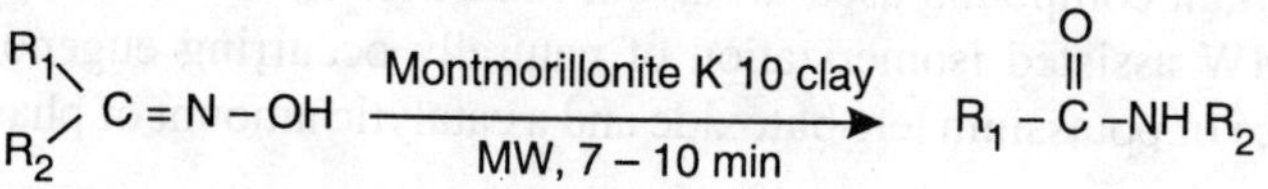

$$\underset{R_2}{\overset{R_1}{>}}C=N-OH \xrightarrow[\text{MW, 7 – 10 min}]{\text{Montmorillonite K 10 clay}} R_1-\underset{\underset{O}{\|}}{C}-NH\,R_2$$

Scheme-39

4.4.4 Benzil-Benzilic Acid Rearrangement

Usually, the above rearrangement has been carried out by heating benzil and alkali metal hydroxides in aqueous organic solvent. It is found that rearrangement proceeds more efficiently and faster in solid state[58] and it takes 0.1 to 6 hr for completion and the yields are 70–93%. The Benzil-Benzilic acid rearrangement could be conducted in solid state by MW irradiation[59] reducing considerably the time of the reaction and good yield (Scheme-40).

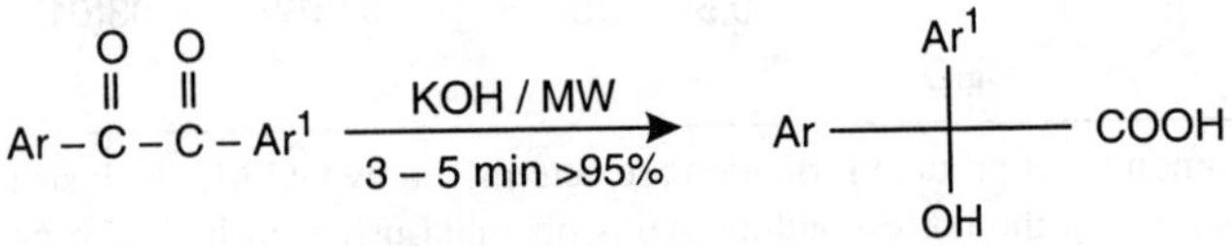

$$Ar-CO-CO-Ar^1 \xrightarrow[3-5 \text{ min} >95\%]{KOH/MW} Ar-C(OH)(Ar^1)-COOH$$

$Ar = Ar^1 = Ph, p\text{-}Cl\ C_6H_4, p\text{-}NO_2\ C_6H_4,$
$Ar = Ph, Ar^1 = p\text{-}Cl\ C_6H_4, p\text{-}NO_2\ C_6H_4, p\text{-}MeOC_6H_4$

Scheme-40

Some other examples of rearrangements reactions include Fries Rearrangement on K10 clay that affords a mixture of ortho and para products[60]; Fries rearrangement that leads to the formation of flavonones[61] and thia-Fries rearrangement of arylsulfonates using aluminium trichloride and zinc chloride on silica gel[62].

4.5 ISOMERISATION REACTIONS

4.5.1 Octylthiocyanate-Octylisothiocyanate Isomerisation

Octylbromide undergoes thiocyanatian reaction[63] with potassium thiocyanide, KSCN in presence of MW. The formed Octylthiocyanates undergo further isomerisation to yield[63] isothiocyanates under mw irradiation (Scheme-41).

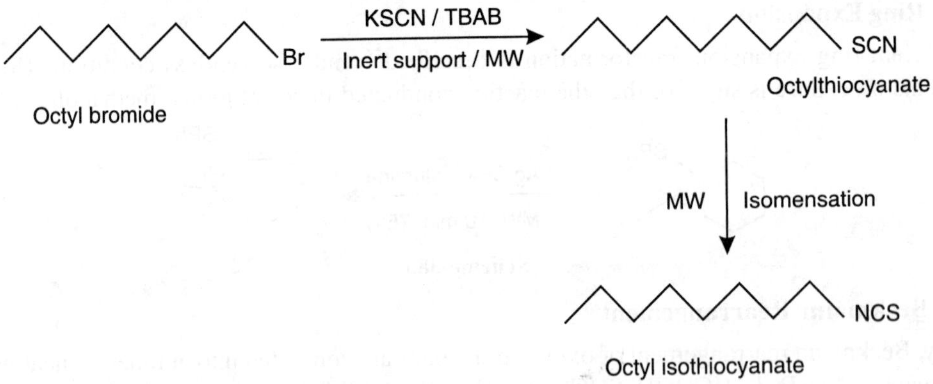

Octyl bromide $\xrightarrow[\text{Inert support / MW}]{KSCN/TBAB}$ Octylthiocyanate (SCN) $\xrightarrow{MW\ \text{Isomensation}}$ Octyl isothiocyanate (NCS)

Scheme-41

4.5.2 Eugenol Isoeugenol Isomerisation

Isoeugenol, an important compound used in flavour industries to manufacture vanillin, is obtained by base catalysed MW assisted isomerization of naturally occurring eugenol under solvent free conditions in presence of potassium tert-butoxide and a catalytic amount of phase transfer reagent[63a] (Scheme-41a).

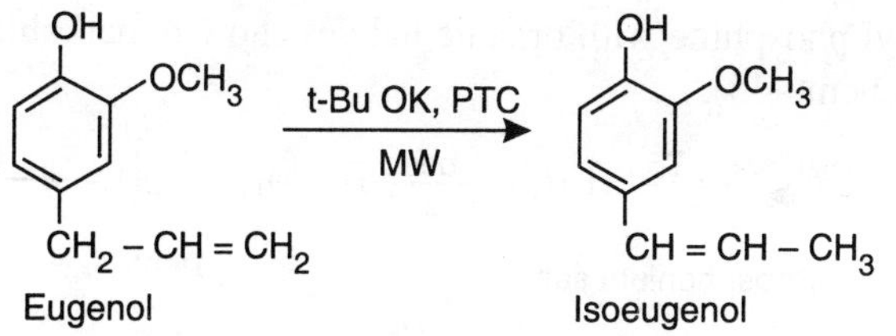

[Scheme-41a: Eugenol → Isoeugenol using t-Bu OK, PTC, MW]

4.6 CONDENSATION REACTIONS

4.6.1 Knoevenagel Condensation

Knoevenagel condensation of aldehydes with creatinine takes place under solvent free reaction conditions at 160–170° using focussed MW irradiation[64] (Scheme-42).

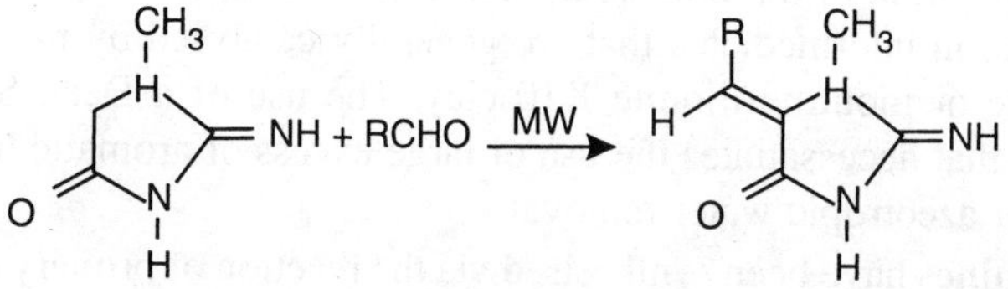

[Scheme-42]

Similarly, the knoevenagel condensation of 5-nitro furaldehyde with active methylene compounds under MW irradiation using K10 and $ZnCl_2$ as catalyst give 5-nitro furfurylidines[65] (Scheme-43).

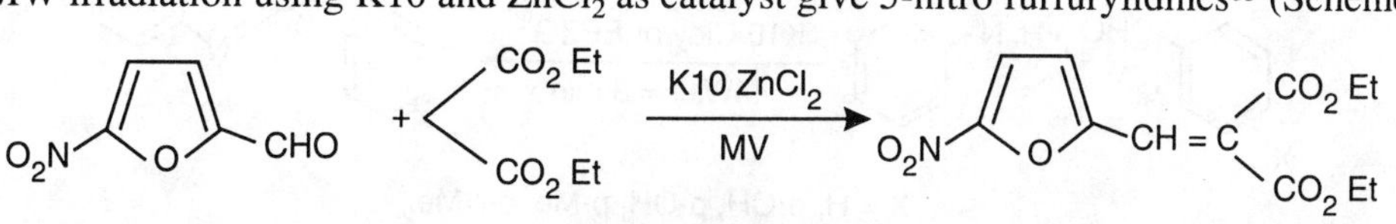

[Scheme-43]

The classical Pechmann approach for the Synthesis of coumarins via the MW assisted reaction[66] has been extended to solvent-free system involving Knoevenagel condensation of salicylaldehydes with a variety of ethyl acetate derivatives under basic conditions (piperidine) to afford coumarins (Scheme-44).

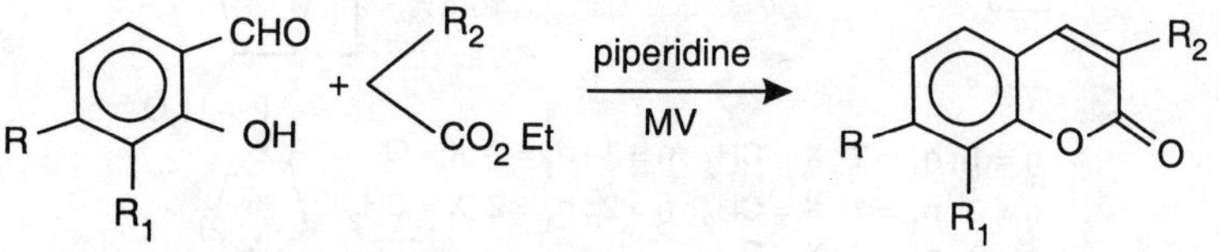

[Scheme-44]

4.6.2 Wittig Olefination Reactions

Stable phosphorous ylides undergo wittig reaction with Ketones in the absence of Solvent on MW irradiation to give improved yields when compared to conventional heating procedures[67]. The preparation of several phosphonium salts has been reported using a domestic MW oven wherein the reaction of neat triphenyl phosphine with organic halides show remarkable rate enhancement in a pressurized vessel[68] (Scheme-45).

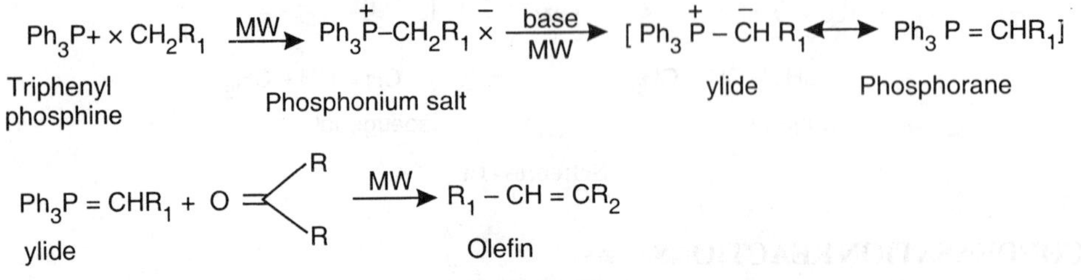
Ph3P+ x CH2R1
MW
Ph3P–CH2R1 x
base
MW
[Ph3 P – CH R1
Ph3 P = CHR1]
Triphenyl phosphine
Phosphonium salt
ylide
Phosphorane
Ph3P = CHR1 + O
R
R
MW
R1 – CH = CR2
ylide
Olefin

Scheme-45

4.6.3 Synthesis of Imines, Enamines, Nitroalkenes and N-Sulfonylimines, Hydrazones and Amides

The conventional preparation of the titile compounds involve the azotropic removal of water from the condensation reaction intermediates that are generally catalysed by p-toluene sulphonic acid, titanium (IV) chloride or montmorillonite K10 clay. The use of a Dean Starks apparatus is an essential requirement that necessatiates the use of large excess of aromatic hydrocaarbons such as benzene or toluene for azeotropic water removal.

Imines and enamines have been synthesised via the reaction of primary and secondary amines with aldehydes and ketones respectively. (Schemes 46 and 47 respectively). Using MW expediated dehydration reactions using montmorillonite K10 clay[44] or Envirocat reagent[45], EPZG®.

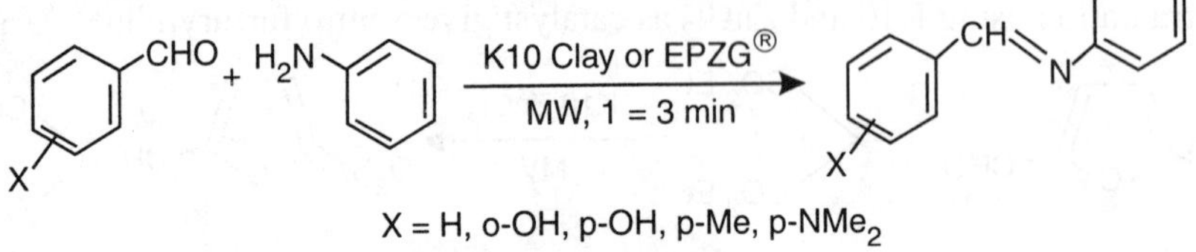
CHO
H2N
K10 Clay or EPZG®
MW, 1 = 3 min
CH
N
X
X
X = H, o-OH, p-OH, p-Me, p-NMe2

Scheme-46 Clay catalysed solvent free MW synthesis of imines

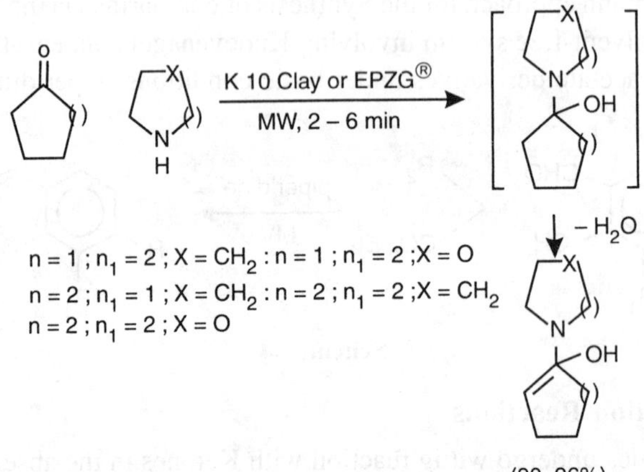
O
X
N
H
K 10 Clay or EPZG®
MW, 2 – 6 min
OH
– H2O
n = 1 ; n1 = 2 ; X = CH2 : n = 1 ; n1 = 2 ;X = O
n = 2 ; n1 = 1 ; X = CH2 : n = 2 ; n1 = 2 ;X = CH2
n = 2 ; n1 = 2 ; X = O
(90–96%)

Scheme-47 Clay catalysed solvent free MW synthesis of enamines

In both the above synthesis, the generation of polar transition state intermediates that couple to microwave is mainly responsible for rapid imine- or enamine forming reactions. The use of a MW oven at lower power levels or intermittent heating has been used to prevent loss of low boiling reactions[44, 45].

α, β-Unsaturated Nitroalkenes have been prepared by the well known **Henry reaction** involving the condensation of nitroalkanes with carbonyl compounds under MV irradiation in presence of catalytic amounts of ammonium acetate[69], thus avoiding the use of large excess of polluting nitrohydrocarbans usually employed in these reactions (Scheme-48).

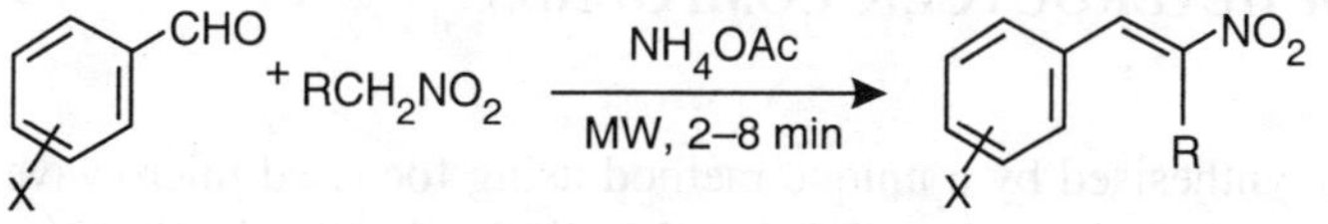

X = H, p-OH, m,p-(OMe)$_2$, m-OMe-p-OH, 1-naphthyl, 2-naphthyl; R = H
X = H, p-OH, p-OMe, m,p-(OMe)$_2$, m-OMe-pOH ; R = Me

Scheme-48 MW-expediated Synthesis of unsaturated alkenes

A number of functional groups such as nitroalkenes, N-substituted hydroxylamines, amines, ketones oximes and α-substituted oximes and ketones[70-72] could be prepared from α, β-unsaturated nitroalkenes.

N-Sulfonylimines have been synthesised by a one pot reaction involving mw heating of alkehydes with sulfonamides using relatively benign reagents, calcium carbonate and montmorillonile K10 clay[73] (Scheme-49).

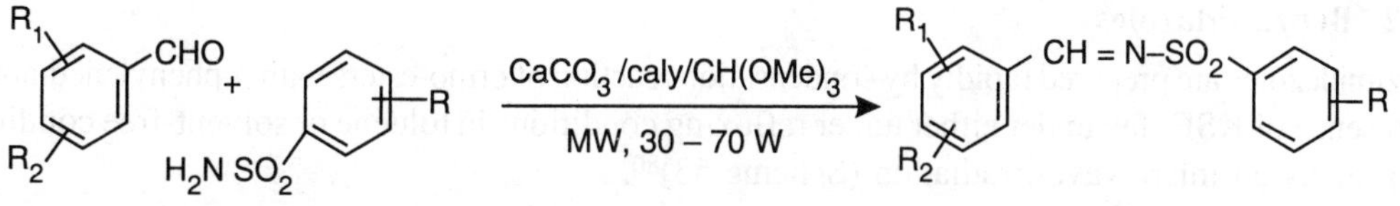

Where R = H, Me, COOMe, Cl; R$_1$ = H, OMe, OCOMe, Br;
R$_2$ =, H, OMe, OCOMe

Scheme-49 One pot solvent free MW preparation of N-sulfonylimines

The formation of **hydrazone derivatives** has been achieved in tolume[74]. Further the reaction of hydrazone with alkali (KOH) accomplishes **Wolff-Kichner reduction** that proceeds in good yield under MW irradiation conditions[75]. Recently, it has been shown that a solvent-free and catalyst free reaction of hydrazines with carbonyl compounds is possible upon MW irradiation in household MW oven (Scheme-50)[76]. The reaction is completed below the melting points of the two reactants possibly via the formation of a eutectic[77].

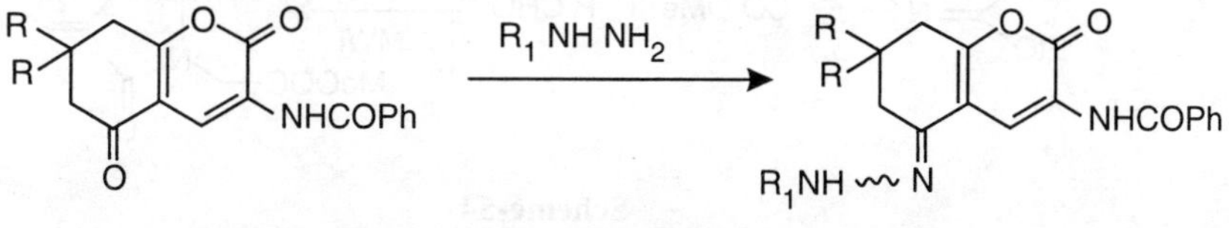

Where R = H or N Me and R$_1$ = aryle or substituted aryl group

Scheme-50 MW assisted Synthesis of hydrazones

Amides have been synthesised by a solid-state synthesis by the reaction of non-enolizable esters and amines in a household MW oven using potassium tert-butoxide[78] (Scheme-51).

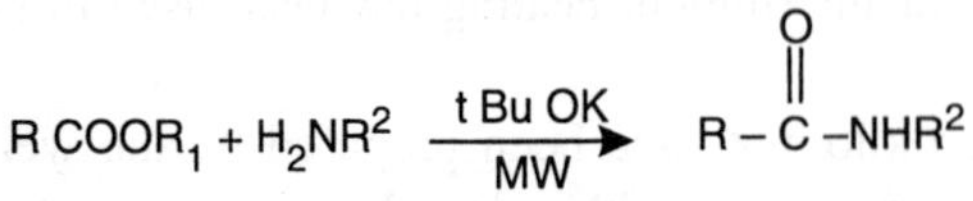

$$R\,COOR_1 + H_2NR^2 \xrightarrow[MW]{t\,Bu\,OK} R-\overset{O}{\underset{\|}{C}}-NHR^2$$

Scheme-51

4.7 SYNTHESIS OF HETEROCYCLIC COMPOUNDS

4.7.1 Aziridines

Aziridines have been synthesised by a unique method using focussed microwave approach under solvent free conditions, where it is observed that the elimination predominates over the Michael addition under MW irradiation unlike classical heating under the same conditions (Scheme-52)[79].

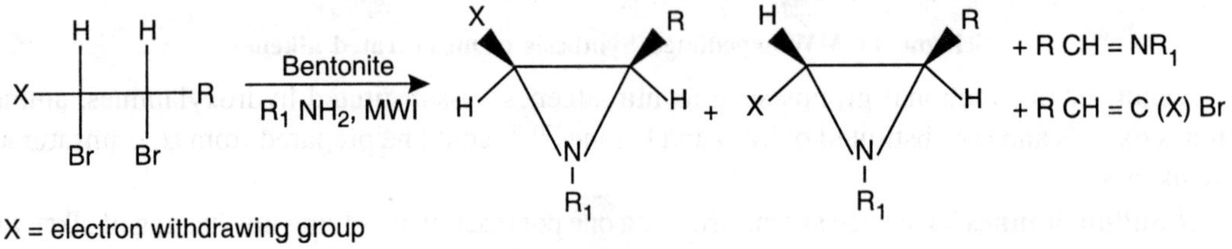

X = electron withdrawing group

Scheme-52

4.7.2 Benzimidazoles

Benzimidazoles are prepared rapidly by condensation reaction of ortho-esters with o-phenylenediamines in presence of KSF clay under either under refluxing conditions in toluene or solvent-free conditions using focussed microwave irradiation (Scheme-53)[80].

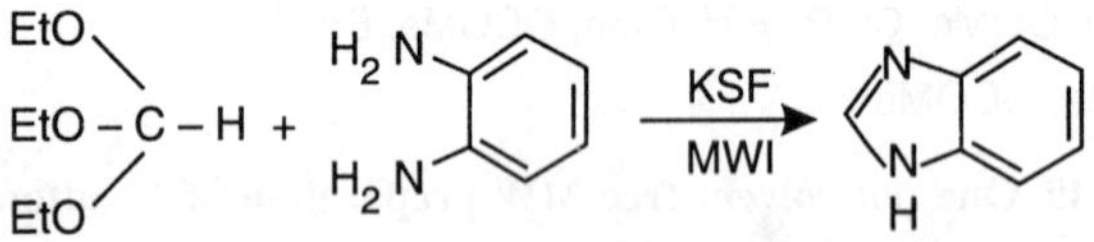

Scheme-53

4-Alkylidine-1H-imidazol-5 (4H) ones are obtained in good to excellent yield by 1, 3-dipolar cydoaddition, in solvent free conditions under focussed microwaves from an activated imidate and aldehydes in the presence of catalytic amounts of anhydrous acetic acid (Scheme-54)[81].

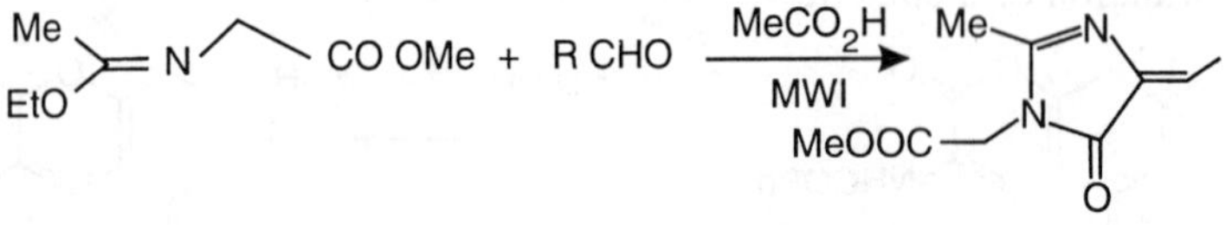

Scheme-54

An expediations solvent-free synthesis of pyrazolino/iminopyrimidino/thiooxopyrimido imidazoline derivatives from oxazolones on solid support using microwave has been described. The reaction time was brought down from hours to minutes as compared to conventional heating (Scheme-55)[82].

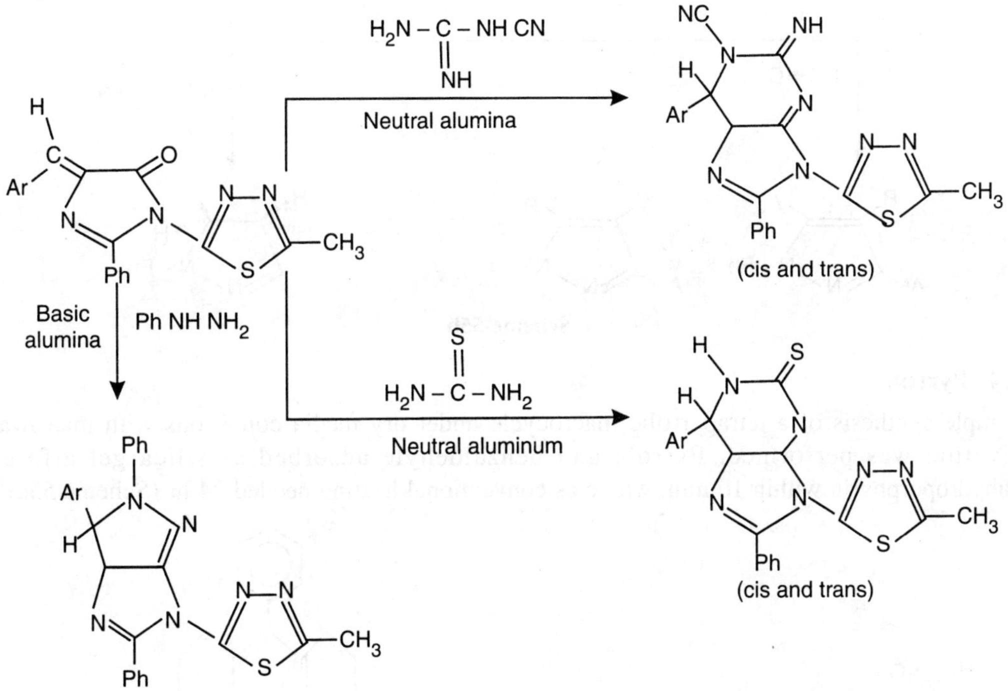

Scheme-55

A general method for solid phase synthesis of N-arylated benzimidazoles, imidazoles, triazoles and pyrazoles has been demonstrated utilizing copper (II) mediated coupling of aryl boronic acid under MMI[82] (Scheme-55a).

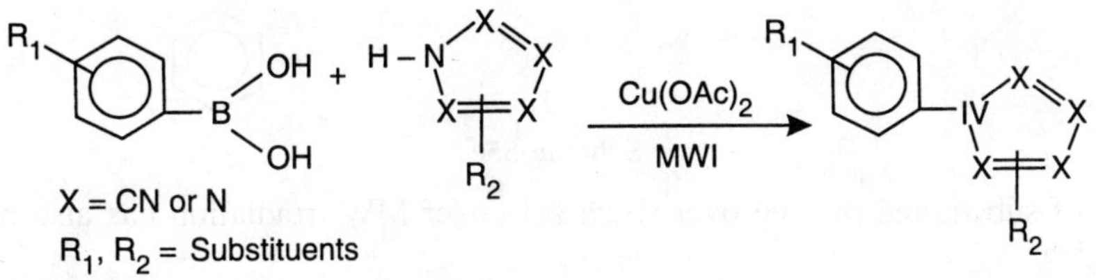

Scheme-55a

4.7.3 Pyrazoles

Pyrazoles are synthesised by a one pot reaction involving cyclocondensation of diarylnitrilimines with alkynes and disubstituted alkenes using irradiation with microwaves[83]. In this synthesis, the nitrilimine was generated in situ (Scheme-55b).

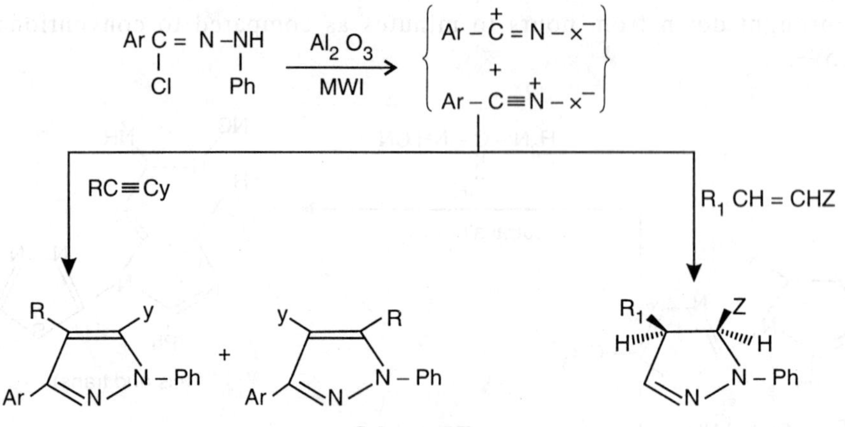

4.24 ■ Alternate Energy Processes in Chemical Synthesis

Scheme-55b

4.7.4 Pyrroles

A simple synthesis of a tetrapyrrolic macrocycle under dry media conditions with microwave activation was performed. Pyrrole and benzaldehyde adsorbed as silica gel afforded tetrahydroporphyrin within 10 min, where as conventional heating needed 24 hr (Scheme 55c)[83a].

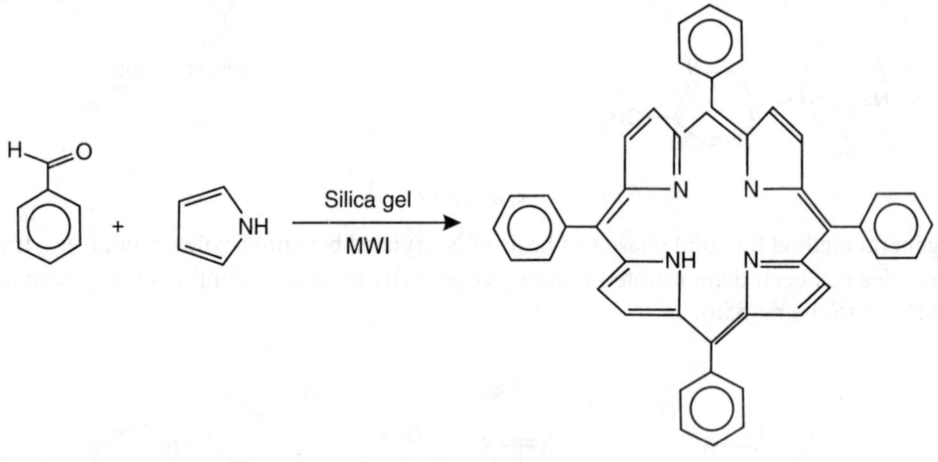

Scheme-55c

Synthesis of substituted pyrrole over silica gel under MW irradiation has also been reported (Scheme-55d)[83b].

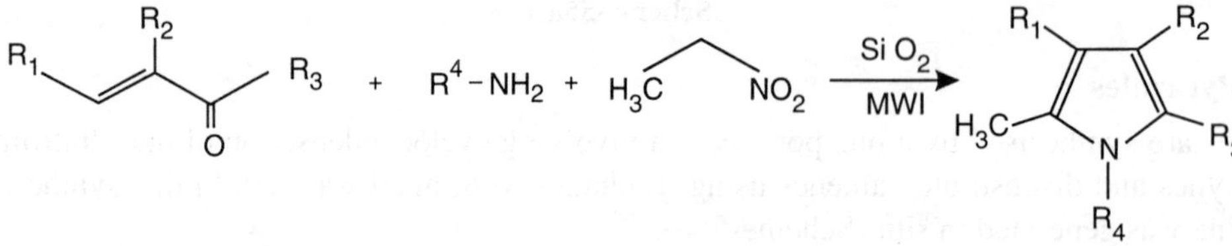

Scheme-55d

4.7.5 Azoles

Oxazolines are readily prepared from carboxylic acids and α, α, α-tri (hydroxymethyl) methylamine under MWI (Scheme-56)[84].

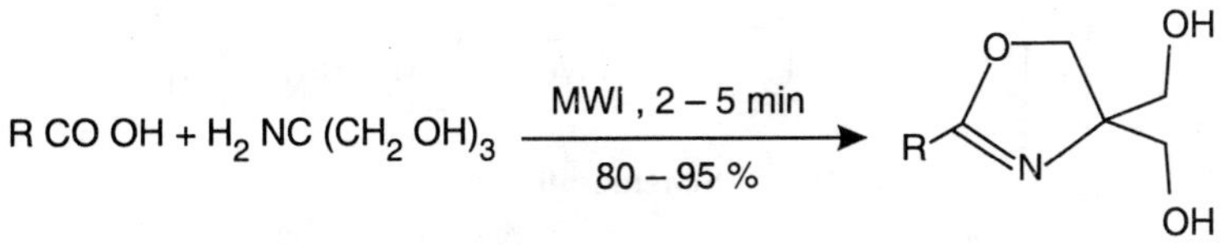

Scheme-56

1, 3-Dipolar cycloaddition of N-methyl-C-Phenylnitrones to methyl acrylate yield isooxazolidines on MW irradiation in presence of solid supports (silica gel, alumina) (Scheme-57)[85].

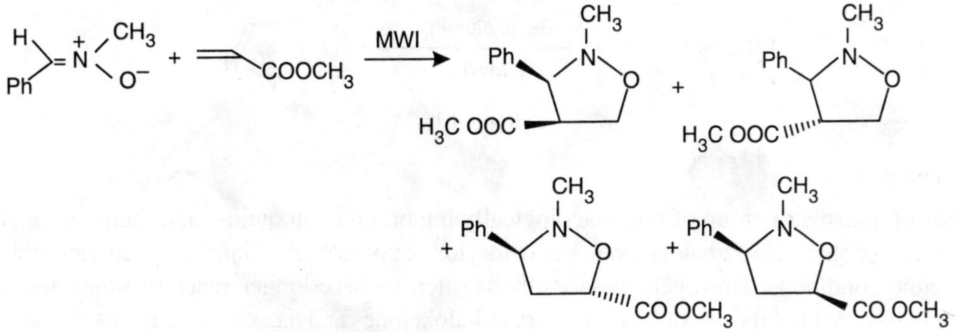

Scheme-57

The solid supported synthesis of azoles and diazines using K_2CO_3 as a solid support has been reported[86] (Scheme 58). This novel technique involves aqueous work up.

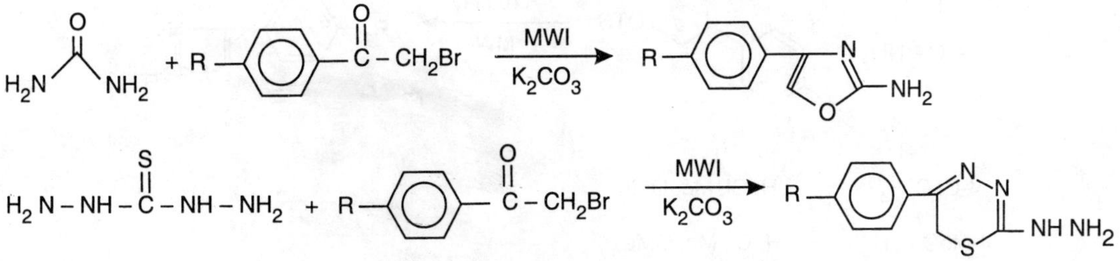

Scheme-58

1, 2, 4-Oxadiazoles are obtained in good yield by the reaction of amidoximes with isopropenyl acetate in presence KSF clay under MWI (Scheme-59)[87].

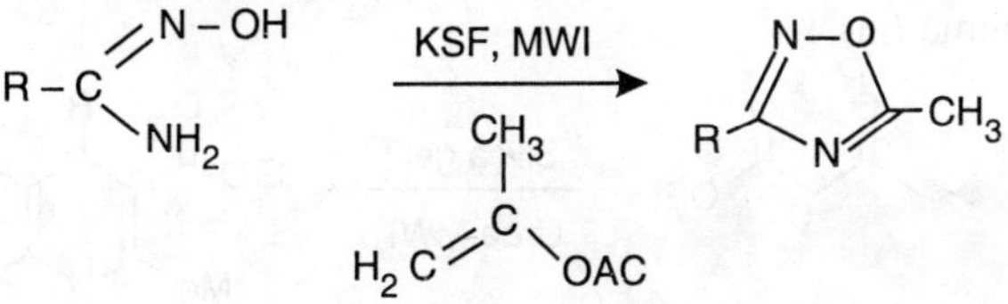

Scheme-59

The 1, 2, 4-oxindiazoles were also obtained[87] by MWI from o-acylamidoximes absorbed on alumina (Scheme-60).

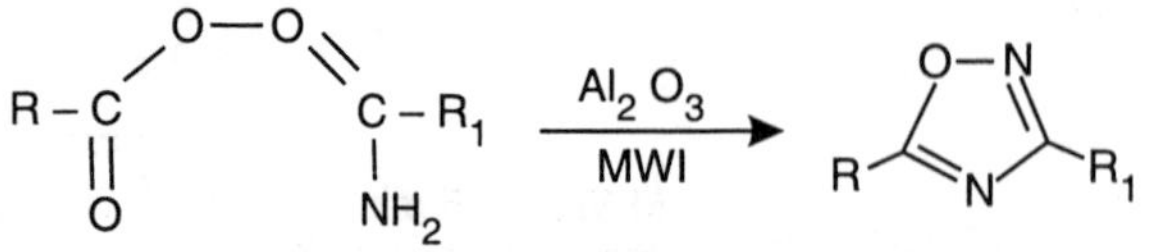

Scheme-60

4.7.6 Isoxazoles

Reaction of aromatic aldehydes with phenyl nitromethane under microwave irradiation (MWI) on basic alumina afforded excellint yields (90–96%) of isoxazoles within 2–3 minutes (Scheme-61)[87a].

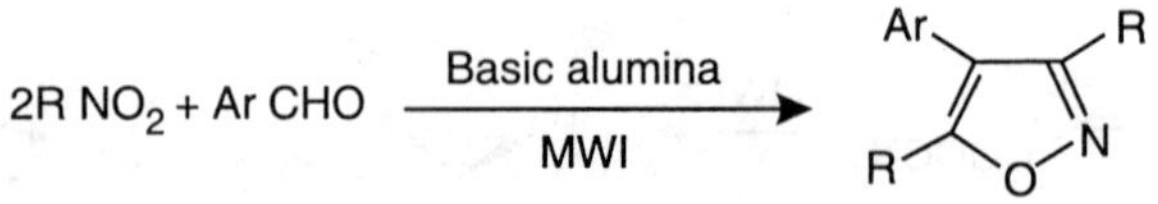

$$2R\,NO_2 + Ar\,CHO \xrightarrow[MWI]{Basic\,alumina}$$

Scheme-61

4.7.7 Thiazoles

Synthesis of thiazole (a group of pharmacologically important compounds) and their derivatives is generally accomplished by α-haloketones[88] or α-tosyloxyketones[89] and thiamides or ethylene thiourea under acidic conditions. However, these methods often involve longer reaction times and suffer from the requirement of the use of lachrymatory α-haloketones and hazardous reagents that generate a waste stream of spent solvent. A rapid solvent-free approach to thiazoles is by simple mixing of thiamides or ethylenethcourea with α-tosyloxykekone in a clay catalysed reaction that accelerated by MW irradiation[90] (Scheme-62).

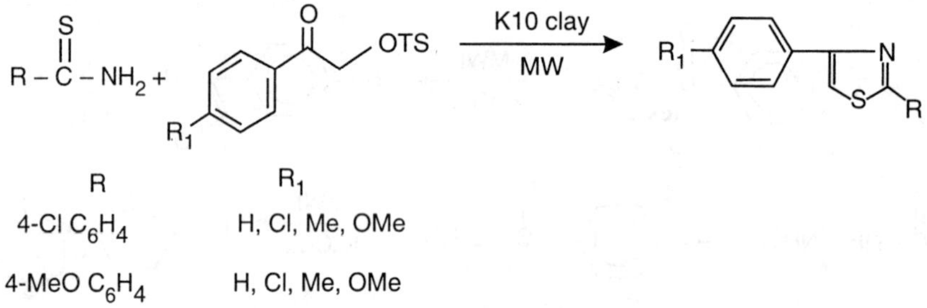

R	R_1
4-Cl C_6H_4	H, Cl, Me, OMe
4-MeO C_6H_4	H, Cl, Me, OMe

Scheme-62

4.7.8 Pyridines

Microwave irradiation of aldehydes and β-ketoesters and urea in presence of silica gave 1, 4-dihydropyridines (Scheme-63)[91].

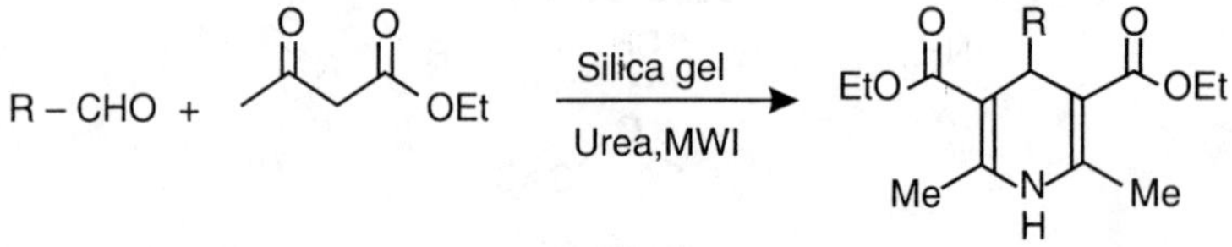

Scheme-63

4.7.9 Quinolines

KSF clay catalysed **Friedlander condensation** of 2-aminoaldehydes or ketones with carbonyl compounds containing α-methylene group has been achieved in solvent free conditions under MWI to give polycyclic quinoline derivative (Scheme-64)[92].

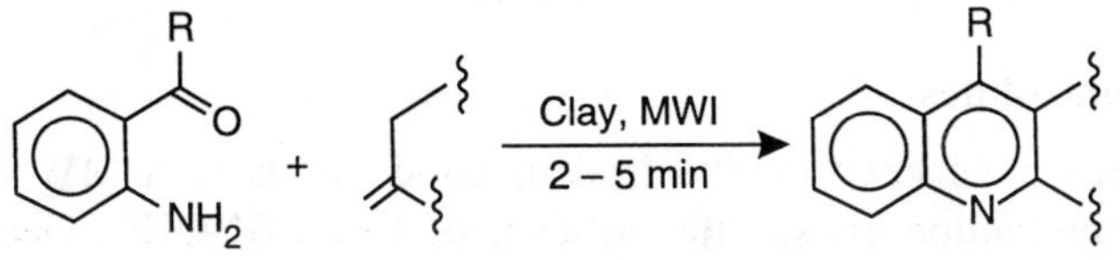

Scheme-64

In an alumina supported synthesis of antibacterial quinolines using microwaves where in the reaction time, was brought down from hours to seconds with improved yields as compared to the conventional heating (Scheme-65)[93].

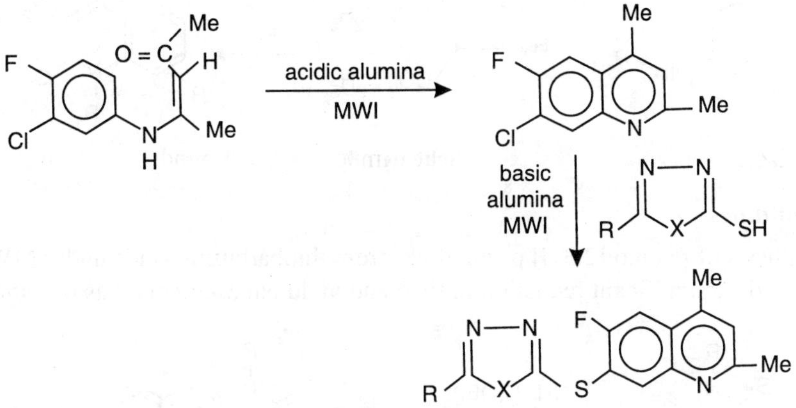

Scheme-65

4.7.10 Quinolones, Quinolinones and Quinazolines

Cyclisation of readily available 2′-aminochalkones using montmorillonite K10 clay under MWI conditions, gave 2-aryl-1, 2, 3, 4-tetrahydro-4-**quinolones** which are valuable precursors for medicinally important quinalones (Scheme-66)[94].

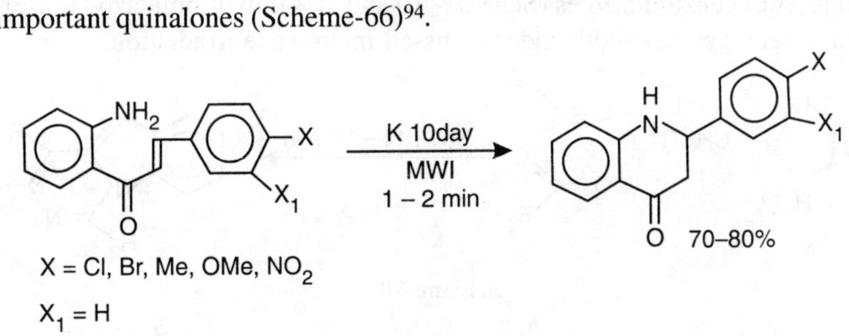

X = Cl, Br, Me, OMe, NO_2
X_1 = H

Scheme-66

2, 3-Disubstituted **quinoxalines** have been synthesised from the aryl or alkyl acyloins and o-phenylenediamine under MW irradiation for 3–6 min (Scheme-67)[95].

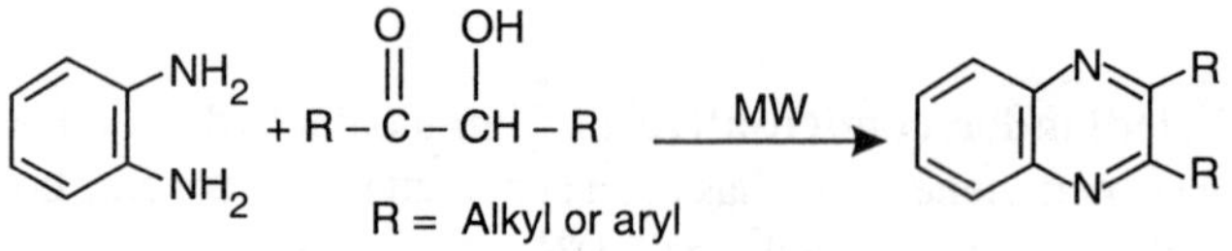

4.28 ■ *Alternate Energy Processes in Chemical Synthesis*

Scheme-67

4.7.11 4-Aminoquinazolines

4-Aminoquinazolines have been synthesised in good yields in a MW oven starting from cyanoaromatics and anthranilonitriles in the presence of 10% t-BUOH[96]. One-pot MW-enhanced synthesis of selective glycine-site NMPA receptor antagonists, 3-aryl-4-hydroxyquinolin-2(IH)-ones have been developed via the amidation of malonic ester derivatives with anilines followed by subsequent cydisation of the intermediate, malondianilides (Scheme-68)[97].

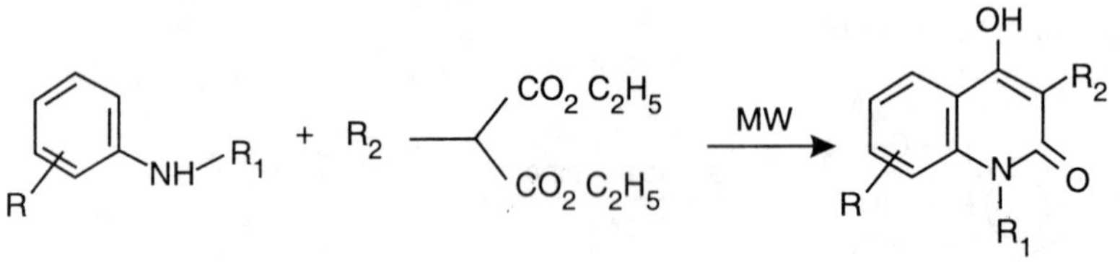

Scheme-68

4.7.12 Pyrimidines

A one pot synthesis of pyrano [2,3-d] pyrimidines from thiobarbituric acids under MW irradiation has been reported. A significant reduction in time and yield enhancement was observed (Scheme-69)[98].

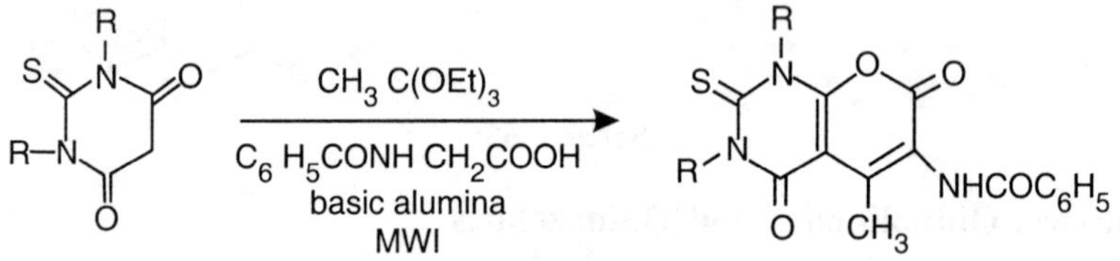

Scheme-69

Pyrimidino [1, 6-a] benzimidazoles (Scheme-70) and 2, 3-dihydroimidazo [1, 2-c] pyrimidines (Scheme-71) have been synthesised[99] under focussed microwave irradiation.

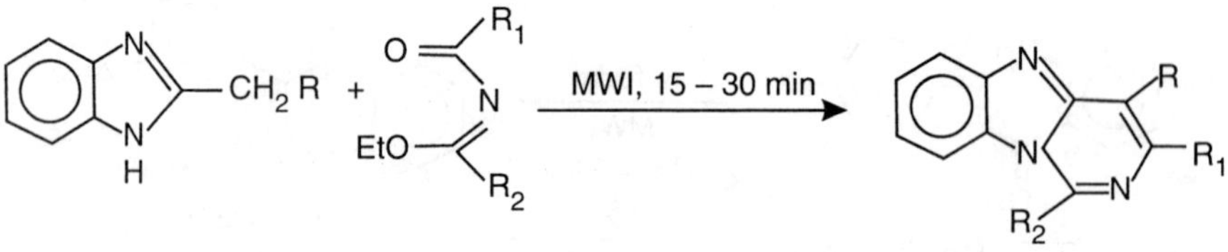

Scheme-70

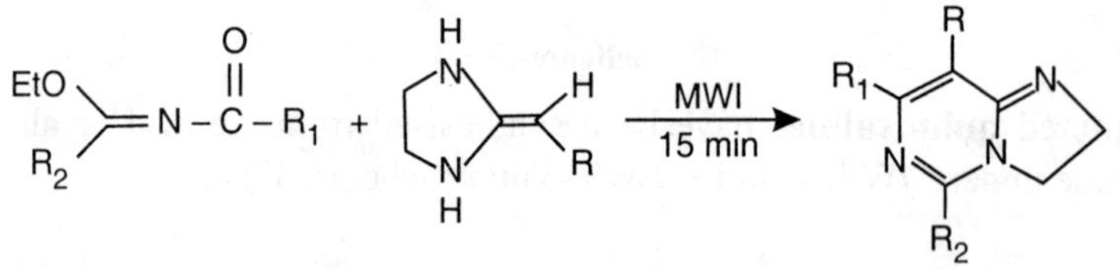

Scheme-71

An efficient synthesis of benzopyranopyrimidenes using three different solid supports viz. acidic alumina montmorillonite, silica gel has been carried out. The products were obtained with improved yields as compared to conventional heating (Scheme-72)[100].

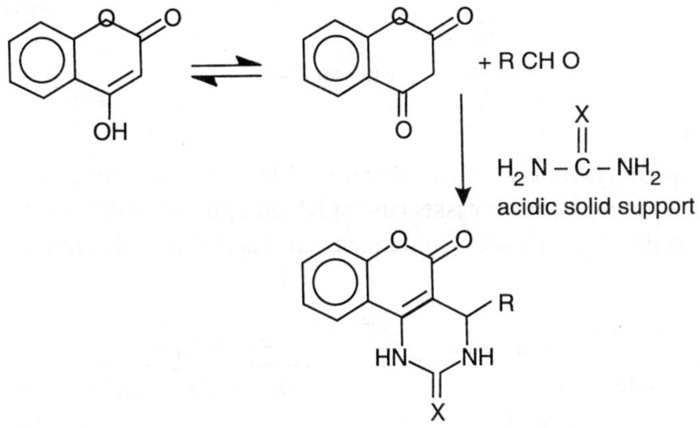

Scheme-72

4.7.13 Oxadiazines

Condensation of N, N'-dimethyl urea, paraformaldehyde and primary amines using montmorillonite K10 clay in dry media under microwave irradiation lead to the formation of triazones. However, condensation of N, N'-dimethyl urea and paraformaldihyde supported on montmorillonite K10 using MWI gave 4-Oxo-Oxadiazinones (Scheme-73)[100a].

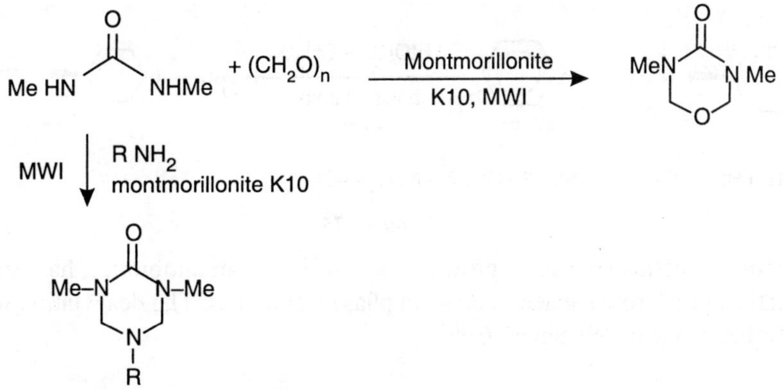

Scheme-73

4.7.14 Thiadiazepines

An environmentally benign synthesis of 1, 2, 4-triazolo [3, 4-b]-1, 3, 4-thiadiazepenes from substituted triazoles and chalkones on basic alumina udner MW irradiation has been achieved (Scheme-74)[101].

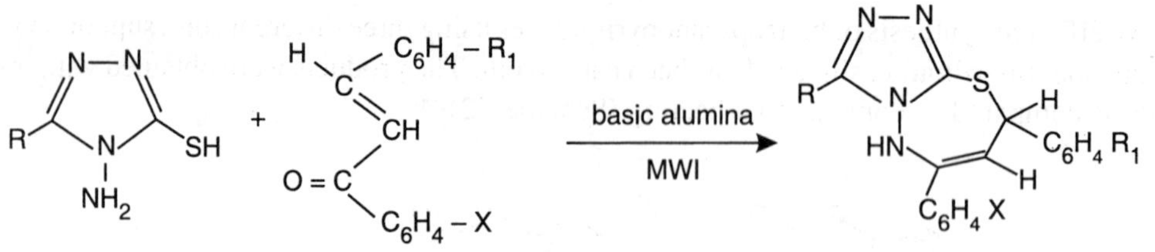

4.30 ■ Alternate Energy Processes in Chemical Synthesis

Scheme-74

4.7.15 β-Lactams

An efficient and rapid synthesis of a number of β-lactoms has been described[102] under MW irradiation. Further in closed Teflon vessels using KF and phase transfer catalyst (PTC), β-lactoms have been synthesised in few minutes from ketene silyl acetal and aldimines (Scheme-74a)[103].

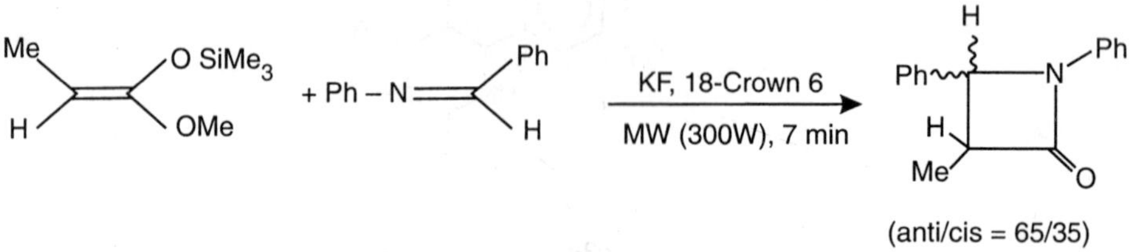

(anti/cis = 65/35)

Scheme-74a

N-(4-Hydoxycyclohexyl)-3-mercapto/cyano-4-aryl-azetidine-2 ones have been synthesised from N-(4-hydroxycyclohexyl)-arylaldimine by reacting with ethyl α-mercapto/α-cyanoacetate on basic alumina under microwave irradiation (Scheme-75), where in not only the reaction time was brought down from hours to minutes in comparison to conventional heating but also the yields were improved[104].

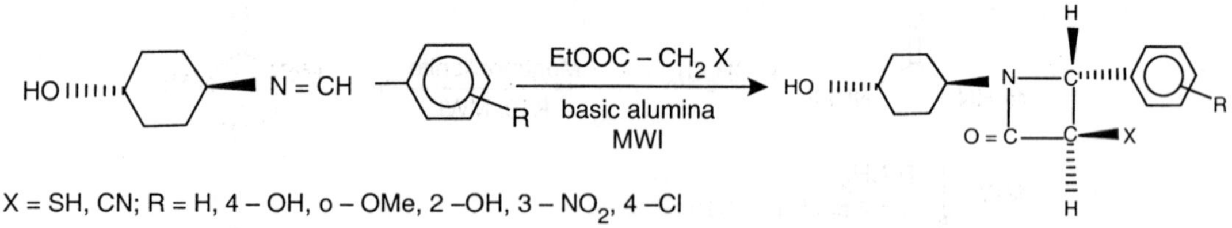

X = SH, CN; R = H, 4 – OH, o – OMe, 2 –OH, 3 – NO_2, 4 –Cl

Scheme-75

Deacylation of cephalosporins, a growing class of β-lactam antibiotics, has been investigated using enzymatic and microwave activated solid phase techniques. The deacylation was achieved in less time with better yields (Scheme-76)[105].

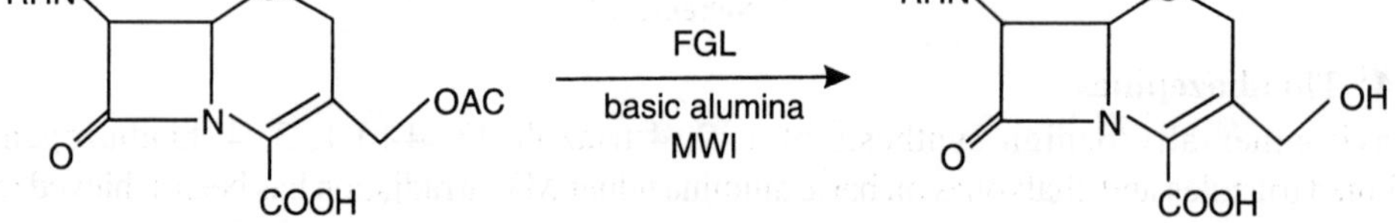

Scheme-76

Reaction of 7-amino-3[5′-methyl-1′, 3′, 4′-thiadiazol-2′-ylthiomethyl] cephalosporanic acid with heterocyclic amines using basic alumina under microwave irradiation afforded new cephalosporin analogs in shorter reaction time with improved yield as compared to conventional heating (Scheme-77)[106].

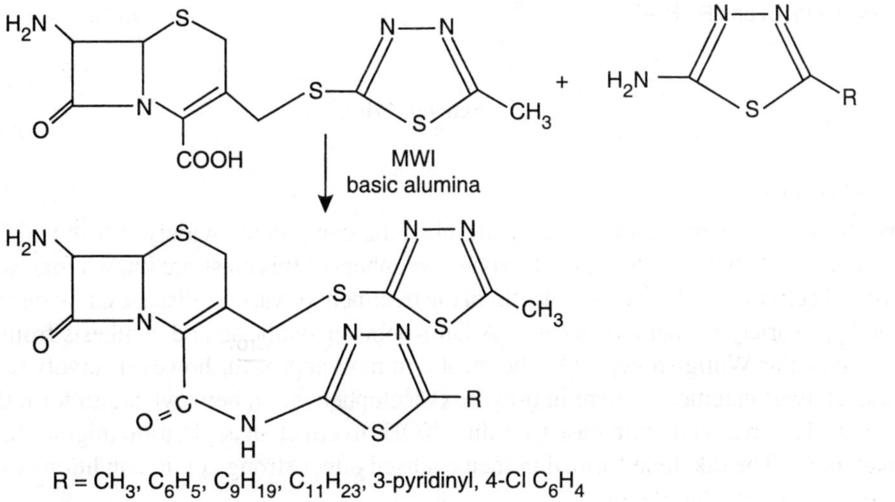
MWI
basic alumina
R = CH3, C6H5, C9H19, C11H23, 3-pyridinyl, 4-Cl C6H4

Scheme-77

An environmentally friendly safe method developed for the preparation of 3-carbomoyl cephalosporin derivatives such as cefuroxime uses o-transcarbomylase, an enzyme of microbial origin for the conversion of 3-hydroxyfunction to the desired 3-carbomyl group. This new synthesis replaces the conventional chemical route, which employs hazardous isocyanates such as dichlorophosphenyl isocyanate or chlorosulfinyl isocyanate to achieve the same conversion (Scheme-77a)[106a].

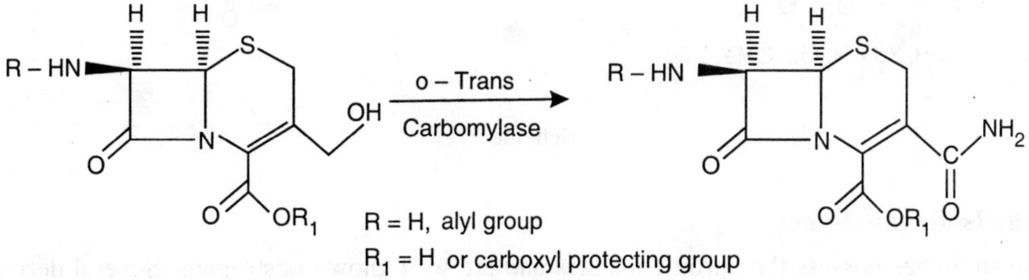
o – Trans
Carbomylase
R = H, alyl group
R1 = H or carboxyl protecting group

Scheme-77a

4.7.16 Furans

Naturally occurring, pharmacologically important 2-aroylbenzofuran are easily obtained under basic solvent-free conditions from α-toyloxyketones and salicylaldehydes in the presence of potassium fluoride doped alumina using MW irradiation (Scheme-77b)[107].

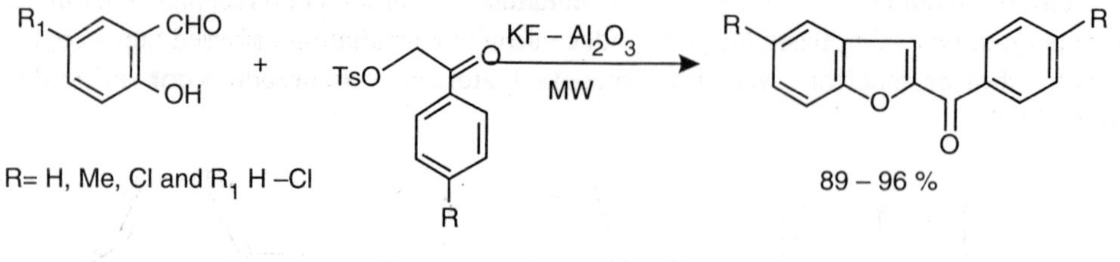

4.32 ■ *Alternate Energy Processes in Chemical Synthesis*

Scheme-77b

4.7.17 Flavones

Flavonoids are a class of naturally occurring phenolic compounds widely distributed in the plant kingdom, the most abundant being the flavones. Members of this class are known for a wide variety of biological activities and have been useful in the treatment of various diseases. Flavones have been prepared by a variety of methods such as Allan Robinson synthesis and synthesis from chalkones via intramolecular Wittig strategy[108]. The most common approach, however, involves the Baker-venkatraman rearrangement, where in o-hydroxyacetophenone in benzoylated to form the benzoyl ester followed by treatment with base (pyridine/KOH) to effect an acyl group migration, forming a 1, 3-diketone[108]. The diketone formed in then cyclised udner strong acidic conditions using H_2SO_4 and acetic acid to give the flavone.

A solvent free synthesis of flavones involves the microwave irradiation of o-hydroxydibenzoylmethanes adsorbed on montmorillonite K10 clay for 1–1.5 min (Scheme-77c)[108].

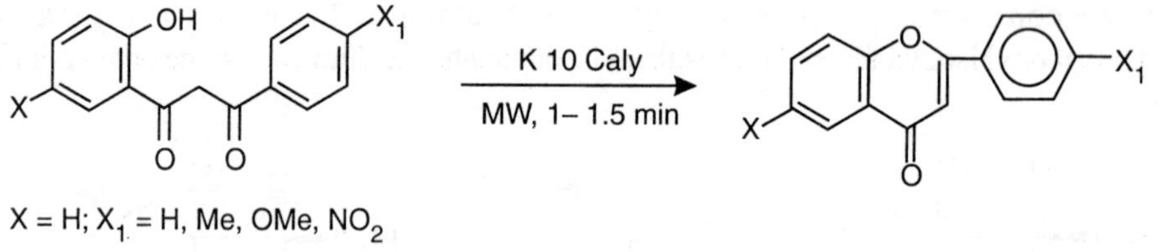

Scheme-77c

4.7.18 Isoflavan-3-enes

Isoflavan-3-enes possess the chrome nuclues and are well known oestrogens. Several derivatives of these oxygen heterocydes have attracted the attention of medical chemists[109–111]. Several methods are available for the synthesis of chromene derivatives.

The MW approach has been extended to a one pot synthesis of 2-aminosubstituted isoflavan-3-enes that involves the generation of the enamine derivative in situ followed by reaction with salicylaldehydes in the same reaction vessel (Scheme-78)[112–114].

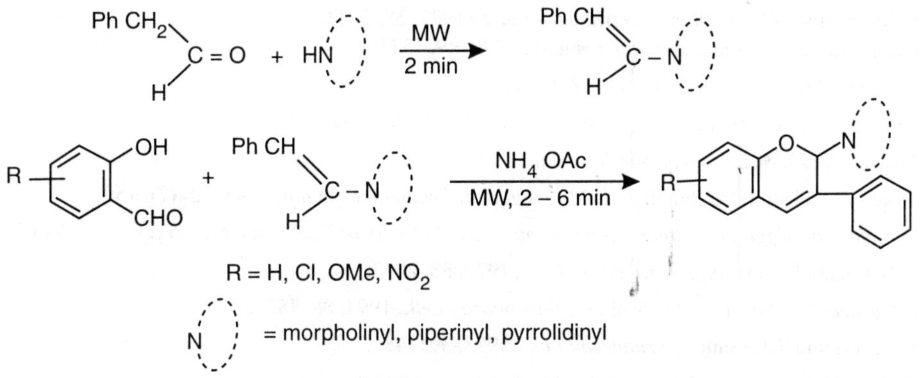

Scheme-78

NOTES AND REFERENCES

1. M. Balogh and P. Laszlo, *Organic Chemistry Using Clays*, Springer, Berlin, 1993.
2. T.W. Greene and P.G.M. Wuts, Protective Groups in Organic Synthesis. 2nd Edn., John Wiley and Sons Inc., New York, N.Y., 1991.
3. M. Csiba, J. Cleophas, A. Loupy, J. Malthete and S.D. Gero, *Tetrahedron Lett.*, 1993, **34**, 1787.
4. D. Villemim, A.B. Alloum and F. Thibault-Starzy's, *Syn. Cammun.*, 1992, **22**, 1359.
5. D. Ruhua, W. Guliang and J. Yao Zhing, *Synth. Commun.*, 1994, **24(i)**, 111.
6. D. Ruhua, W. Yuliang and Y. Yaozhong, *Synth. Commun.*, 1994, **24(13)**, 1917.
7. B. Perio, M.J. Dozias, P. Jacquault and J. Hamilin, *Tetraedron Lett.*, 1997, **38**, 7867.
8. D. Bogdal, J. Pielichowki and A. Boron, *Synlett.*, 1996, 873.
9. A. Vass, J. Toth and E. Pallai Varsanyi, Effect of Inorganic Solid Support for microwave assisted organic reactions, presented at the international conference on Microwave Chemistry, Prague, Czech. Republic, Sept. 6–11, 1998.
10. R.S. Varma, M. Varma and A.K. Chatterjee, *J. Chem., Soc.*, Perkin Trans. I, 1993, 999.
11. R.S. Varma, A.K. Chatterjee and M. Varma, *Tetrahedron Lett.*, 1993, **34**, 3207.
12. P. de la Cruz, A. de la Hoz, L.M. Font, F. Langa and M.C. Perez, -Kodriguez, *Tetrahedron Lett.*, 1998, **39**, 6053.
    (a) D. Jinchang, G. Henjic, W. Jinzhu and L. Caizhen, *Synth. Commun.*, 1994, **24**, 301.
13. W.F. Huffimann, R.F. Hall, J.A. Grant and H.D. Holden, *J.Med. Chem.* 1978, **21**, 413.
14. T. Tsuji, T. Kataoka, M. Yoskioka, Y. Sendo, S. Nishitant, S. Hirai, T. Maeda and W. Nagata, *Tetrahedron Lett.*, 1979, 2793.
15. C.W. Roberts, *J. Am. Chem. Soc.*, 1954, **76**, 6203.
16. R.S. Varma, A.K. Chatterjee and M. Varma, *Tetrahedron Lett.*, 1993, **34**, 4603.
17. R.S. Varma, A.K. Chatterjee and M. Varma, *Tetrahedron Lett.*, 1993, **34**, 4603.
18. L. Polombi, F. Bonadies and A. Scettri, *Tetrahedron*, 1997, **53**, 15867.
19. R.S. Varma, *Pure Appl. Chem.*, 2001, **73**, 193.
20. R.S. Varma, and H.M. Meshram, *Tetrahedron Lett.*, 1997, **38**, 5427.
21. A.I. Bosch, P. da la Cruez, E. Diez-Barra, A. Loupy and F. Langa, *Synlett.*, 1995, 1259.
22. R.S. Varma, R. Dahiya and R.K. Saini, *Tetrahedron Lett.*, 1997, **38**, 8819.

23. R.S. Varma and H.K. Meshram, *Tetrahedron Lett.*, 1997, **38**, 7973.
    (a) R.S. Varma, D. Kumar, *Synth. Commun.*, 1999, **29**, 1333.
24. R.S. Varma, D. Kumar, *Org. Lett.*, 1999, **1**, 697.
25. A Loupy, M. Pedoussaut and J. Sanoulet, *J. Org. Chem.*, 1986, **51**, 740.
26. A Dietrich and J.M. Lehn, *Tetrahedron Lett.*, 1973, 1225.
27. L. Loupsy, P. Pigeon, M. Ramdani and P. Jaequault, *Synthetic Commun.*, 1994, **24(2)**, 159.
28. *Comprehensive Organic Synthesis (Oxidation)*, (Ed.) B.M. Trost Pergaman, New York 1991, Vol 7.
29. R.S. Varma and R. Dahiya, *Tetrahedron Lett.*, 1977, **38**, 2043.
30. R.S. Varma, R.K. Sarine and R. Dahiya, *Tetrahedron Lett.*, 1997, **38**, 7823.
31. R.S. Varma and R.K. Saini, *Tetrahedron Lett.*, 1998, **39**, 1481.
32. R. Ballini, Y. Bosica and M. Parrini, *Tetrahedron, Lett.*, 1998, **39**, 7963.
33. R.S. Varma, R. Dahiya and R.K. Saini, *Tetrahedron Lett.*, 1997, **38**, 7029.
34. R.S. Varma and R. Dahiya, *Tetrahedron Lett.*, 1998, **39**, 1307.
35. R.S. Varma, D. Kumar and R. Dahiya, *J. Chem. Res (s)*, 1998, 324.
36. R.S. Varma, R. Dahiya and D. Kumar, *Molecules on line*, 1998, **3**, 82.
    (a) R.S. Varma, R.K. Saini and H.M. Mishram, *Tetrahedron Lett.*, 1997, **38**, 6525.
37. R.S. Varma, R.K. Saini and R. Dahiya, *J. Chem. Res (s)*, 1998, 120.
38. R.S. Varma and R. Dahiya, *Synth. Commun.*, 1998, **28**, 4087.
39. H. Benhaliliba, A. Derdour, J.-P Bazurean, F. Texier-Boullet and J. Hamelin, *Tetrahedron Lett.*, 1998, **39**, 541.
    (a) O. Garefa, F. Dolgado, A.C. Cano and C. Alverez, *Tetrahedron Lett.*, 1993, **34**, 623; C. Alvarez, F. Delgado, O. Garefa, S. Medina and C. Marquez, *Synth. Commun.*, 1991, **21**, 619.
    (b) F. Delago, C. Alvarez, O. Garcia, G. Penieres, and C. Marquiz, *Synth. Commun.*, 1991, **21**, 2137.
40. R.S. Varma and R.K. Saini, *Tetrahedron Lett.*, 1997, **38**, 4337.
41. R.F. Borch, M.D. Berstein and H.D. Durst, *J. Am. Chem. Soc.*, 1971, **93**, 289.
42. A.F. Abdel-Magid, K.G. Carson, B.D. Harris, C.A. Maryanoff and R.D. Shah, *J. Org. Chem.*, 1996, **61**, 3849 and references cited therein.
43. V. Giancarlo, A.G. Giumanini, P. Strazzoline and M. Poiana Synthesis, 1993, 121.
44. R.S. Varma, R. Dahiya and S. Kumar *Tetrahedron Lett.*, 1997, **38**, 2039
45. R.S. Varma and R. Dahiya, *Synlett.*, 1997, 1245.
46. R.S. Varma and R. Dahiya, *Tetrahedron*, 1998, **54**, 6293.
47. G.H. Posner, A.W. Runquist and M.J. Chapdelani, *J. Org. Chem.*, 1972, **42**, 1202 and references cited therein.
48. D. Barbry and S. Torchy, *Tetrahedron Lett.*, 1997, **38**, 2959.
49. S. Cannizzano, *Ann.*, 1853, **88**, 129.
50. T.A. Geisssman, *Organic Reactions*, 1944, II, 94.
51. C.G. Swain, A.I. Powell, W.A. Sheppard and C.R. Morgan, *J. Am. Che. Soc.*, 1979, **101**, 3576.
52. E.C. Ashby, D.T. Coleman III, M.P. Gamasa, *Tetrahedron Lett.*, 1983, **24**, 851.
53. R.S. Varma, K.P. Naicker and P.J. Liesen, *Tetrahedron Lett.*, 1998, **39**, 8437.
54. E. Gutirrez, A. Loupy, G. Bram and E. Ruiz-Hitzky, *Tetrahedron Lett.*, 1989, **30**, 945.
55. D. Villemin and B. Labiad, *Synth. Commun.*, 1922, **22**, 2043.
56. I. Almena, A. Diaz-ortiz, E. Diez-Barra, A. Hos and A. *Loupy Chem. Lett.*, 1996, 333; S. Caddick, *Tetrahedron*, 1995, 10400.

57. A.I. Bosch, P. de la Cruez, E. Diez-Barra, A. Luppy and F. Langa, *Synlett.*, 1995, 1259
58. F. Toda, Y. Tamaka, Y. Kagawa and Y. Sakaino, *Chem. Lett.,* 1990, 373.
59. H.-M. Yu, S.-T. Chen, M.-J. Tseng and K.-T. Wang, *J. Chem., Res(s),* 1999, 62.
60. G.L. Kad, IR Trehan, J. Kaur, S. Nayyar, A. Arora and J.S. Brar, *Ind. J. Chem.*, 1999, **35B**, 734.
61. F.M. Moghaddam, M. Ghaffarzadeh, S.M.Abdi Oskoui, *J. Chem. Res. (S)*, 1999, 574.
62. F.M. Moghaddam, M.G. Dakamin, *Tetrahedron Lett.*, 2000, **41**, 3479.
63. A. Vass, J. Toth and E. Pallai-Varsanyi, Abstract No. OR 19, International Conference on Microwave Chemistry, Pregue, Czech Republic, Sept 6–11 (1998).

    (a) A. Loupy and Le Ngoc Thach, *Synth. Commun.*, 1993, **23**, 2571.
64. D. Villemin and B. Martin, *Synth. Commun.*, 1995, **25**, 3135.
65. D. Villemin and B. Martin, *J. Chem., Res. (S)*, 1994, 146.
66. V. Singh, J. Singh, P. Kaur and G.L. Kad, *J. Chem. Res (S),* 1997, 58.
67. A. Spinella, T. Fortunati and R. Sorienti, *Synlett.*, 1997, 93.
68. J.J. Kiddle, *Tetrahedron Lett.*, 2000, **41**, 1339.
69. R.S. Varma, R. Dahiya and S. Kumar, *Tetrahedron Lett.*, 1997, **38**, 5131.
70. R.S. Varma and G.W. Kabalka, *Heterocycles*, 1986, **24**, 2645.
71. G.W. Kabalka and R.S. Varma, *Org. Prep. Proc. Internl.* 1987, **19**, 283.
72. G.W. Kabalka, L.H.M. Guindi and R.S. Varma, *Tetrahedron*, 1990, **46**, 7443.
73. A. Vass, J. Dudas and R.S. Varma, *Tetrahedron Lett.*, 1999, **40**, 4951.
74. S. Gadhwal, M. Boruah and J.S. Sandhu, *Synlett.*, 1999, 1573.
75. E. Parquet and Q. Lin. *J. Chem.*, Eeu., 1997, 74.
76. M. Jeselnik, R.S. Varma, S. Polanc and M. Kocevar. *Chem. Commun.*, 2001, 1716.
77. M. Jeselnik, R.S. Varma, S. Polanc and M. Kocevar, *Green Chemistry*, 2002, **4**, 35.
78. R.S. Varma and K.P. Naicker, *Tetrahedron Lett.,* 1999, **40**, 6177.
79. A Saoidi, J. Hamelin and H. Benhaoua, *J. Chem, Res. (S)*, 1996, 492.
80. D. Villemin, M. Hammadi and B. Martin, *Synth. Commun.*, 1996, **26**, 2895.
81. G. Kerneur, J.M. Lerestif, J.P. Bazureau and J. Hamelin, *Synthesis,* 1997, 287.
82. M. Kidwai, P. Sapra, K.R. Bhushan and P. Misra, *Synthesis*, 2001, **10**, 1509.

    (a) A.P. Combs, S. Saubern, M. Rafalski and P.Y.S. Lam, *Tetrahedron Lett.*, 1999, **40**, 1623.
83. K. Bougrin, M. Soufiaoui, A. Loupy and P. Jacquault, *New J. Chem.*, 1995, **19**, 213.

    (a) A. Petit, A. Loupy, P. Maillard and M. Momentau, *Syn. Commun.,* 1992, **23**, 1137.
84. A.L. Marrera-Terrero and A. Loupy, *Synlett.*, 1996, 245.
85. A. Padwa, L. Fisera, K.F. Koehler, R. Rodriguez and Y.S.K. Wong, *J. Org. Chem.*, 1984, **49**, 276.
86. M. Kidwai, R. Venkataramann and B. Dave, *J. Heterocycl, Chem.,* 2002, **39**, 1-3.
87. B. Oussaid, L. Moeini, B. Martin, D. Villemin and B. Garrigues, *Synth. Commun.*, 1995, **25(10)**, 1451.

    (a) M. Kidwai and P. Sapra, *Org. Prep. Proced. Int.*, 2001, **33**, 381.
88. M. Fefer and L.C. King, *J. Org. Chem.*, 1961, **26**, 828.
89. O. Prakash, N. Rani and S. Goyal, *J. Chem. Soc.*, Perkin Trans, 1992, **1**, 707.
90. R.S. Varma, D. Kumar and P.J. Liesen, *J. Chem. Res.*, Perkin Trans I, 1999, 4093.

91. J.S. Yadav, B.V.S. Reddy and P.T. Reddy, *Synth. Commun.*, 2001, **31(3)**, 425.
92. G. Sabitha, R.S. Babu, B.V.S. Reddy and J.S. Yadav, *Synth. Commun.*, 1999, **29(24)**, 4403.
93. M. Kidwai, K.R. Bhushan, P. Sapra, R.K. Saxena and R. Gupta, *Bioorg. Med. Chem.*, 2000, **8**, 69.
94. R.S. Varma and R.K. Saini, *Synlett.*, 1997, 857.
95. F. Juncai, L. Yang, M. Quinghua and L. Bin, *Synth. Commun.*, 1998, **28**, 193.
96. J.A. Seijas, M.P. Vazquez-Tato and M.M. Martinzer, *Tetrahedron Lett.*, 2000, **41**, 2215.
97. J.H.M. Lange, P.C. Verveer, S.J.M. Osnabrug and G.M. Visser, *Tetrahedron Lett.*, 2001, **42**, 1367.
98. M. Kidwai, K. Venkataramanan, R.K. Garg and K.R. Bhushan, *J. Chem. Res. (S)*, 2000, 586.
99. M. Rahmouni, A. Derdour, J.P. Bazureau and J. Hamelin, *Tetrahedron Lett.*, 1994, **35**, 4563.
100. M. Kidwai and P. Sapra, *Synth. Commun.*, 2002, **32(11)**, 1693.
    (a) S. Balalaie, M.S. Hashtroudi and A. Sharifi, *J. Chem., Res (S)*, 1999, 392.
101. M. Kidwai, P. Sapra, P. Misra, R.K. Saxena and M. Singh, *Bioorg. Med. Chemistry*, 2001, **9**, 217-220.
102. A.K. Bose, M. Jayaraman, S.S. Okawa, S.S. Beri, E.W. Robb and M.S. Manhas, *Tetrahedron Lett.*, 1996, **37(39)**, 6989.
103. F. Texier Boullet, R. Latouche and J. Hamelin, *Tetrahedron Lett.*, 1993, **34**, 2123.
104. M. Kidwai, R. Venkatramanan and S. Kohli, *Synth. Commun.*, 2000, 30(6), 989.
105. M. Kidwai, B. Dave, K.R. Bhushan, P. Misra, R.K. Saxena, R. Gupta, R. Gulati and M. Singh, *Biocatalyst and Biotransformations*, 2002, **20(5)**, 377-379.
106. M. Kidwai, P. Sapra, K.R. Bhushan, P. Misra, R.K. Saxena, R. Gupta and M. Singh, *Bioorg. Chemistry*, 2001, **29**, 380.
    (a) I.D. Flemeng, M.K. Turner and S.J. Brewer, US. 4075061, 1978; *Chem. Abstr.* 1978, **88**, 168488.
107. R.S. Varma, D. Kumar and P.J. Liesen, *J. Chem. Soc.* Perkin Trans I, 1998, 4093.
108. R.S. Varma, R.K. Saini and D. Kumar, *J. Chem. Res. (S)*, 1998, 348.
109. W. Lawson, *J. Chem. Soc.*, 1954, 4448.
110. S. Durani, A.K. Agarwal, R. Saxena, B.S. Setty, R.C. Gupta, P.L. Koli, S. Ray and N. Anand, *J. Steroid Biochem.*, 1979, **11**, 67.
111. T.A. Grese and L.D. Pennington, *Tetrahedron Lett.*, 1995, **36**, 8913.
112. F.M.Dean and R.S. Verma, *Tetrahedron Lett.*, 1981, **22**, 2113.
113. F.M. Dean and R.S. Varma, *J. Chem. Soc. Perkin Trans*, 1982, **1**, 1193 and 2271.
114. R.S. Varma and R. Dahiya, *J. Org. Chem.*, 1998, **63**, 8038.

5

MISCELLANEOUS REACTIONS

5.1 A SINGLE STEP CONVERSION OF ARYL ALDEHYDES TO AROMATIC NITRILES

The conversion of aldehydes to the corresponding nitriles is an important transformation[1]. In most of the cases the aldoxime is first prepared and subsequently dehydrated using a wide variety of reagents like O, N-bis (trifluoroacetyl) hydroxylamine or trifluoroacetohydroximic acid[2], chloramine/base[3], (H_2SO_4/SiO_2)[4], p-chlorophenyl chlorothionoformate/pyridine[5], triethylamine/dialkyl hydrogen phosphinates[6], $TiCl_4$/pyridine[7], triethylamine/phosphonitrilic chloride[8] and 1, 1′-dicarbonylbiimidazoles[9]. These methods involve dehydration of oxime which is a time demanding process for one pot reactions[10]. It is found that aldehydes are rapidly converted into nitriles in good yields (89–95%) with hydroxylamine hydrochloride supported on montmorillonite K10 clay in the absence of solvent[11, 12].

The above mentioned reaction is a general one as exemplified by a variety of aldehydes (Scheme-1) that undergo this facile conversion to afford high yields of the corresponding nitriles (89-95%) within a short MW irradiation[11, 12].

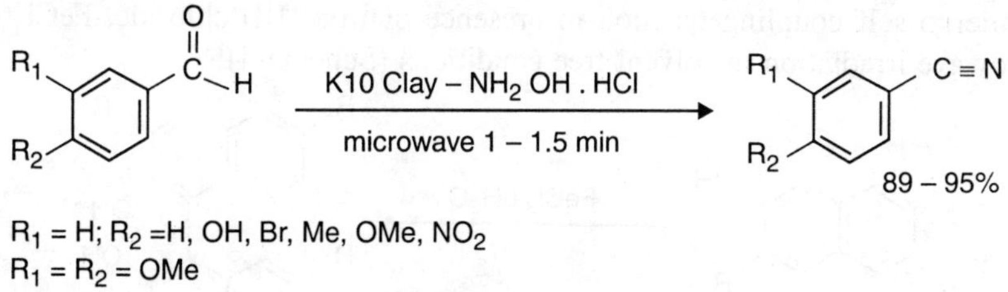
R1
R2
O
C
H
K10 Clay – NH2 OH . HCl
microwave 1 – 1.5 min
R1
R2
C ≡ N
89 – 95%
R1 = H; R2 =H, OH, Br, Me, OMe, NO2
R1 = R2 = OMe

Scheme-1

5.2 SYNTHESIS OF ANHYDRIDES FROM DICARBOXYLIC ACIDS

Dicarboxylic acids can be converted into anhydrides (Scheme-2) in the presence of isopropenyl acetate (which acts as a water scavenger) under microwave irradiation[13] using montmorillonite-KSF. The driving force is the formation of acetone.

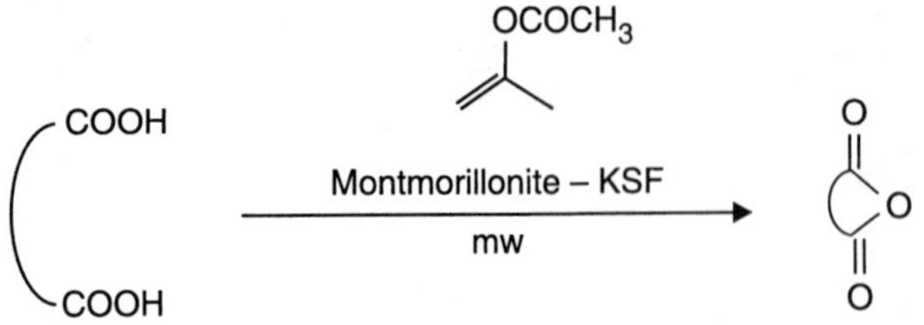

5.2 ■ Alternate Energy Processes in Chemical Synthesis

Scheme-2

This method is rapid and convenient and avoids using corrosive reagents like CH_3COCl, $SOCl_2$, $(CH_3CO)_2O$.

5.2a SIDE CHAIN NITRATION OF STYRENE TO β-NITROSTYRENE

A facile solid slate synthesis of β-nitrostyrene from readily available feedstock, styrene and its substituted derivatives using inexpensive 'doped' clay reagents, Clayfen and clayan (Scheme-3) has been described[14]. In this procedure, the neat reagent, styrene and clayfen or clayan are mixed in a glass container and the solid mixture heated in an oil bath (~100–110°C, 15 min) or irradiated in a microwave oven (~100–110°, 3 min). In the latter case, intermittent warming for 30s intervals to maintain temperature below 60–70° gives good results. The major product is β-nitrostyrene.

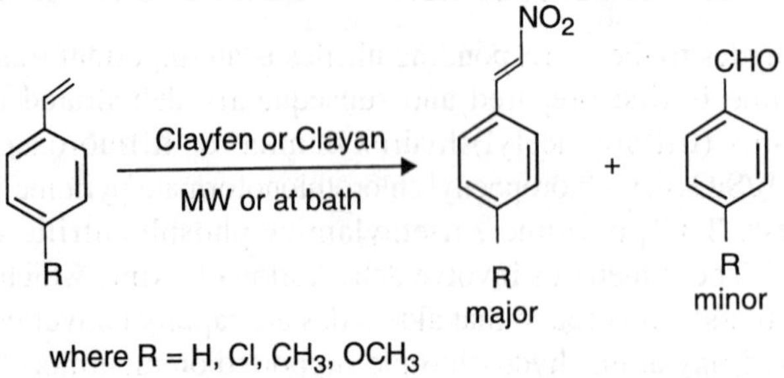

where R = H, Cl, CH_3, OCH_3

Scheme-3

In case the reaction is conducted in solution phase, polymeric products are formed.

5.3 OXIDATIVE COUPLING OF β-NAPTHOLS

β-Napthols undergo self coupling reaction in presence of iron (III) chloride, $FeCl_3 6H_2O$ under focussed microwave irradiation in solvent free conditions (Scheme-4)[15].

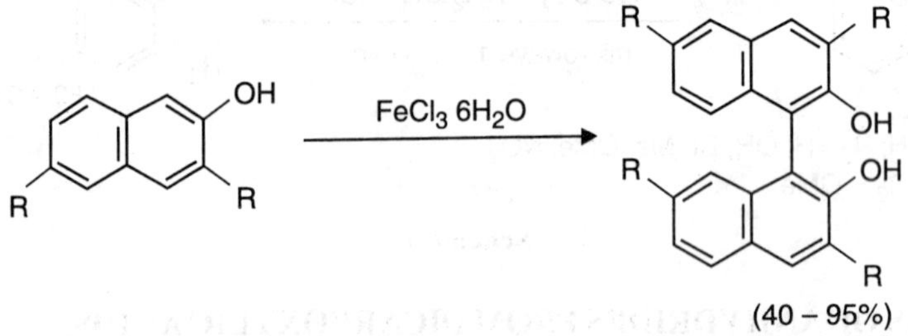

(40 – 95%)

Scheme-4

5.4 METHYLENATION OF 3, 4-DIHYDROXYBENZALDEHYDE

Methylenation of 3, 4-dihydroxybenzaldehyde takes place rapidly in presence of a phase transfer catalyst on a benign calcium carbonate surface. In this case, the bonding of the vicinal hydroxyl groups is low thereby enhancing the reaction with the alkylating agent under solvent-free microwave irradiation (Scheme-5)[16].

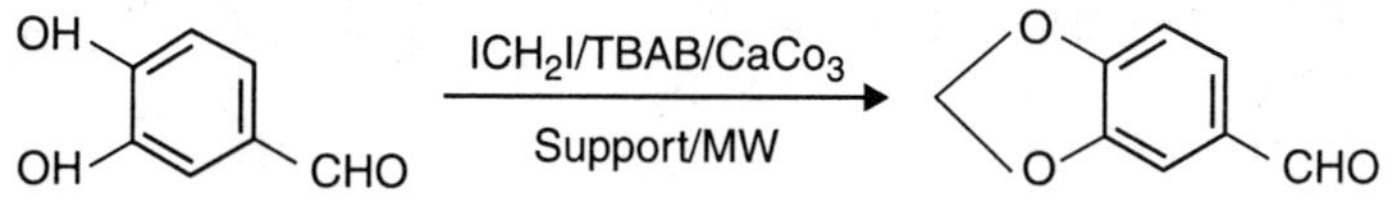

Scheme-5

5.5 MICHAEL ADDITION

A number of 2'-hydroxy, 4',6'-dimethylchalcones undergo a solid state intramolecular Michael type addition to yield[17] the corresponding flavanones (Scheme-6).

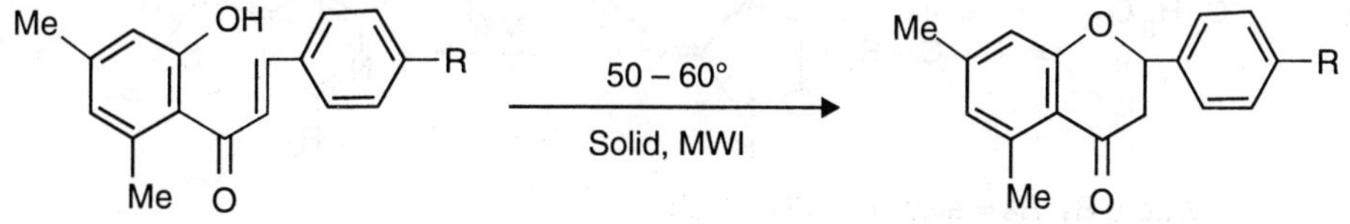

2' Hydroxy-4', 6'-dimethylchalcones
R = H, Cl or Br

5, 7-dimethyl flavanones

Scheme-6

The Mechael addition of chalcone to 2-phenylcyclohexonone under PTC conditions[18] gave 2, 6-disubstituted cyclohexanone derivatives in high disteroselectivity (99% ee) (Scheme-7).

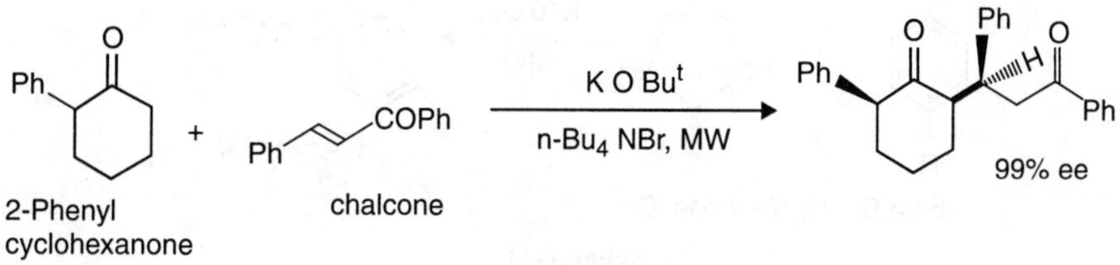

Scheme-7

An interesting example in the solvent free Michael addition reaction of nitromethane to chalcone in the presence of alumina under microwave irradiation conditions that gives the adduct in 90% yield (Scheme-8)[19].

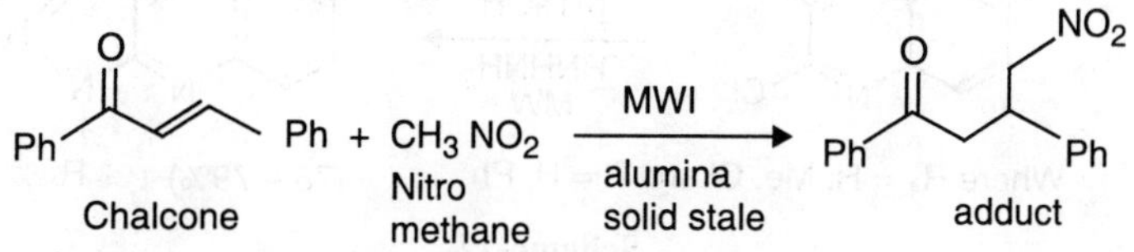

Scheme-8

5.6 SYNTHESIS OF BRIDGEHEAD NITROGEN HETEROCYCLIC COMPOUNDS

Microwave assisted synthesis of bridgehead nitrogen heterocycles has been achieved. Thus, pyrimidino [1, 6-a] benzimidazoles are synthesised from N-acylimidates and activated 2-benzimidazoles (Scheme-9)[20].

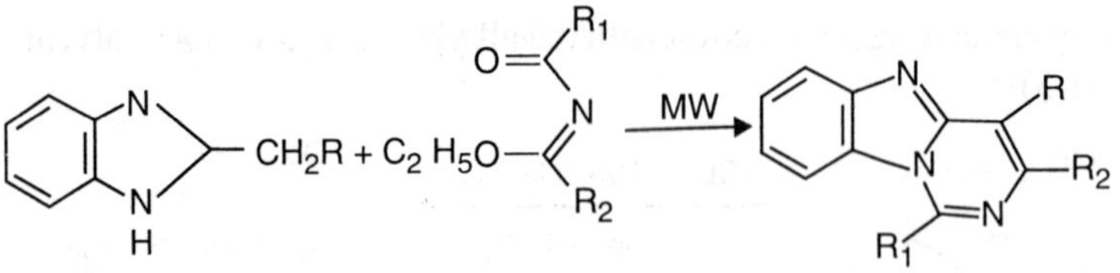

5.4 ■ Alternate Energy Processes in Chemical Synthesis

Where R = CN, CO_2 Me, CU_2 Et

Scheme-9

In a similar way MW irradiation of N-acylimidates and imidazoline ketene aminals give 2, 3-dihydroimidazo [1, 2-c] pyrimidines (Scheme-10)[21].

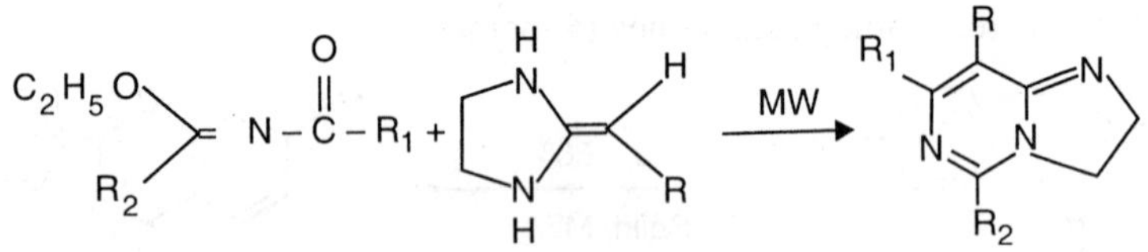

Where R_1, R_2 = alkyl, R = CN, CO_2Et

Scheme-10

Microwave-accelerated synthesis of the corresponding bridgehead heterocycles gets completed in a short time by the reaction of α-tosyloxyketones with ethylenethioureas (Scheme-11)[22]. In the conventional heating in an oil bath the reaction remains incomplete[22, 23].

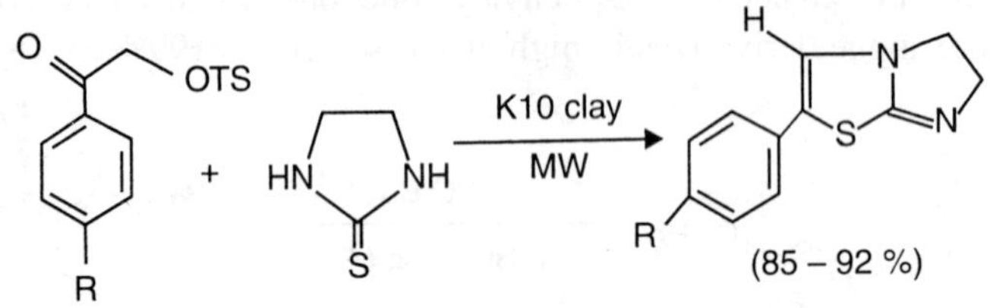

where R = H, Me, OMe, Cl

(85 – 92 %)

Scheme-11

Pyrazolo [3, 4-b] quinolines and pyrazolo [3, 4-c] pyrazoles have been synthesised using MW irradiation from β-chlorovinylaldehydes and hydrazines in presence of p-toluene sulfonic acid (p-Ts OH) (Scheme-12)[24].

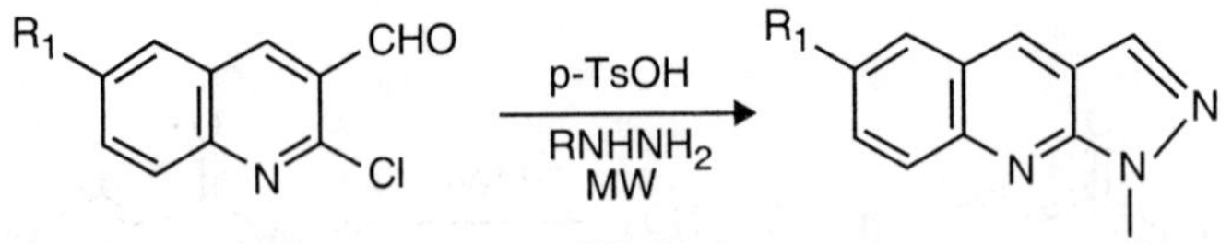

Where R_1 = H, Me, Cl and R = H, Ph

(78 – 79%)

Scheme-12

Imidazo [1, 2-a] pyridines, imidazo [1, 2-a] pyrazines and imidazo [1, 2-a] pyridimenes under solvent free condition using MW irradiation[25] by the **ugi reaction**. The procedure consist in mixing aldehydes (aliphatic, aromatic and vinylic) and the corresponding 2-amino-pyridine, pyrazine or pyrimidine in presence of catalystic amount of clay (50 mg) to generate the iminium intermediate. Subsequently isocyanate (aliphatic, aromatic or cyclic) is added to the same container and the reactants are further exposed to MW to afford the corresponding imidazo [1, 2-a] pyridines, imidazo [1, 2-a] pyrazines and imidazole [1, 2-a] pyrimidines respectively (Scheme-13).

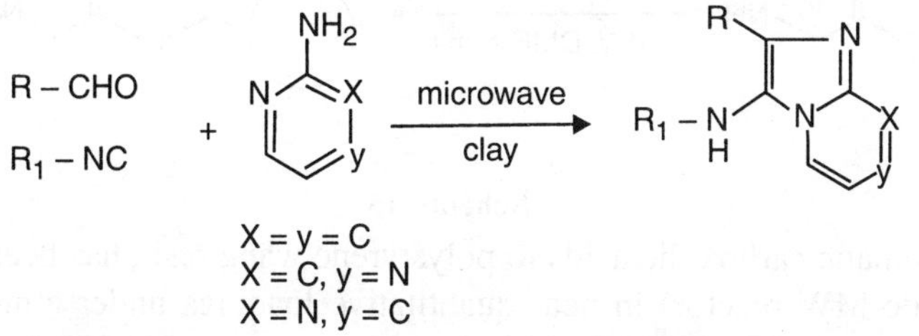

$$R-CHO + R_1-NC + \underset{\substack{X=y=C \\ X=C, y=N \\ X=N, y=C}}{\text{(aminoheterocycle)}} \xrightarrow[\text{clay}]{\text{microwave}} R_1-\underset{H}{N}-\text{(imidazo-fused product)}$$

Scheme-13

5.7 ENZYME CATALYSED REACTIONS

Normally in synthetic transformations, enzymes are used in aqueous or organic solvent at moderate temperature in order to preserve the activity of the enzymes. However, in view of the newer development that immobilize enzymes on solid supports[26], they are now amenable to operate at relatively higher reaction temperature with adequate pH control. The application of MW irradiation have been investigated with the enyme systems. Pseudomonas lipase (dispersed in Hyflo super cell that essentially consists of diatomeous silica around pH 8.5–9.0) and commercially available SP 435 NOVOZYM (*Candida antarctica lipase* grafted on an acrylic resin)[27].

Using the above ensymes, the solvent-free resolution of recemic 1-phenylethanol has been achieved under MW irradiation conditions by transesterification (Scheme-14).

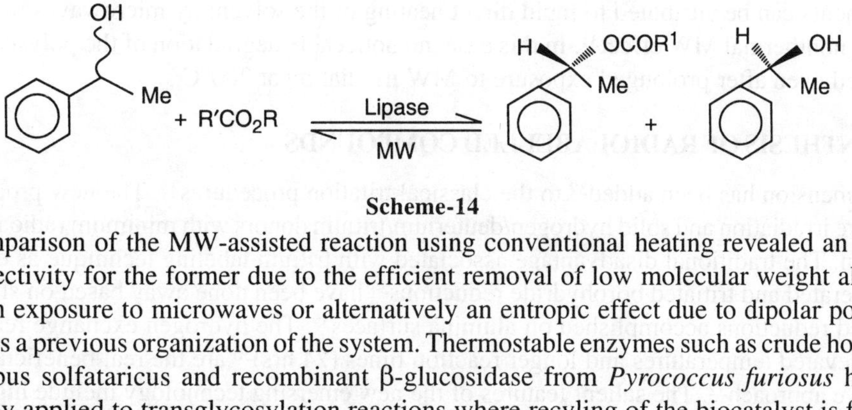

Scheme-14

A comparison of the MW-assisted reaction using conventional heating revealed an enhanced enantioselectivity for the former due to the efficient removal of low molecular weight alcohols or water upon exposure to microwaves or alternatively an entropic effect due to dipolar polarization that induces a previous organization of the system. Thermostable enzymes such as crude homogenate of sulfolobus solfataricus and recombinant β-glucosidase from *Pyrococcus furiosus* have been successfully applied to transglycosylation reactions where recyling of the biocatalyst is feasible[28].

5.8 SOLID-PHASE ORGANIC SYNTHESIS (SPOS)

Using an unmodified MW oven and a custom-made solid phase reaction vessel, solid-phase organic synthesis has been carried out[29]. Using this approach several Fmoc-protected amino acids and peptide fragments have been coupled with glycine pre-loaded polystyrene Wang resin (PS-Wang) in DMF, using either the symmetric anhydride or pre-formed precursors as N-hydroxybenzotriazole active esters (HOBt) under atmospheric pressure (Scheme-15).

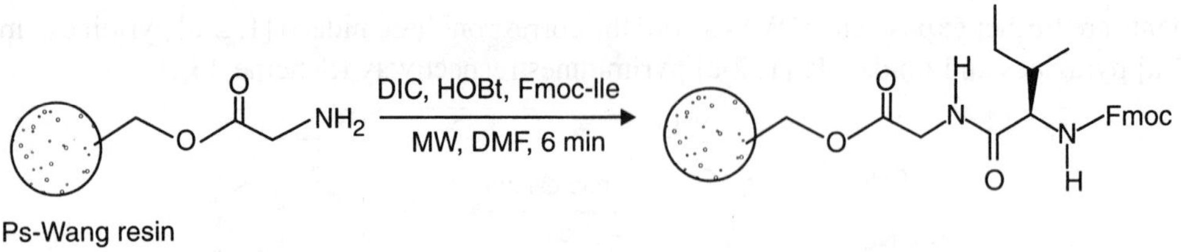

5.6 ■ Alternate Energy Processes in Chemical Synthesis

Scheme-15

Coupling of aromatic carboxylic acids to polystyrene wang resin has been reported[30] (using a dedicated multimode MW reactor) in near quantitative loadings under atmospheric pressure in 1-methyl-2-pyrrolidone (NMP) at 200°C within 10 min, as compared to 2–3 days using conventional coupling protocols at room temperature. In a related study, the attachment of carboxylic acid to chloromethylated polystyrene resins via cesium carbonate method has also been achieved[31] with significant rate enhancements as compared to the conventional thermal scheme (Scheme-16).

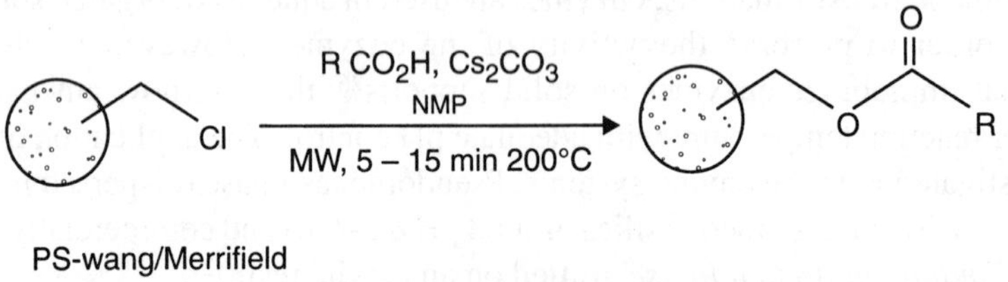

Scheme-16 MW-assisted coupling of carboxylic acid to merrifield resins

In the above procedure, reaction times are reduced from 12–48 hr with heating at 80°C to 5–15 min with MW flash heating at temperatures up to 200°C. It is believed that the observed rate enhancements can be attributed to rapid direct heating of the solvent by microwaves rather than to a specific nonthermal MW effect[31]. In this case, no noticeable degradation of the polystyrene resins is observed even after prolonged exposure to MW irradiation at 200°C.

5.9 SYNTHESIS OF RADIOLABELLED COMPOUNDS

A new dimension has been added[32] to the classical tritation procedures[33]. The new procedure uses microwave irradiation and solid hydrogen/deuterium/tritium donors with minimum radioactive waste generation. The traditional disadvantage associated with tritium labeling technique as exemplified with deuterated and tritiated borohydride reductions[34] have been done away based on similar MW-expediated reductions accomplished on alumina surfaces[35]. The hydrogen exchange reactions that require elevated temperatures and longer reaction times (24 hrs)[36] are the real beneficiaries of this microwave approach[32]. The salient features of the new emersing technology include high purity of labelled materials, efficient insersion and excellent regio-selectivity (Scheme-17).

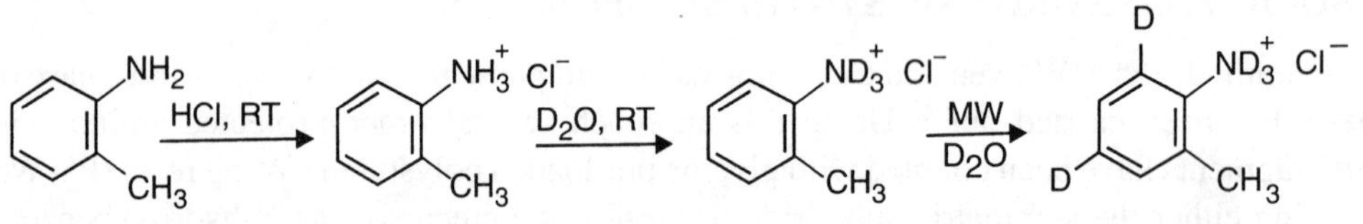

Scheme-17

Selective, faster and cleaner applications of MW-accelerated reactions have been demonstrated in the synthesis of a variety of radiolabelled (^{3}H, ^{11}C and ^{19}F) organic compounds via the nucleophilic aromatic and aliphatic sybstitution reactions, esterifications, condensations, hydrolysis and complexation reactions using monomodal MW cavities on microscale[37]. Thus a substaintially reduced level of radioactive waste is generated in these reactions.

Hydrogenation reactions in which $H_2/D_2/T_2$ gases are replaced by various formates proceed rapidly under MW irradiation canditions (Scheme-18)[38]. The pattern of labelling can be easily modified and the advantages are especially noteworthy in the case of tritium where high specific activity tritiated water is hazardous to use.

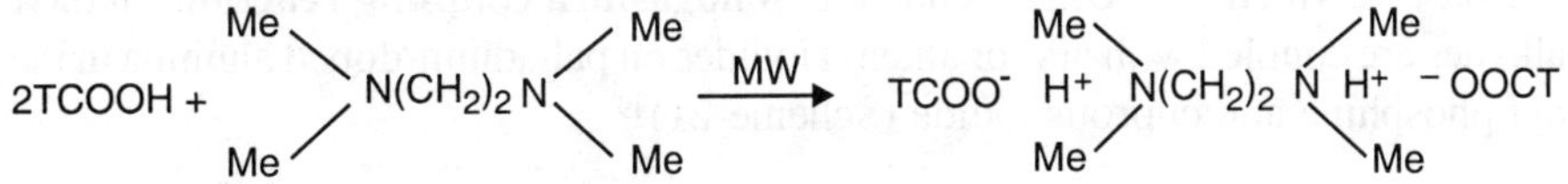

$$2\text{TCOOH} + \text{Me}_2\text{N(CH}_2)_2\text{NMe}_2 \xrightarrow{\text{MW}} \text{TCOO}^- \text{H}^+ \text{Me}_2\text{N(CH}_2)_2\text{NMe}_2 \text{H}^+ {}^-\text{OOCT}$$

Scheme-18

5.10 ORGANOMETALLIC REACTIONS (REACTIONS INVOLVING C-C BOND FORMATION)

C-C Bond forming reactions, viz. Suzuki, Heck and Stille reactions[39–41] are found to be transition-metal catalysed MW assisted reactions. Suzuki cross coupling of aryl halides with arylboronic acid (Suzuki cross coupling reaction) using $PdCl_2$, KF and poly ethylene glycol (PEG) take place under microwave irradiation to give biaryls. This environmentally friendly process offers easy access to biaryls[39] (Scheme-19).

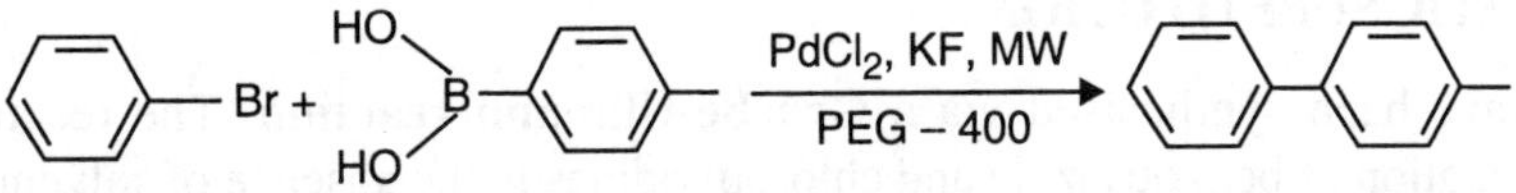

Scheme-19

In the above protocol, the catalyst can be recycled[40]. Recently, a ligand free palladium-catalyused Suzuki reaction has been reported in water[41], which uses low palladium loadings (0.4 mol%).

The original molybedenum-catalysed asymmetric allylic alkylation developed by Trost *et al* has now been transformed into a fast and efficient reaction under non-inert conditions using MW acceleration[42].

This modification has enabled the reaction to be performed in one-step employing stable precatalyst [Mo(CO)$_6$] in low concentration and under air-stable environment. The scope of the flourous stille coupling with respect to both the tin and triflate/halide components has been extended using the MW protocol that gets completed in 2 min. compared to 1day in the conventional thermal reaction[43].

A number of reactions have been described earlier in solution phase chemistry mostly palladium catalysed reactions of aryl halides and olefins[44] for C-C bond formation reactions[45]. Palladium acetate is the commonly used catalyst, although other palladium complexes have also been used. A solvent free **Heck reaction** has been conducted in excellent yields using a household MW oven and palladium acetate as catalyst and triethylamine as base (Scheme-20)[46]. The reaction takes much shorter time compared to classical heating methods.

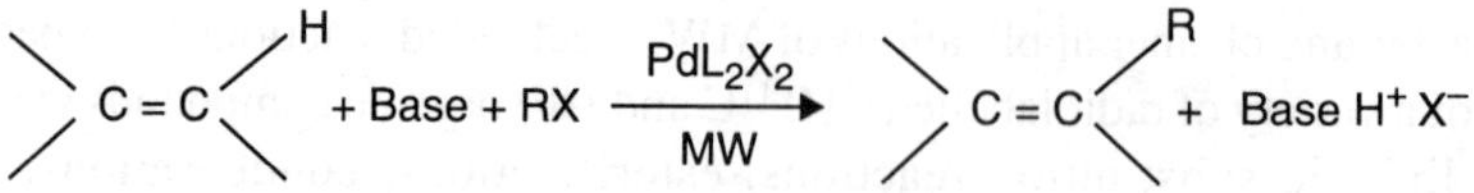

5.8 ■ Alternate Energy Processes in Chemical Synthesis

$$\text{>C=C<} \text{H} + \text{Base} + \text{RX} \xrightarrow[\text{MW}]{\text{PdL}_2\text{X}_2} \text{>C=C<} \text{R} + \text{Base H}^+ \text{X}^-$$

Scheme-20

A MW-assisted palladium catalysed coupling of heteroaryl and aryl boronic acids with iodo and bromo-substituted benzoic acids, anchored on Tenta Gel has been achieved[47]. An environmentally friendly Suzuki cross-coupling reaction has been developed which uses small amounts of recyclabe poly ethylene glycol (PEG) as the reaction medium and palladium chloride as catalyst[40]. A solvent-free Suzuki coupling has also been reported an palladium-doped alumina in the presence of potassium fluoride as a base[48], which has been extended to **Sonogashira coupling reaction.** In the latter case terminal alkynes are coupled with aryl or alkenyl iodides on palladium-doped alumina in the presence of triphenyl phosphine and cuprous iodide (Scheme-21)[49].

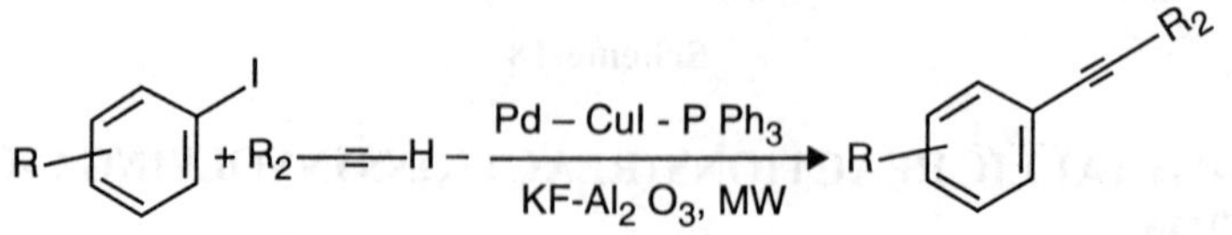

$$R\text{-Ar-I} + R_2\text{-}\equiv\text{-H} \xrightarrow[\text{KF-Al}_2\text{O}_3, \text{MW}]{\text{Pd-CuI-PPh}_3} R\text{-Ar-}\equiv\text{-}R_2$$

Scheme-21

Intramolecular hydroacylation of 1-alkenes with aldehydes is a greener alternative to classical approach using homogeneous catalyst in toluene. This reaction is a general reaction and uses Rh(I) complex (**Wilkinson catalyst**) uder solvent-free conditions and there is considerable rate enhancment of polar transition intermediates[50].

5.11 AROMATIC SUBSTITUTION

β-carbolines have been synthesised via a **Graebe-Ullmann reaction**. The reaction involved[51] microwave irradiation of benzotriazole and chloropyridines in the absence of solvent in presence of pyrophosphoric acid. The formed product (2) is obtained in 13–16 minutes (Scheme-22). Attempts to support the reactants on silica gel or montmorillonite gave only moderate yield of the intermediate pyridyl benzotriazole (1). If the reaction is carried out under thermal condition without Microwaves, the required product (2) is obtained in similar yields but required much longer reaction times.

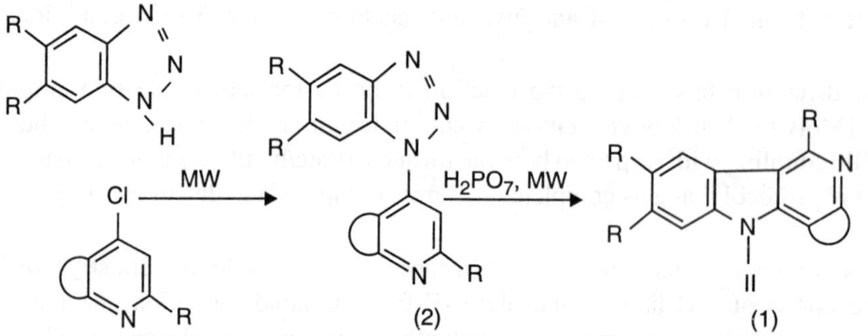

Scheme-22

Bisnaphthols have been prepared in moderate to good yields by microwave irradiationn of dry solid reagents, $FeCl_3.6H_2O$ and the appropriate naphthol is a resonance cavity (See section 5.3).

5.12 PERICYCLIC REACTIONS

Some Pericyclic reactions have already been discussed in section 3.29 under microwave irradiation.

An intra molecular Diels-Alder reaction has been reported by adsorbing the furan derivative (1) on silica gel, saturating with water and irradiating with microwaves. Cycloadducts (2) and (3) were obtained in 64% yield[52]. In this reaction (Scheme-23) water is crucial for the success of this transformation, acts as a source of heat, accelerates the cydoaddition by the hydrophobic effect and favours the hemiacetalketone equilibrium. Conventional thermolysts of the Furan (1) gives poor yields of cycloaddition due to substantial decomposition.

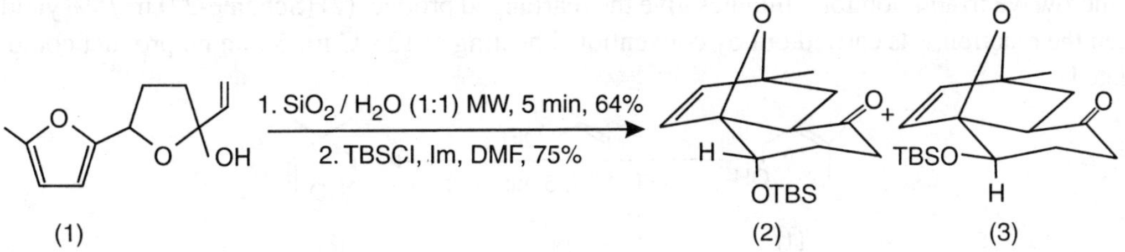

Scheme-23

Methyl nitroacetate and dimethyl acetylene dicarboxylate in presence of acid (SiO_2 or molecular sieve) in an open vessel on microwave irradiation gave the cycloadduct in moderate to good yield[53] (Scheme-24).

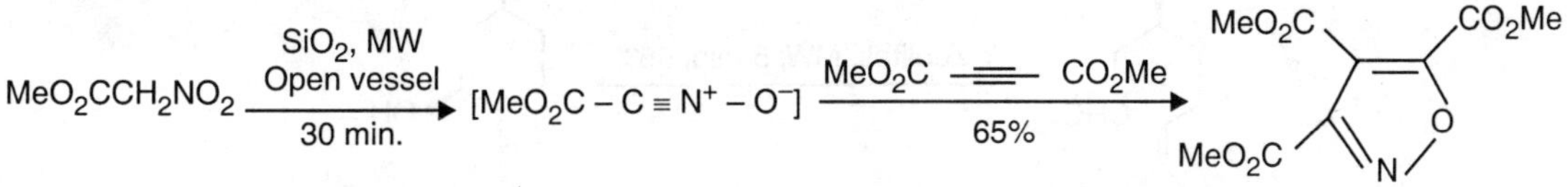

Scheme-24

Cycloheten-3-ol (1) on microwave irradiation with triethyl orthoacetate and KSF clay in DMF underwent **ortho-ester claisen rearrangement** to give the rearranged product (2) in good yield[54] (Scheme-25) (See also section 3.9). Under the conventional conditions in which propionic acid was used as a catalyst (as oppose to KSF clay), the reaction required 12.5 hours heating to give a 68% yield to the product (2).

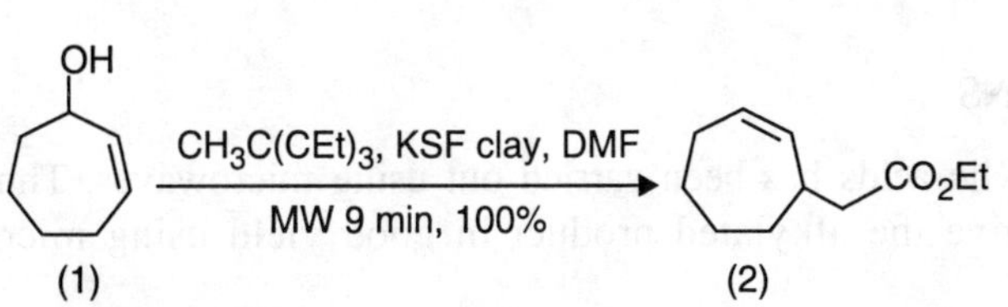

Scheme-25

Fishcher Indole Synthesis has be carried out[55] using cydohexanone and phenyl hydrazine in presence of montmorillonite KSF clay (Scheme-26). Using 'dry conditions' for a closely related cyclisation gave only traces of the product[56].

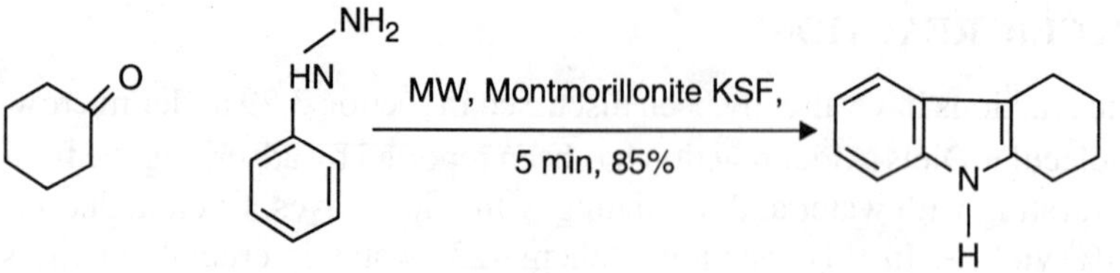

5.10 ■ Alternate Energy Processes in Chemical Synthesis

Scheme-26

Montmorillonite clays has also been used for other organic transformations including [2, 3] sigmatropic rearrangements. Thus, adsorption of 3-methyl but-2-en-1-ol (1) onto KSF clay followed by microwave irradiation for 5 minutes give the rearranged product (2) (Scheme-27) in 75% yield[57]. When the reaction was carried out by conventional heating at 135°C for 5 min no product could be isolated.

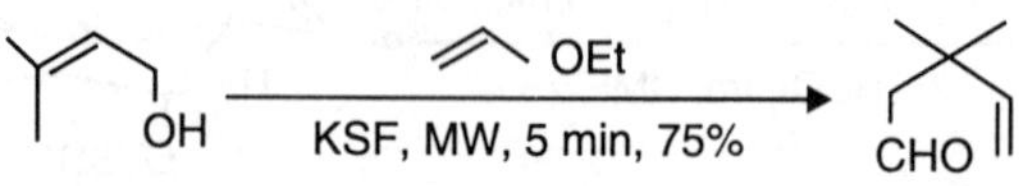

Scheme-27

A number of other sigmatropic rearrangements can be carried out by adsorption of the substractes on y zeolites. Two of such rearrangements[58] are given in Scheme-28.

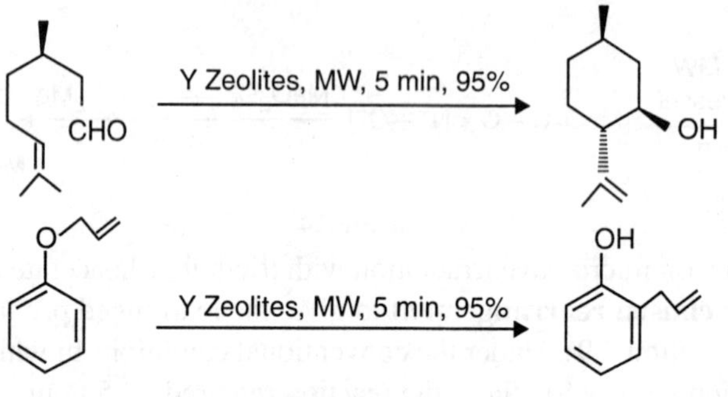

Scheme-28

5.13 ALKYLATIONS

Alkylation of carboxylic acids has been carried out using microwaves. Thus, potassium acetate and octyl bromide give the alkylated product in good yield using microwave irradiation[59] (Scheme-29).

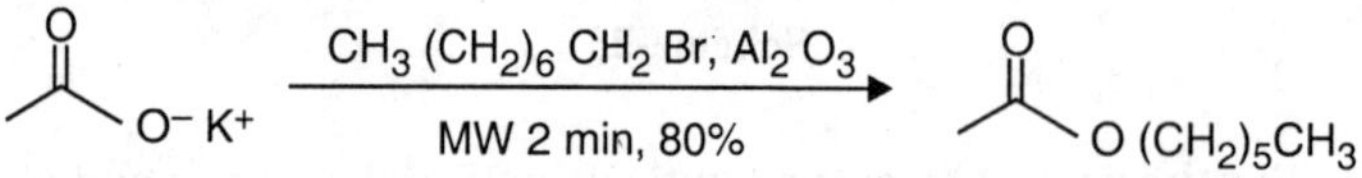

Scheme-29

On larger scale (up to 0.1 mole) the above alkylation could be carried out by irradiation in open vessels. An alternative procedure involving alkylation of carboxylic acids with alkyl halides in presence of phase transfer catalyst (PTC) has been developed. It gives high yields of esters with 10 minutes irradiation. Thus, the reaction of hexanoic acid with benzyl bromide and PTC in a sealed tube gives the ester in good yield (72%)[60].

Sulphones have been prepared by microwave assisted alkylation of sodium phenylsulphinate. Thus, irradiation of a mixture of the sulphinate and benzyl halide adsorbed onto alumina gave the desired sulphone in 40–99% yield[61] (Scheme-30).

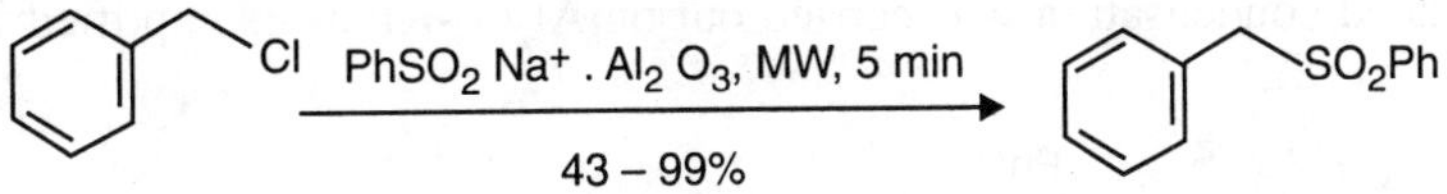

PhCH$_2$Cl $\xrightarrow{\text{PhSO}_2\text{ Na}^+ \cdot \text{Al}_2\text{O}_3, \text{ MW, 5 min}}$ PhCH$_2$SO$_2$Ph
43 – 99%

Scheme-30

Allyldiphenylphosphine oxide has been obtained via phosphinic allylation. Though the reaction proceeds satisfactorily by stirring the neat reagents at room temperature overnight, but the reaction time can be reduced to 1–3 minutes using microwave irradiation (Scheme-31)[62].

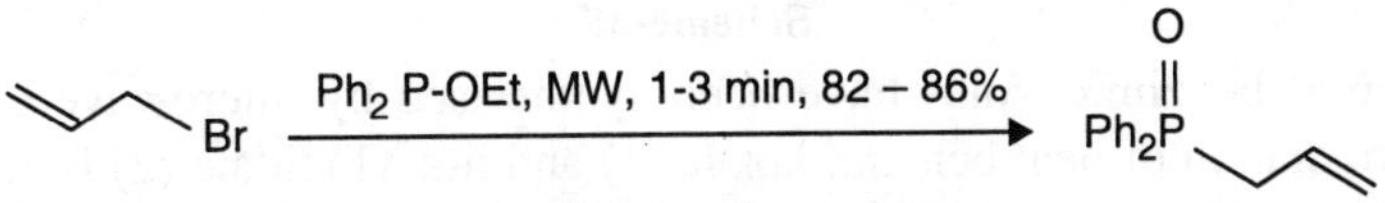

CH$_2$=CHCH$_2$Br $\xrightarrow{\text{Ph}_2\text{P-OEt, MW, 1-3 min, 82 – 86\%}}$ Ph$_2$P(=O)CH$_2$CH=CH$_2$

Schemes 31

Active methylene groups can be alkylated using microwave irradiation. Thus, treatment of phenylsulphonylacetate with alkyl halide in presence of potassium carbonate and phase transfer catalyst leads to the alkylated product (Scheme-32) in good yield after 2–3 minutes of microwave irradiation[63].

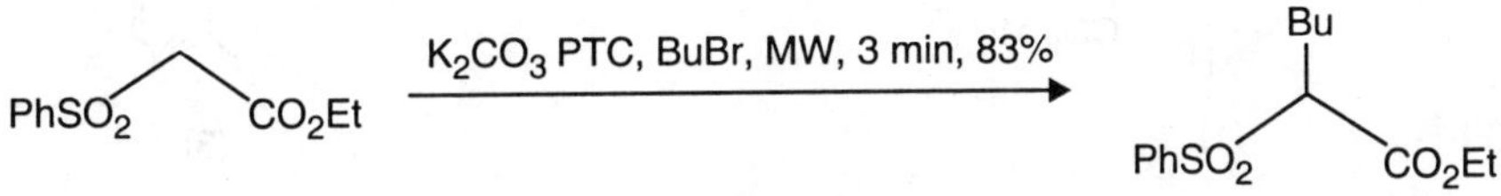

PhSO$_2$CH$_2$CO$_2$Et $\xrightarrow{\text{K}_2\text{CO}_3 \text{ PTC, BuBr, MW, 3 min, 83\%}}$ PhSO$_2$CH(Bu)CO$_2$Et

Scheme-32

5.14 CONDENSATIONS

Some condensation reactions have already been described (see section 4.6). Following are given some other condensation reactions of importance.

Condensation of cyclohexane 1, 2-dicarboxylic acid to the corresponding anhydride has been reported[64] (Scheme-33).

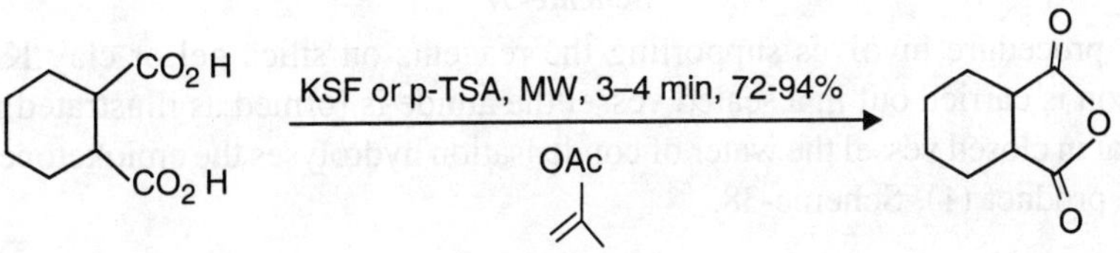

cyclohexane-1,2-(CO$_2$H)$_2$ $\xrightarrow[\text{OAc}]{\text{KSF or p-TSA, MW, 3–4 min, 72-94\%}}$ cyclohexane-1,2-dicarboxylic anhydride

Scheme-33

Active methylene compounds have been condensed with aldehydes using microwaves. Thus, condensation of 5-nitro-2-furaldehyde with active methylenes (like ethyl cyanoacetate) was achieved in high yields by adsorption of the reagents on the lewis acid K10, ZnCl$_2$ and irradiation[65] with microwaves (Scheme-34).

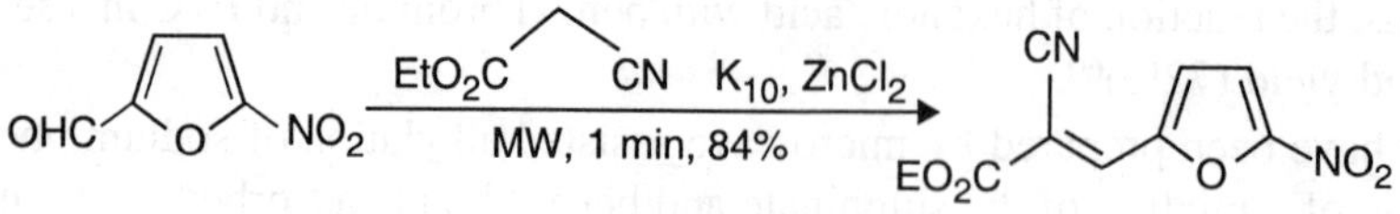

5.12 ■ Alternate Energy Processes in Chemical Synthesis

Scheme-34

A closely related condensation was carried out on Al_2O_3-KF as examptified in the following Scheme-35[66].

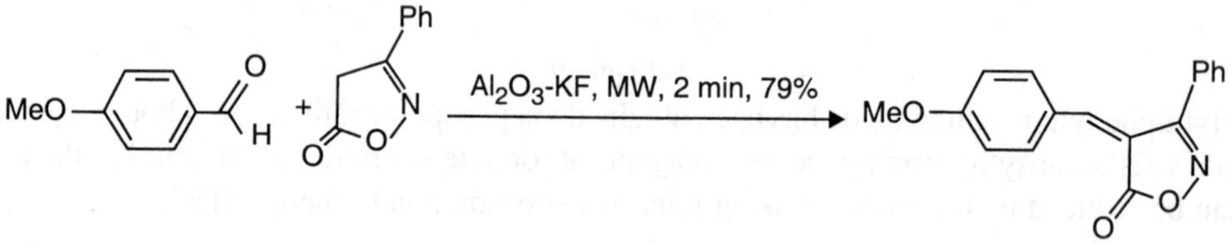

Scheme-35

Pyrimido [1, 6-a] benzimidazoles have been synthesised by microwave cyclocondensation approach. Thus, irradiation of neat benzimidazole (1) and n-acylimidate (2) is an open vessel in a microwave oven gave a good yield of the benzimidazole (3) (Scheme-36). The reaction involves a conjugate addition-elimination, condensation sequence. The use of open vessel permits the vaporisation of ethonol and water produced in the reaction[67].

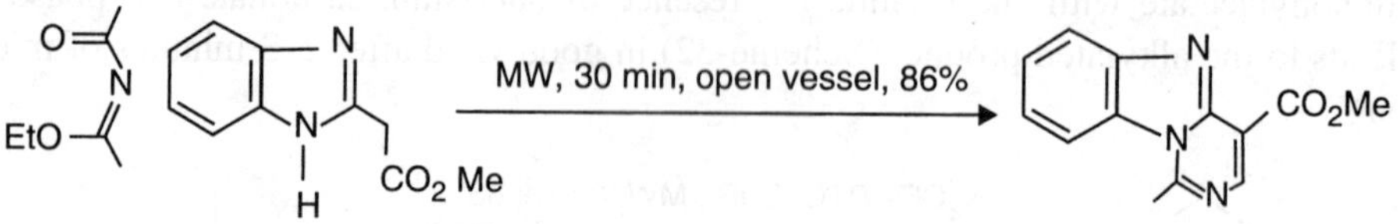

Scheme-36

Enaminoketones are obtained[68] by the microwave irradiation of a β-diketone with an amine as examplified in Scheme-37.

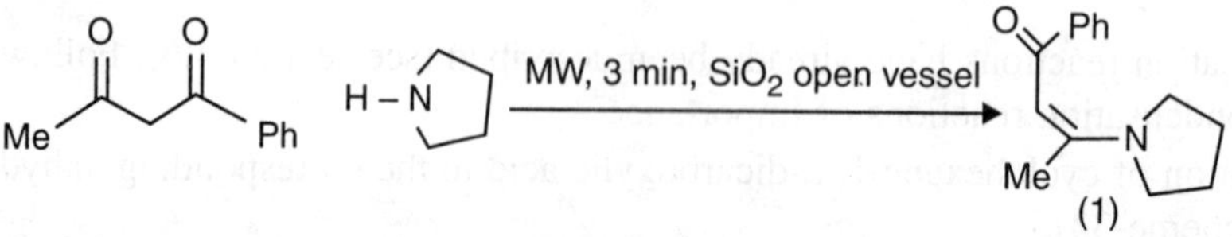

Scheme-37

The above procedure involves supporting the reagents on silica gel or clay K10. If is some cases the reaction is carried out in a sealed vessel the amide is formed as illustrated in Scheme-38. It is believed that in closed vessel the water of condensation hydrolyses the emioketone [(1), Scheme-37] to give the product (1), Scheme-38.

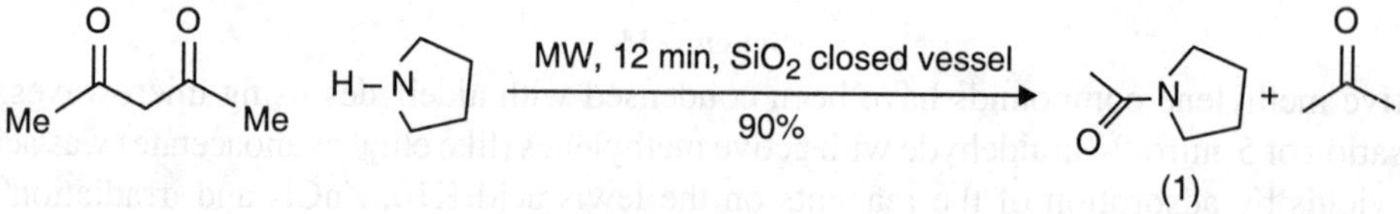

Scheme-38

Microwave assisted condensation has also been used to prepare heterobicycles as shwon below in Scheme-39. Thus, irradiation of the diamine (1) with Ketoester (2) leads to the formation of the product (3) in 86% yield. The reaction was performed by supporting the reagents on the alumina followed by irradiation in an open vessel[69].

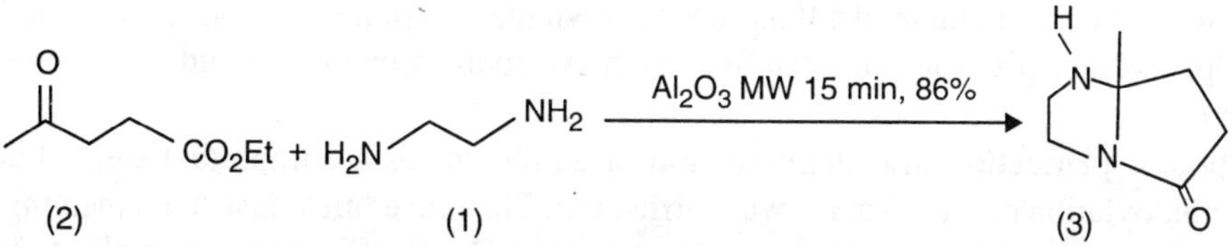

Scheme-39

An industrially important raw material anthraquinone has been prepared[70] by the use of microwaves. Thus, the microwave assisted acid catalyst cyclodehydration of benzophenone o-carboxylic acid (1) could be achieved using the same batch of catalyst without reduction in yields (12 reactions) (Scheme-40). Using conventional heating the yield was around 50% after four reactions using the batch of catalyst.

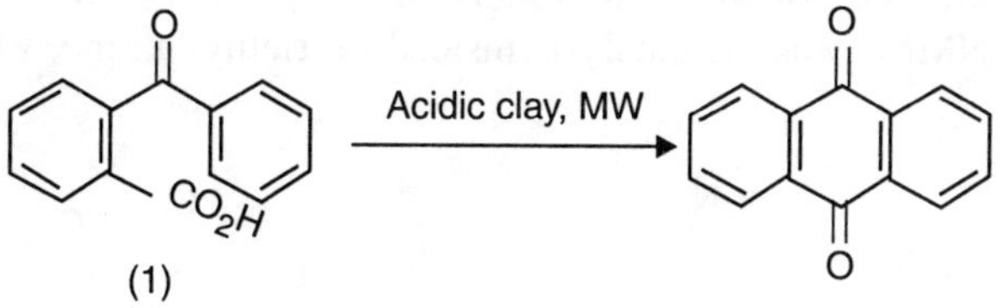

Scheme-40

An interesting application of microwave assisted cyclocondensation is in the preparation of tetrapyrrole as shown in Scheme-41. Although the yield is about 10 per cent, the isolation and purification of the product is simple[71].

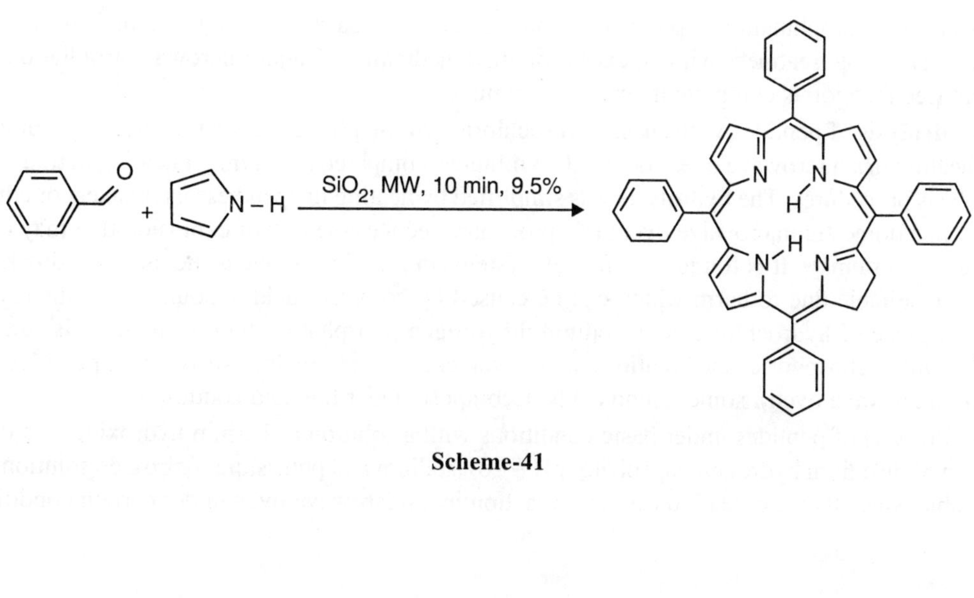

Scheme-41

5.15 SYNTHESIS OF PEPTIDES

Microwave assisted solid phase peptide synthesis[72] has also been achieved. Thus, coupling of Fmoc-protected amino acids using either symmetrical anhydride or active ester methodology can be carried out using microwave conditions with a significant reduction in reaction time (2–6 minutes) compared to conventional methods. It has been shown that coupling reactions proceed in high yield with no apparent racemisation. This can find extensive application in the rapid synthires of peptides and proteins.

In peptides, **protection and deprotection** of amino acides are important steps. Earlier large scale N-phthaloylation of amino acids was carried out. The procedure consisted in healing an amino acid with phthalic anhydride in refluxing toluene using triethylamine as a catalyst[73]. A flask with a reflux condenser and a Dean-stark water separator was used as the reaction vessel. For reactions making 0.1 M product, the reaction took about 5 hrs for completion. It is now possible to carry out the reaction in open vessel using DMF as the reaction medium[74]. Phthaloylation was complete in few minutes. No racemization was observed when an optically active amino acide such as L-phenylalanine was used. An improvement in the method is to use tetrachlorophthalic anhydride in place of phthalic anhydride[75]. By this process 2-5g of N-tetrachloro phthaloyl glycine in 94% yield in 90 seconds was obtained. In this reaction DMF was used as the reaction medium and N-methylmorpholine (NMM) was the catalyst (instead to triethyl amine, which has a lower b.p.) (Scheme-42).

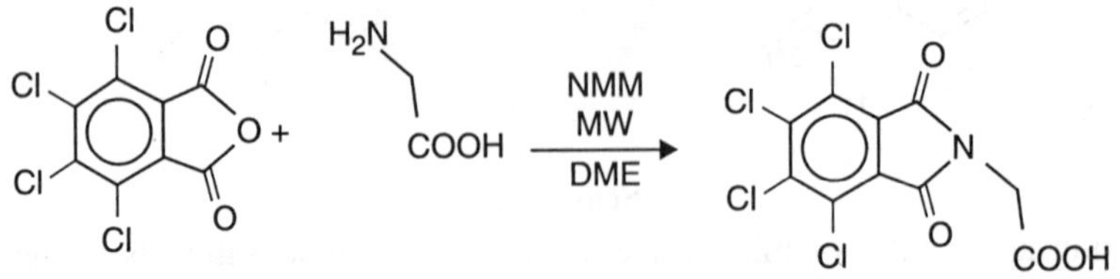

Scheme-42

The above reactions (Scheme-42) could be scaled up to use 1 mol of tetrachlorophthalic anhydride, the N-protected glycine was formed in 90% yield after 8 min of irradiation.

The tetrachlorophthalimido protective group can be removed at room temperature, in the absence of any solvent, by treatment with an excess of ethylenediamine. Under microwave irradiation, this solvent free reaction is complete in only 1–2 minutes.

Hydrolysis of peptides with strong hydrochloric acid in a sealed tube under argon is complete after healing in a microwave oven for about 15 minutes, compared to overnight heating in traditional hydrolysis procedures. The hydrolysis was simplified by heating in open vessels (beakers or conical flasks) in a domestic microwave oven; this procedure reduces the risk of explosion that may occur in a sealed system[75a]. It is found that in open systems in microwave ovens the use of hydrochloric acid is unsuitable due to harm which can be caused by corrosive acid vapours to the microwave oven. In place of hydrochloric acid, sodium dihydrogen phosphate solution (pH~2.8) is suitable. Though dilute phosphoric acid solutions are convenient for rapid hydrolysis of small peptides (1–2 min in microwave oven), some amino acids decompose under hot acid conditions.

Hydrolysis of peptides under basic conditions with a solution of barium hydroxide of a dilute solution of disodium hydrogen phosphate (pH ~ 10), sodium and potassium hydroxide solutions are unsuitable, since they can lead to runaway reaction in a microwave oven under certain conditions.

Di, tri and tetrapeptides were completely hydrolysed in 3–15 min when saturated barium hydroxide solution was used.

For determing C-terminal of a peptide **Akabori reaction (hydrazinolysis)** is usually carried out ordinarily, the heating is done in a sealed tube for over 12 hours at a high temperatures. The Akabari reaction has been successfully conducted in a few minutes in a domestic microwave oven in open glass vessels (Scheme-42a).

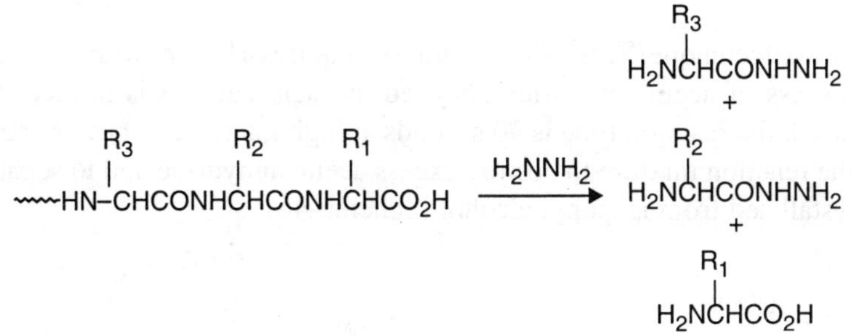

Scheme-42a

Hydrazinolysis of different di and tripeptides could be completed in 3–10 min.

5.16 REACTIONS INVOLVING SILICON REAGENTS

The reaction of silyl ketene actals with imines under microwave condition[76] give the addition product (1) (Scheme-43). It is of great advantage to adsorb the reagents to K10 montmorillonite clay.

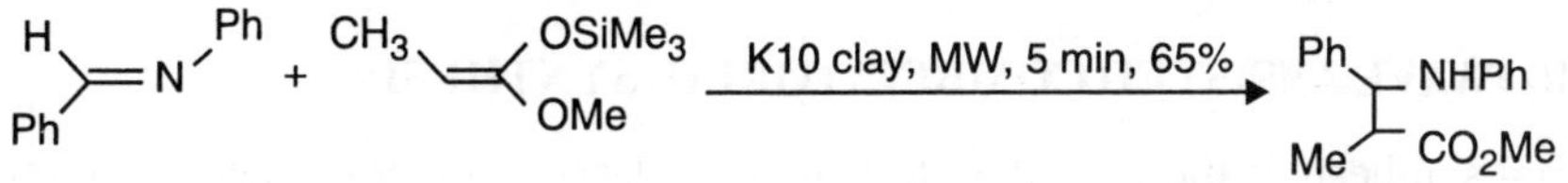

Scheme-43

When the above reaction (Scheme-43) is carried out by mixing neat reagents with 18-crown-6 and irradiation is a closed vessel, β-lactams (1) are isolated in good yield (Scheme-44).

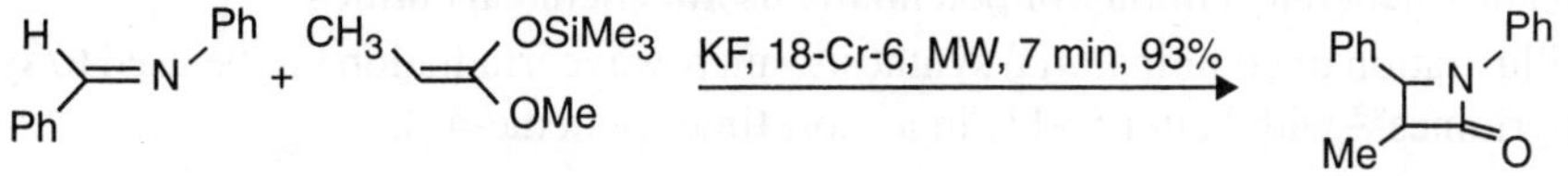

Scheme-44

Silyl-Reformatsky process has been developed using silicon containing nucleophile. It has been found that a range of both supported and unsupported reaction conditions (alkali metal fluorides/ Al_2O_3, MgO or clay) can be used to prepare useful addition products as shwon below[77, 78] in Scheme-45.

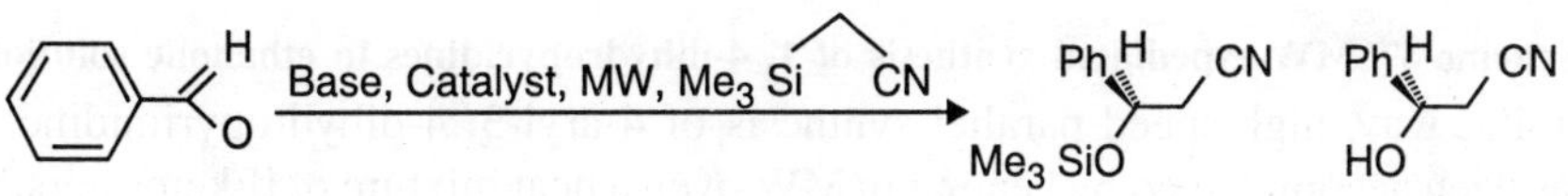

Scheme-45

5.16 ■ *Alternate Energy Processes in Chemical Synthesis*

5.17 SYNTHESIS OF ASPIRIN

Aspirin is a single drug which is produced and used in maximum amount on a global scale. It is conventiaonally prepared from salicylic acid and acetic anhydride in presence of small amount of sulphuric or phosphoric acid (about 30 min. heating at 85–95°). About 4M equiv. of acetic anhydride are required for each mole of salicylic acid. Only one mole of acetic anhydride is the reactant and the rest is a reaction medium. After the reaction, excess acetic anhydride is hydroloped by ice water to acetic acid.

Using microwave technique[75a], aspirin preparation requires a beaker with a loose cover. Salicylic acid and slight excess of acetic anhydride are used, no acid catalyst is needed. When 2–20g of salicylic acid is used, the reaction time is 90 seconds at high microwave power (800–1000W). Ice-water-added to the reaction mixture to destroy excess acetic anhydride and to separate aspirin. It is filtered and recrystallised from isopopyl alcohol (Scheme-46).

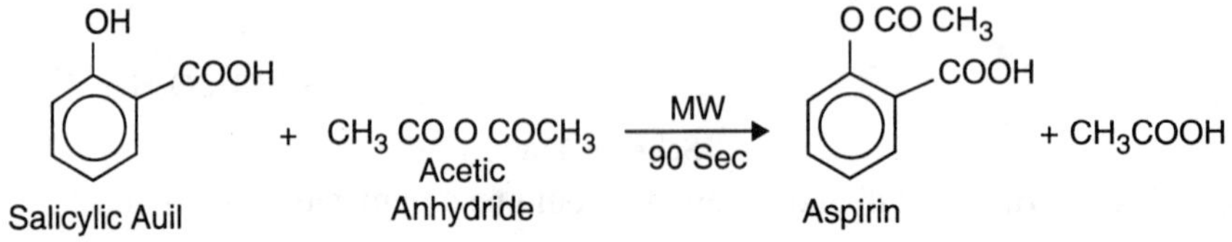

$$\text{Salicylic Auil} + CH_3 CO\ O\ COCH_3 \text{ (Acetic Anhydride)} \xrightarrow[90\ \text{Sec}]{MW} \text{Aspirin} + CH_3COOH$$

Scheme-46

Several hundred grams of salicylic acid can be used; the reaction mixture, which is a paste is spread out in a uniform layer in an inexpensive glasses plate or baking disk. Reaction time is 8–10 minutes in a microwave oven.

5.18 MICROWAVE ASSISTED COMBINATORIAL SYNTHESIS

Combinatorial synthesis is the procedure to generate a large number of chemical compounds in a much shorter time compared to traditional synthesis which involves slow and painstaking work of making one compound at a time in a single reaction. At present there is lot of interest to accelerate the technologies to identify novel drug targets via high-throughput synthesis and combinatorial chemistry[79-80]. Several combinatorial techniques are available where in the required products could be made as mixtures or in a parallel fashion, using either solution or solid phase techniques. Such synthesis readily generate a library of potentially useful chemical entities[81].

As an illustration of combinatorial synthesis, microwave irradiation has be used to synthesise 1, 4-dihydropyridines[82] with better yields in a short time (Scheme-47).

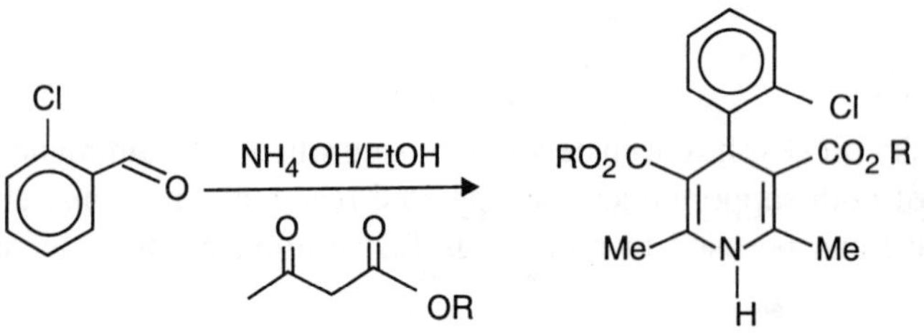

Scheme-47 MW-expediated synthesis of 1, 4-dihydropyridines in ethanolic solution

In a similar way, high-speed parallel synthesis of 4-aryl-3, 4-dihydropyrimidine 2(1H)-ones (DHPM) has been accomplished by heating in MW oven a neat mixture of β-ketoesters, aryl halides and urea derivatives with polyphosphate esters (PPE) as the reaction mediator[83]. In view of the readily available aromatic aldehydes, β-keloesters, and urea derivatives, a large collection of DH PMS can be potentially prepared (Scheme-48) using the recently automated, high throughput robotic technologies for performing microwave assisted combinatiorial synthesis[84].

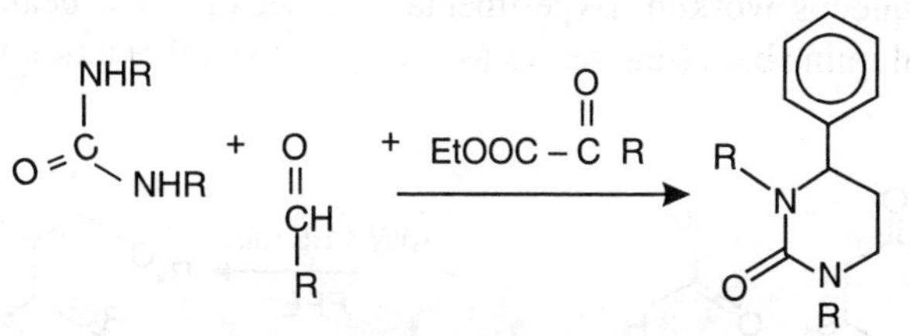

Scheme-48 Synthesis of DHPM

In combinatorial synthesis, the parallel processing of synthetic sequence has been one of the central themes[79–80]. In this sequence, the assembly of library of compounds occurs using an ordered array of spatially separated small reaction vessels. The location of the compound in the array provides the structure of the compound. The most commonly used format for parallel synthesis, is the 96-well microtier plate that enables, often in an automated fashion, the assembly of hundreds of compounds by parallel synthesis[79, 80].

The MW-assisted approach seems to be ideally suited for rapid product generation in high yield under these uniform conditions. Thus, the nucleophilic substitution of an alkyl iodide with 60 piperidine or piperazine derivatives (Scheme-49) in individual sealed polypropylene vials was the first reported parallel reactions conducted under multimode MW irradiation conditions[85].

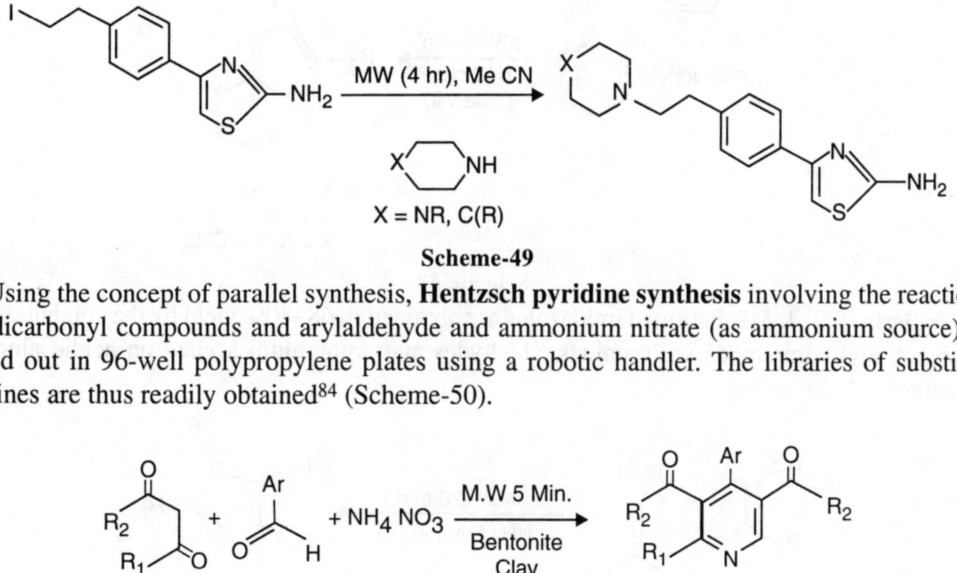

Scheme-49

Using the concept of parallel synthesis, **Hentzsch pyridine synthesis** involving the reaction of 1, 3-dicarbonyl compounds and arylaldehyde and ammonium nitrate (as ammonium source) was carried out in 96-well polypropylene plates using a robotic handler. The libraries of substituted pyridines are thus readily obtained[84] (Scheme-50).

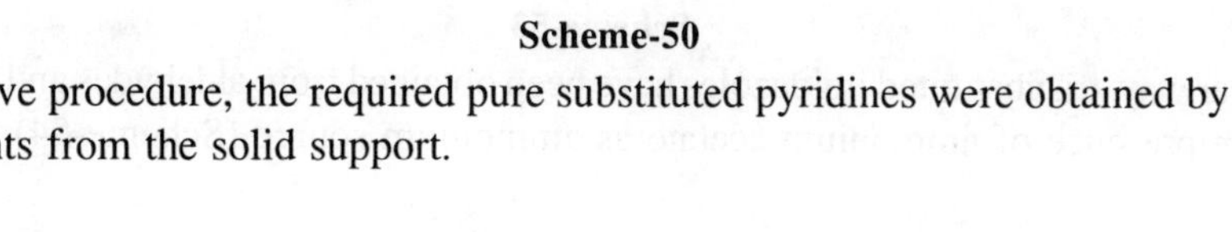

Scheme-50

In the above procedure, the required pure substituted pyridines were obtained by extraction by organic solvents from the solid support.

5.18 ■ Alternate Energy Processes in Chemical Synthesis

In a siimilar way, the classical **Biginelli condensation** products (the dihydropyrimidines) were obtained in a MW-accelerated approach[83] (Scheme-51). In this procedure, the neat mixture of β-ketoesters, aryl aldehyde and (thio) ureas with polyphosphate ester (PPE) as reaction medium are irradiated in a domestic MW oven for 1.5 min and the formed dihydropyrimidines were obtained in 61–95% yield after aqueous workup. Experimentally, a set of small beakers containing specific building blocks in an alumina bath (that serves as a holder of small reaction vessel and also as a heat sink).

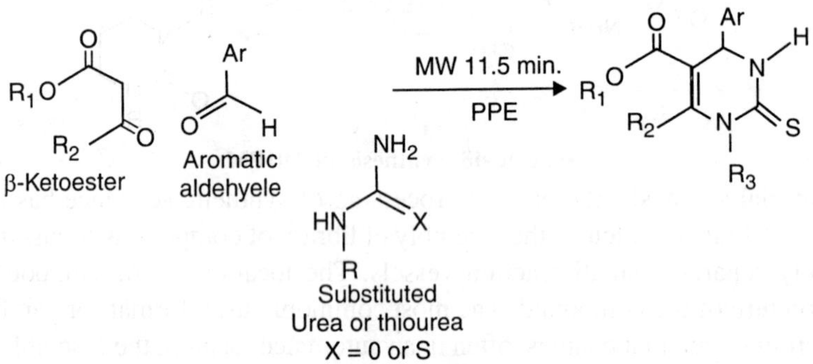

Scheme-51

Several other solvent free MW-accelerated process for parallel synthesis have been reported. As an example, imidazo-annulated pyridines, pyrazines and pyrimidines have been obtained in high yield by an **Ugi-type reaction** using montmorrilonite K10 clay as catalyst[86] (Scheme-52).

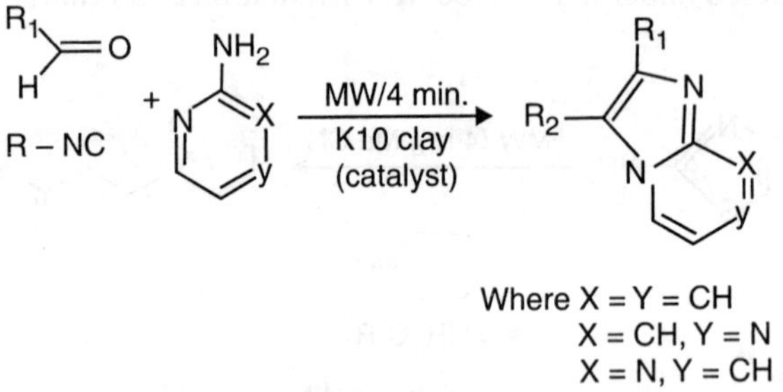

Where X = Y = CH
X = CH, Y = N
X = N, Y = CH

Scheme-52

Similarly, 1, 2, 4, 5-Substituted imidazoles are obtained in 68–80% yield by the condensation of 1, 2-dicarbonyl compounds with amines, aldehydes and ammonium acetate on acidic alumina[87] (Scheme-53).

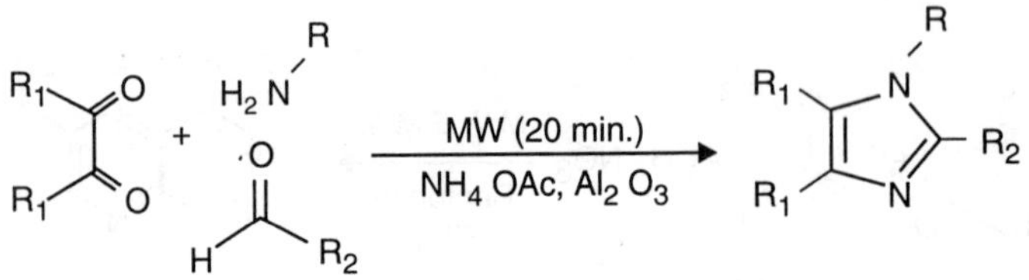

Scheme-53

In a similar way, trisubstituted imidazoles have been obtained from aldehydes and 1, 2-dicarbonyl compounds in presence of ammonium acetate as ammonium source (Scheme-54).

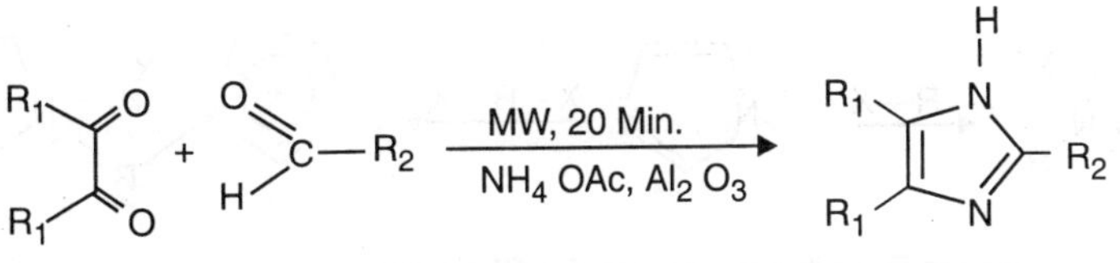

Scheme-54

A number of thioamides were obtained[88] by the reaction of amide derivatives with Lawesson's reagent (MW-irradiation, 8 min) (Scheme-55).

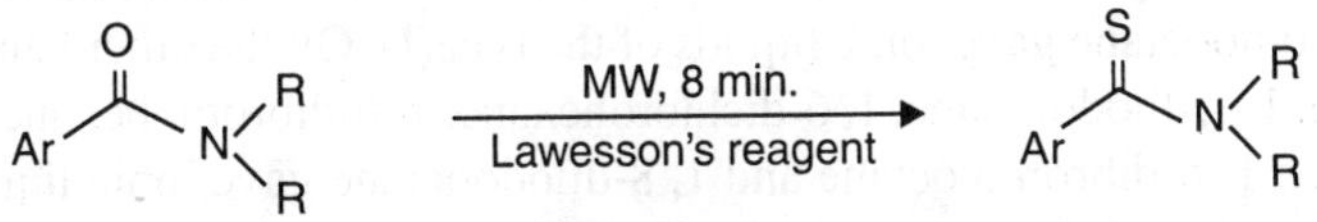

Scheme-55

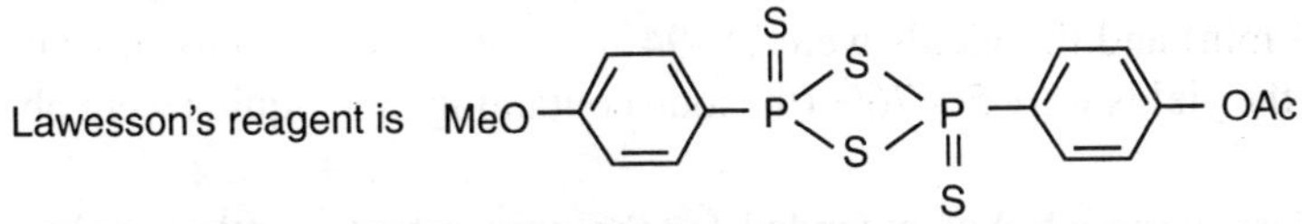

Lawesson's reagent is MeO—C₆H₄—P(=S)(S)—S—P(=S)(S)—C₆H₄—OAc

(structure: MeO-aryl-P(=S)(μ-S)₂P(=S)-aryl-OAc)

5.19 SYNTHESIS OF IONIC LIQUIDS USING MICROWAVES

The industrial chemistry is adopting the concept of green chemistry to meet the fundamental challenge of protecting the human health and environment while maintaining commercial viability. The most important aspect is the replacement of volatite organic solvents from the reaction medium and substitution by nonvolatile or recyclable alternatives. A very important development in the last few years has ben the discovery and use of ionic liquids as solvent in synthetic organic chemistry[89]. Ionic liquids comprise a wide range of anions that form low melting salts with several organic cations like imidazolium, pyridinium, phosphonium and ammonium etc. The most widely studied examples of ionic liquids contain 1, 3-dialkylimidazolium cations. The anions include the halide (Cl, Br, I), tetrafluoroborate, hexafluorophosphate, CF_3-or similar fluoninaled alkyl groups[90]. The ionic liquids have become the solvents of choice due to their non-volatile nature with barely measurable vapour pressure. These ionic liquids were originally used in electrochemical applications[91], but are currently being explored as environmentally benegn solvent substitutes for conventional volatile solvents in a variety of chemical synthesis[89].

Originally the ionic liquids were conventionally prepared by heating the reactants (the imidiazole and the alkyl halide) in refluxing solvents. The procedure used large excess of aklyl halide/organic solvent[92]. An elegant method[93] of preparation of ionic liquids involves heating the reactants in an microwave oven at reduced power level[94] (240 watt). It was found that at elevated power levels evaporation of alkyl halide and partial decomposition or charring occurred. This was attributed to localized overheating of ionic liquids resulting in low yields. The problem of decomposition, charring etc. could be avoided by heating the reactants with intermittent heating and mixing at moderate power level. This modification gave better yields and cleaner ionic liquid formation. The solventless approach[93] required only few minutes of reaction time compared to several hours required by conventional heating procedures, which uses excess of reactants. The preparation[94] of mono (I) and dicationic (II), 1, 3-dialkyimidazolium halides is given in (Scheme-56).

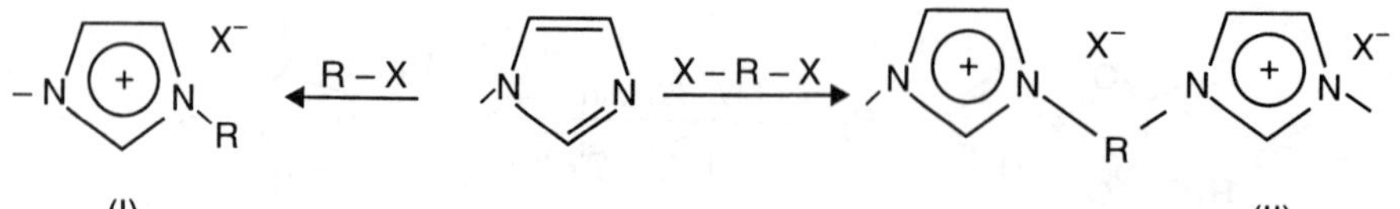

Scheme-56 MW-assisted preparation of ionic liquids

The alkyl halides vig. 1-bromobutane, 1-chlorobutane, 1-bromohexane, 1-iodohexane, 1-iodoheptane-1-bromooctane gave ionic liquids of the type(I). On the other hand alkyl halides viz., 1, 4-dibromobutane, 1-4, diiodobutane, 1, 6-dichlorohexane, 1, 6-dibromohexane, 1, 6-diiodohexane, 1, 8-dichloro-octane, 1, 8-dibromo octane and 1, 8-diiodooctane gave ionic liquids of the type (II). The ratio of reactants viz. alkyl halide and 1-methylimidazole used was 1 : 1 incase of I and 1 : 2 in case of II. The reaction took 1–1.25 minutes for completion (for iodohalidies the reaction was complete in 0.83 min) and the yields were 78–94%. However conventional heating (80° oil bath) took 3–5 hrs and the yields were 56–76% (the iodo containing ione liquids were obtained in 93–97% yield)[95, 96].

The solventless approach was extended for the preparation of other ionic salts (III) bearing tetrafluoroborate anions[97]. The method involves exposing N, N'-dialkyl imidazolium chloride (IV) and ammonium tetrafluoroborate salt to MW irradiation (Scheme-57). The procedured used was similar to that used earlier, viz. intermittant heating and mixing at moderate power level.

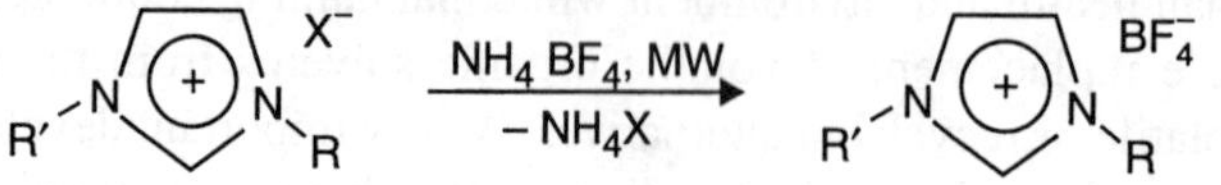

Scheme-57 MW-assisted preparation of dialkyl imidazolium tetrafluoroborates

Thus, by using 1-butyl-3-methylimidazolium chloride (I, R = butyl, R' = methyl, X⁻ = Cl⁻) and ammonium tetrafluoroborate, the ionic liquid 1-butyl-3-methyl-imidazolium tetrafluoroborate III, R = butyl, R' = methyl) was obtained. Using the above procedure following 1, 3-dialkyl imidazolium tetrafluoroborates were prepared (MW heating for 2–3.5 min using 240–360 MW power): III, R' = CH_3 and R= butyl, Isobutyl, hexyl, heptyl and octyl.

It is appropriate to state that water and air-stable ionc liquids comprising 1, 3-dialkylimidazolium cation and tetrafluorobrate anion was earlier prepared via the anian metathesis method from 1, 3-dialkyl imidazolium halides and sodium tetrafluorobarate in organic solvents[98]. In a similar way, the preparation of hydrophobic (Water immiscible) ionic liquids containing hexafluorophosphate anions was also prepared by anion metathesis in organic solvent or water[90].

The ionic liquids are polar and ionic in character and so couple to MW irradiation very effecently and so are ideal microwave absorbing candidates[99] for expediting chemical reactions[100].

A general use of these ionic liquids is the protection and deprotection of alcohols. This has been demonstrated in a facite tetrahydrophyranylation reaction[101]. Also rate enhancements have been reported[102] in 1, 3-dipolar cycloaddition reactions including the use of covalently grafted dipolarophiles on the ionic liquids.

Sonochemical synthesis of ionic liquids

It has already been mentioned that though the preparation of MW assisted 1, 3-dialklylimidazolium halides and tetrafluoroborates has reduced the reaction time from several hours to a few minutes and also avoiding a large excess of alkyl halids/organic solvents as the reaction medium. However, continuous microwave heating may result in overheating due to exothermic nature of the reaction that leads to charring and formation of coloured products.

A convenient procedure for the preparation of ionic liquid involves to use of ultrasound. Ultrasound -promoted chemical reactions are well known and proceed via the formation and adiabatic collapse of the transient cavitation bubbles[103]. In this procedure[104] the neat reactants in a closed container were exposed to ultrasound as the energy source using a sonication bath. Using this procedure a number of commonly used 1, 3-dialkylimidazolium halides (V), and (VI) were prepared (Scheme 58).

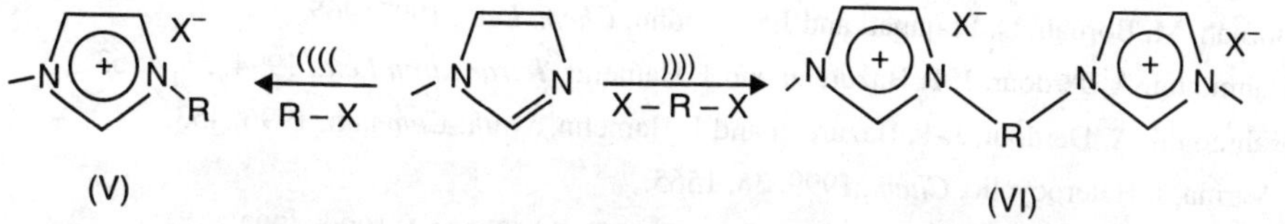

Scheme-58 Ultrasound-acceletrated synthesis of Ionic liquids

In the above procedure, the use of alkyl halides viz. 1-bromopropane, 1-chloropropane, 1-bromobutane, 1-iodobutane, 2-bromobutane, 1-chlorohexane, 1-bromohexane, 1-iodohexane, 1-iodoheptane, 1-chlorooctane, 1-bromooctane, and 1-iodooctane gave the ionic liquids of the type (V). However, the use of 1, 4-diiodohexane, 1, 8-dichlorooctane, 1, 8-dibromooctane and 1, 8-diiodooctane gave ionic liquids of the type (VI). The reaction time was 2-6 hrs and yield 90–95% in most of the cases.

5.20 CONCLUSION

The combination of supported reagents and microwave irradiation can be used to carry out a wide range of reactions in short times and with high conversions and selectivity. The reactions could be conducted in water, in organic solvents and without the need of solvents in the solid state. The later approach can prove beneficial since the recovery of solvents from conventional reaction systems always results in some losses. Recovery of both products and inorganic supports/catalysts in generally possible, leading to an efficient and low waste routes to a wide range of products.

NOTES AND REFERENCES

1. M. Miller and Y. Loudon, *J. Org. Chem.*, 1975, **40**, 126.
2. J.H. Pomerong and C.A. Craig, *J. Am. Chem. Soc.*, 1959, **81**, 6340.
3. D.T. Mowrg, *Chem., Rev.*, 1948, **42**, 250.
4. H.M. Sampath Kumar, P.K. Mohanty, M. Suresh Kumar and J.S. Yadav, *Syn.Commun*, 1997, **27**, 1327.
5. D.L. Clive, *Chem. Commun.*, 1970, 1014.
6. P.J. Foley, *J. Org. Chem.*, 1969, **34**, 2805.
7. W. Lehnert, *Tetrahedron Lett.*, 1971, **6**, 559.
8. G. Rosini, G. Baccolini and S. Cacchi, *J. Org. Chem.*, 1973, **38**, 1060.

9. H.G. Folly and D.R. Dalton, *J. Chem. Soc, Chem. Commun.*, 1973, 628.
10. D. Villemin, M. Lalaui and A.B. Alloum, *Chem. Ind.* (London). 1991, 176.
11. R.S. Varma and K.P. Naicker, *Molecules on line*, 1998, **2**, 94.
12. R.S. Varma, K.P. Naicker, D. Kumar, R. Dahiya and P.J. Liesen, *J. Microwave Power Electromag. Energy*, 1999, **34**, 113.
13. D. Villemin, B. Labiad and A. Loupy, *Synthetic Commun.*, 1993, **23(4)**, 419.
14. R.S. Varma, K.P. Naicker and P.J. Liesen, *Tetrahedron Lett.*, 1998, **39**, 3977.
15. D. Villemin and F. Sauvaget, *Synlett.*, 1994, 435.
16. A. Vass, J. Toth and E. Pallai-Varsanyi, OR, 19, presented at international conf. in microwave chem., Prague, Czech Republic, Sept. 6–11, 1998.
17. B. Satish, K. Panneersel-Vam, D.Zacharids and G.R. Desivaju, *J. Chem, Soc. Perkin Trans.*, 1995, **2**, 325.
18. E. Diez-Barra, A. de la Hoz, S. Merino and P. Sanchez-Verdu, *Tetrahedron Lett.*, 1997, **38**, 2359.
19. A. Boruah, M. Boruah, D. Prajapati and J.S. Sandhu, *Chem. Lett.*, 1997, 965.
20. M. Rahmouni, A Derdour, J.-P. Bazureau and J. Hamelin, *Tetrahedron Lett.*, 1994, **35**, 4563.
21. M. Rahmouni, A. Derdour, J.-P, Bazureau and J. Hamelin, *Synth. Commun*. 1996, 453.
22. R.S. Varma, J. Heterocyclic, *Chem.* 1999, **36**, 1565.
23. R.S. Varma, D. Kumar and J.P. Liesen, *J. Chem. Res.*, Perkin Trans. I, 1999, 4093.
24. S. Paul, M. Gupta, R. Gupta and A. Loupy, *Tetrahedron Lett.*, 2001, **42**, 3827.
25. R.S. Varma and D. Kumar, *Tetrahedron Lett.*, 1999, **40**, 7665.
26. E. Guibe-Jampel and G. Rousseau, *Tetrahedron Lett.*, 1987, **28**, 3563.
27. J.-R. Carrillo-Munoz, D. Bouvert, E.Guibe-Jampl, A. Loupy, A Petit, *J. Org. Chem.*, 1996, **61**, 7746.
28. M. Gelo-Pujic, E. Guibe-Jampel, A. Loupy and A. Trincone, *J. Chem. Soc. Perking Trans I*, 1997, 1001.
29. H.-M. Yu, S.-T. Chen and K.-T. Wang, *J. Org. Chem.*, 1992, **57**, 4781.
30. A. Stadler and C.O. Kappe, *Tetrahedron*, 2001, **57**, 3915.
31. A. Stadler and C.O. Kappe, Eur. *J. Org. Chem.*, 2001, 919.
32. S. Anto, J.R. Jones and S.Y. Lu., Microwave enhanced acid-catalysed hydrogen/deuterium exchange reactions. p. 34, Presented at the International Conference on Microwave Chemistry, Prague, Czech Republic, Sept. 6–11, 1998.
33. K.E. Wilzbach, *J. Am. Chem. Soc.* 1957, **79**, 1013.
34. J.R. Jones, Development of new microwave enhanced tritiation procedures, OR 9, presented at the international conference on microwave Chemistry, Prague, Czech Republic, Sept. 6–11, 1998.
35. R.S. Varma and R.K. Saini, *Tetrahedron Lett.*, 1997, **38**, 4337.
36. N.H.Werstiuk, in Isoiopes in the physical and biological sciences, (Ed.) E. Buncell and J.R. Jones, Elasevier, Amsterdam, 1987, Vol. 1A, pp. 124–155.
37. N.Elander, J.R. Jones, S.Y. Lu and S. Slone-Elander, *Chem. Soc, Rev.* 2000, **29**, 239.
38. M.H. Al-Qahtani, N. Cleator, T.N. Donks, R.N. Garman, J.R. Jones, S. Stefaniak, A.D. Morgan and A.J. Simmonds, *J. Chem. Res. (S)*, 1998, 400.
39. M. Larhed and A. Hallberg, *J. Org. Chem.*, 1996, **61**, 9582.
40. V.V. Namboodiri and R.S. Varma, *Green Chem.*, 2001, **3**, 146.
41. N.E. Leadbeater and M. Marco, *Org. Lett.*, 2002, **4**, 2973.
42. N.-F.K. Kaiser, U. Bremberg, M. Larhed, C. Moberg and A. Hallberg, *Angew Chem. Int.* Ed. Engl., 2000, **39**, 3596.

43. M. Larhed, M. Hoshino, S. Hadida, D.P. Curran and A. Hallberg, *J. Org. Chem.*, 1977, **62**, 5583.
44. R.F. Heck, *Org. React.*, 1982, **27**, 345.
45. R.S. Varma, K.P. Naicker and P.J. Liesen, *Tetrahedron Lett.*, 1999, **40**, 2075.
46. A.D. Oritz, P. Prieto and E. Vazquez, *Synlett.*, 1997, 269.
47. M. Larhed, G. Linderberg and A. Hallberg, *Tetrahedron Lett.*, 1996, **378**, 8219.
48. G.W. Kabalka, R.M. Pagni, V.V. Namboodri and C.M. Hair, *Green Chem.*, 2000, **2**, 120.
49. G.W. Kabalka, L. Wang, V.V. Namboodri and R.M. Pagni, *Tetrahedron Lett.*, 2000, **41**, 5151.
50. C.-H. Juss, J.-H. Chring, D.-Y. Lee, A. Loupy and S. Chatti, *Tetrahedron Lett.* 2001, **42**, 4803.
51. A. Mohna, J.I. Vaguero, J.I. Garefa and J. Alvarez-Builla, *Tetrahedron Lett.*, 1993, **34**, 2673.
52. W.B. Wang and E.J. Roskamp, *Tetrahedron Lett.*, 1992, **33**, 7631.
53. H. Tonaux, H. Klein, F. Texier-Boullet and J. Hamelin, *J. Chem. Research(S)*, 1994, 116.
54. G.B. Jones, R.S. Hurber and S. Chau, Tetrahedron, 1993, **49**, 369; R.S. Hurber, G.B. Jones, *J. Org Chem.* 1992, **57**, 5778.
55. D. Villemin, B. Labiad and Y. Ouhilal, *Chem & Ind.*, 1989, 607.
56. R.A. Abramovitch and A. Bulman, *Synlett.*, 1992, 795.
57. A.B. Alloum, B. Labiad and D. Villemin, *J. Chem. Soc., Chem. Commun.* 1989, 386.
58. J. Ipaktschi and M. Bruck, *Chem. Ber.*, 1990, **123**, 1591,
59. G. Bram, A. Loupy, M. Majdoub, E. Gutierrez and E. Ruiz-Hitzky, *Tetrahedron*, 1990, **46**, 5167.
60. Y. Yuncheng and J. Yulin, *Synth. Commun.*, 1992, **22**, 3109.
61. D. Villemin and A.B. Alloum, *Synth. Commun.*, 1990, **20**, 925.
62. R.J. Giguere and B. Herberich, *Synth. Commun.*, 1991, **21**, 2197.
63. W. Yuliang and J. Yaozhong, *Synth. Commun.* 1992, **22**, 2287.
64. D. Villemin, B. Labiad and A. Loupy, *Synth. Commun.*, 1993, **23**, 419.
65. D. Villemin, and B. Martin, *J. Chem. Research (S)*, 1994, 146.
66. D. Villemin and A.B. Alloum, *Synt. Commun*, 1990, **20**, 3325; D. Villemin and B. Labiad, Synth. Commun., 1990, **20**, 3207; D. Villemin and B. Labiad, Synth. Commun., 1990, **20**, 3213; D. Villemin and B. Labiad, *Synth. Commun.*, 1990, **20**, 3333.
67. M. Rahmouni, A. Derdour, J.P. Bazureau, and J. Hamelin, *Tetrahedron, Lett.*, 1994, **35**, 4563.
68. B. Rechsteiner, F. Texier-Boullet and J. Hamelin, *Tetrahedron Lett.*, 1993, **34**, 5071.
69. J.F. Pilard, B. Klien, F. Texier-Boullet and J. Hamelin, *Synlett.*, 1992, 219.
70. G. Bram, A. Loupy M. Majdaub and A. Petit, *Chem. & Ind.*, 1991, 396.
71. A. Petit, A. Loupy, Ph Maillard and M. Momenteau, *Synth. Commun.*, 1992, **22**, 1137.
72. H-M. Yu, S-T. Chem and K-T. Wang, *J. Org. Chem.*, 1992, **57**, 4781.
73. A.K. Bose and F. Greer, *J. Org. Chem.*, 1958, **23**, 1335.
74. A.K. Bose, M.S. Manhas, M.Ghosh, V.S. Raju and K. Tabei, *Heterocycles*, 1990, **30**, 741; A.K. Bose, M.S. Manhas, M. Ghosh, M. Shah, V.S. Raju and S.S. Bari, *J. Org. Chem.*, 1991, **56**, 6968.
75. A.K. Bose, M. Jayaraman, A. Okawa, S.S. Bari, E.W. Robb and M.S. Manhas, *Tetrahedron Lett.*, 1996, **39**, 6989.
    (a) A.K. Bose, B.K. Banik, N. Lavlinskaia, M. Jayaraman and M.S. Manhas, *Chemtech*, Sept. 1997, 18.
76. F. Texier-Boullet, P. Latouche and J. Hamelin, *Tetrahedron Lett.*, 1993, **34**, 2123.
77. R. Latouche, F. Texier-Boullet and J. Hamelin, *Bull. Soc. Chem. Fr.*, 1993, **130**, 535.
78. R. Latouche, F. Texier-Boullet and J. Hamelin, *Tetrahedron Lett.*, 1991, **32**, 1179.

5.24 ■ Alternate Energy Processes in Chemical Synthesis

79. F. Dorwald Zaragoza, *Organic Synthesis on Solid Phase*, Wiley-VCH, Weinheim (2000).
80. I. Sucholeiki, *High-throughput synthesis, Principles and Practices*, Marcel Dekker, New York (2001).
81. G.A. Stronhmeier and C.O. Kappe, *J. Comb. Chem.*, 2002, **4**, 154.
82. A. Alajarin, J.J. Vaquero, J.L. Garcia Navio and *J. Alvarez-Builla, Synlett.*, 1992, 297.
83. C.O. Kappe, D. Kumar and R.S. Varma, *Synthesis*, 1999, 1979.
84. I.C. Cottrill, A.Y. Usyatinsky, J.M. Arnold, D.S. Clark, T.S. Dornick, P.C. Michels and Y.L. Khmelnitsky, *Tetrahedron Lett.*, 1998, **39**, 1117.
85. C.N. Selway and N.K. Terret, *Bioorg. Med. Chem.*, 1996, **4**, 645.
86. R.S. Varma and D. Kumar, *Tetrahedron Lett.*, 1999, **40**, 7665.
87. A. Ya. Usyatinsky and Y.L. Khmelnitsky, *Tetrahedron Lett.*, 2000, **41**, 5031.
88. R. Olsson, H.C. Hansen, C.-M. Andersson, *Tetrahedron Lett.*, 2000, **41**, 7947.
89. (a) J. Dupont, R. de Sozua, P.A.Z. Suarez, *Chem. Rev.*, 2002, **102**, 3667;
    (b) D.Zhao, M. Wu, Y. Kou, E. Min., *Catalysis Today* 2002, **74**, 157;
    (c) J.F. Brennecke, Al chE Journal, 2002, **47**, 2384;
    (d) P. Wasserscheid, W. Keim, Angew, *Chem. Int.* Ed. Engl. 2000, **39**, 3772;
    (e) J. Welton, *Chem. Rev.*, 1999, **99**, 2071;
    (f) H. Olivier, *J. Mol. Cat. A. Chem.*, 1999, **146**, 285;
    (g) J.D. Holbrey, K.R. Seddon, *Clean Prod. Proc.* 1999, **1**, 223.
90. P. Bonhôte, A.P. Dias, N. Papageorgiou, K. Kalyanasundram, M. Grätzel, *Inorg. Chem.*, 1996, **35**, 1168.
91. (a) C.L. Hussey, *Molten Salt Chem.*, 1983, **5**, 185;
    (b) J.S. Wilkes, J. Levisky, R.A. Wilson, C.L. Hussey, *Inorg. Chem.*, 1982, **21**, 1263.
92. P. Volker, W. Bohm, W.A. Herrmann, *Chem., Eur. J.*, 2000, **6**, 1017.
93. (a) R.S. Varma, V.V. Namboodiri, *Chem. Commun.*, 2001, 643;
    (b) R.S. Varma, V.V. Namboodiri, *Pune Appl. Chem.*, 2001, **73**, 1309.
94. A recently introduced household microwave oven (Parasonic) equipped with invertor technology provides a good control of the microwave power to a desirable level.
95. J.D. Holberg, K.R. Seddon, *J. Chem. Soc. Dalton*, Trans. 1999, 2133.
96. R.D. Rogers and K.R. Seddon, Ionic Liquids as Green Solvents ACS Symposium 856 and the references cited therein.
97. V.N. Namboodiri, R.S. Varma, *Tetrahedron Lett.*, 2002, **43**, 5381.
98. J.S. Wilkes, M.J. Zaworokto, *J. Chem. Soc. Chem. Commun.*, 1992, 965.
99. N.E. Leadbeater and H.M. Trenius, *J. Org. Chem.*, 2002, **67**, 3145.
100. E. Vander Eyckon, P. Appkkuttan, W. De, Borggraeve, W. Dehaen, D. Dallinger and C.O. Kappe, *J. Org. Chem.*, 2002, **67**, 7904.
101. V.N. Namboodiri and R.S. Varma, *Chem. Commun.*, 2002, 342.
102. J.F. Dubreuil and J.P. Bazureau, *Tetrahedron Lett.*, 2000, B41, 7351.
103. (a) Synthetic organic chemistry, J.L. Luche (Ed.), Plenum Press, New York, 1998; A. Gaplovsky, M. Gaplovsky, S. Toma, J.L. Luche, *J. Org. Chem.*, 2000, **65**, 84444; R.R. Deshmukh, R. Rajagopal, K.V. Srinivasan, *Chem. Commun.*, 2001, 1544; J.M. Leveque, J.L. Luche, C. Petrier, R. Roux, W. Bonvath, *Green Chem.*, 2002, **4**, 357.
104. V.N. Namboodiri and R.S. Varma, *Org. Lett.*, 2002, **4**, 3161.

Part-II

ULTRASOUND ASSISTED ORGANIC SYNTHESIS

Part-II

ULTRASOUND ASSISTED ORGANIC SYNTHESIS

6

INTRODUCTION

Ultrasound has been used in the service of mankind since 1950. It is more familiar in the context of animal communications (bats, dog whistles, etc.), medical diagnosis (foetal scanning), materials testing (Flow detection), under-water ranging (depth gauges) and cleaning ultrasonic baths.

The name ultrasound is given to sound waves having frequencies higher than those to which the human ear can respond (i.e. > 16 KHz) (Hz = Hertz = cycles per second). It is considered to lie between 5 MHz (for gases) and 500 MHz (for liquids and solids). For many years ultrasound has found a variety of uses in engineering, science and medicine. However, its application to chemistry has received attention only in the recent past. A large number of publications have appeared in the past few years which describe various applications of ultrasound to chemical synthesis. The term 'sonochemistry' is used to describe the effect of ultrasonic sound waves to chemical reactivity. A number of reviews on the chemical applications of ultrasound have been published[1-7].

6.1 INSTRUMENTATION

The instrumentation for the generation of ultrasound requires an ultrasonic transducer, a device by which electrical or mechanical energy can be converted into sound energy. The most commonly used are the electromechanical transducers which convert energy into sound—they are most commonly based on piezo-electric effect. The piezo-electric effect is the production of a potential difference across opposite faces of a crystal of a material when it is subjected to sudden compression and is found in some crystalline materials like quartz. The inverse effect is produced when a rapidly alternating potential is placed across the faces of piezo-electric crystal. This will induce dimentional changes in the crystal and, thus, generate vibrational (sound) energy[1-4].

The ultrasonic equipments are normally of four types[2] as explained below:

(i) *Whistle Reactor* is a mechanical transducer device and is predominantly used for homogenization and emulsification.

(ii) *Ultrasonic Cleaning Bath* is the most accessible and simplest instrument available. In this, the amount of power reaching the 'reaction' immersed in the bath is not readily quantifiable because

6.4 ■ Alternate Energy Processes in Chemical Synthesis

it will depend on the size of the bath and the type of the reaction vessel. Also, the temperature control is not easy in this system. A drawback in this system is that they do not all operate at the same frequency, and so do not give reproducible results.

(iii) *Direct Immersion Sonic Horn* is the most efficient method of transmitting ultrasonic energy into a reaction. In this case, an ultrasonic probe is placed directly in the reacting system and such a equipment is used for biological cell disruption. The advantage of such a system is that much higher ultrasonic powers can be used since energy losses during the transfer of ultrasound through the bath media and reaction vessel walls are eliminated. These devices can be tuned to give optimum performance in the reaction mixture over a range of powers. However, in such a system there are difficulties of temperature control and operation at fixed frequency. Direct sonication results also in generation of radical species by the action of the probe tip on the solvent.

(iv) *The Cup Horn* was originally designed for all disruption, but is more controllable than a cleaning bath in terms of power and temperature and less drastic in action than a sonic horn. Such a system allows more quantitative and reproducible studies. The frequency is fixed and the power is tuneable.

Two commonly used ultrasound instruments as shown in the figures below:

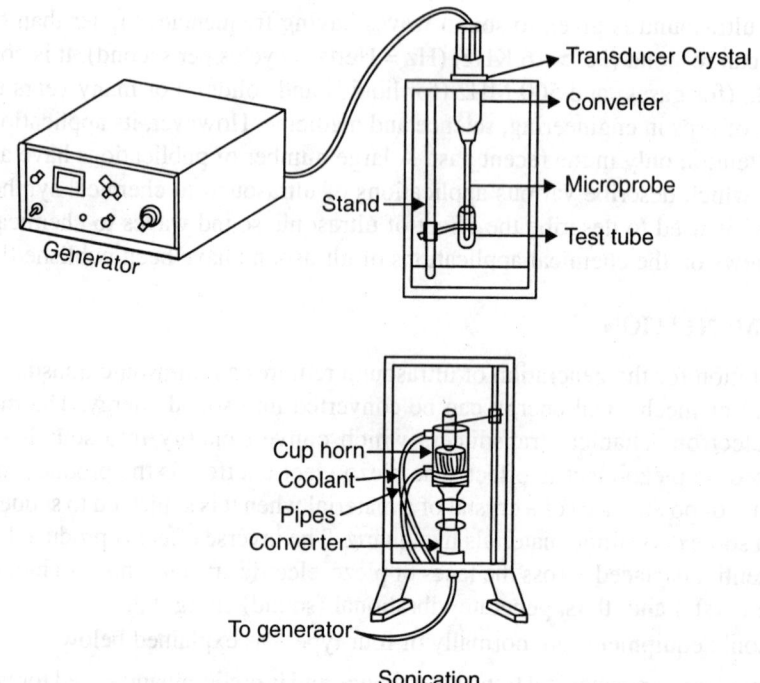

Sonication

6.2 THE PHYSICAL ASPECTS[2]

Ultrasound being a sound wave is transmitted through any substance, solid, liquid or gas, which posses elastic properties. When transmitted through a medium, it produces alternate compression and rarefraction (stretching cycle of waves). The use of ultrasound may be divided into two areas. *First*, low amplitude (high frequency) propagation, which is concerned with the effect of the medium on the wave. Generally speaking, the low amplitude waves are used to measure the velocity and absorption coefficient of the wave in the medium. *Secondly,* high energy (low frequency) propagation, which is concerned with the effect of the wave in the medium. Examples of the second type are ultrasonic cleaning, drilling, soldering, emulsification of chemical processes, *etc*. These processes are the result of either the mechanical agitation caused by the wave or are of consequence of cavitation (tiny bubbles) produced in the liquid.

It is only after the understanding of the phenomenon of cavitation with significant development in the transducer design that a rapid expansion in the application of ultrasound to chemical processes occurred. It is the subsequent fate of some of the cativation bubbles, as they oscillate in the applied sinusoidal acoustic field, which is the origin of sonchemical effects. There are two forms of cativation—stable and transient. The stable cavities are those which oscillate, often non-linearly, about some equilibrium size—such bubbles have an existence of many cycles. These are also capable of being transformed into transient cavities. The transient cavities, on the other hand, generally exist for less than a single acoustic cycle during which time they expand to at least double their initial size before collapsing violently into smaller bubbles generating high energies the bubbles involving pressures of hundreds of atmospheres and temperature of thousands of degrees. An estimate of temperature and pressures involved in the final phase of the implosion of a bubble containing nitrogen in water at ambient temperature (20°) and ambient pressure (1 bar), gives 4200 K and 975 bar. It is the existence of these very high temperatures within bubble that have formed the basis for the radical production and sonoluminescene. On the other hand, the release of pressure as a shock wave is a factor which has been used to account for increased chemical reactivity (due to increased molecular collision) and polymer degradation.

The mechanical and chemical effects of the collapsing bubble will be felt in two distinct regions (a) within the bubble itself, which can be though of as a high temperature and pressure microreactor. It is the existence of these very high temperatures within the bubble that has formed the basis for the radical production (homogeneous sonochemistry) and sonoluminiscence, and (b) in the immediate vicinity of the bubble where shock wave is produced in collapse as a result of release of pressure. This shock wave is believed to account for increased chemical reactivity due to increased molecular collisions (heterogeneous sonochemistry). The phenomenon of cavitation is dependent upon local environment and is effected by change in frequency, solvent, system vapour pressure and hence temperature, external pressure, etc.

The types of ultrasound which are used in chemistry are basically divided into 'Power Ultrasound' between 20 and 100 KHz, which is used for cleaning, plastic welding and to affect chemical reactivity, and 'High Frequency Ultrasound' in the range 2-10 MHz, which is used in metal scanning, chemical analysis, in the study of relaxation phenomena, etc. The first type, i.e., power ultrasound provides a form of energy for the modification of chemical reactivity which is different from the normally used, i.e., heat, light and pressure and it is this type which is of interest to synthetic chemists.

6.3 TYPES OF SONOCHEMICAL REACTION

The synthetic chemist is mainly concerned with reactions in solution. The effect of ultrasound on various reactions is summarised in the following types of reactions.

6.3.1 Homogeneous Reactions

The chemical effects of ultrasound on liquids have been studied for many years. If water is sonicated, cavitation induces the homolytic cleavage of water—the primary products are H_2 and H_2O_2; other high-energy intermidiates, viz., $HO_2^{\bullet}$, $H^{\bullet}$, $OH^{\bullet}$ and e^- (aq) have been suggested. This could have been an interesting synthetic method for generation of hydrogen, but it is not so. The formation of ammonia by the sonification of aqueous solution of nitrogen and hydrogen is worth mentioning. The main problem in non-aqueous sonochemistry is that organic solvents have high vapour pressure and, therefore, low temperatures are required to achieve effective cavitation.

6.3.2 Heterogeneous Liquid-Liquid Reactions

The use of phase transfer catalysts in organic aqueous biphasic system is well known to catalyse these reactions. However, ultrasound is much more effective in these reactions because ultrasonic waves generate extremely fine emulsions which result in very large interfacial contact areas between the liquids—the result is a dramatic increase in the reactivity between species dissolved in the separate liquids and, therefore, should react much faster than the conventional phase-transfer conditions. In some cases, it has been found that a combination of sonication and PTC has a better overall effect than either of the two techniques alone.

6.3.3 Heterogeneous Solid-Liquid Reactions

These are of two types: Type I—those in which the solid serves as one of the reagents, and is consumed during the reaction. The Type II are those in which the solid often a metal functions either as a catalyst or is consumed. The Type I reactions have been used with success to improve yields—this is due to the dispersing and microstreaming effects of ultrasound.

In Type II reactions, the cavitational erosion is the major effect which is observed when ultrasonic waves propagate towards, or in the vicinity of a solid. The erosion of the metal follows the sequence Pb > Mg > Zn > Sn > Cu. Alkali metals have been submitted to sonochemical conditions in a variety of reactions.

NOTES AND REFERENCES

1. J.P. Lorimer and T.J. Mason, *Chem. Soc. Rev.,* 1987, **16**, 239–274.
2. J.L. Lindly and T.J. Mason, *Chem. Soc. Rev.* 1987, **16**, 275–311.
3. C. Einhorn, J. Einhorn and J.L. Luche, *Synthesis*, 1989, 787–813.
4. J.M. Khurana; *Chemistry Education*, 1990, 24–29.
5. K.S. Suslick, *Modern Synthetic Methods*, 1986, **4**, 1.
6. K.S. Suslick, *Adv. Organomet. Chem.*, 1986, **25**, 73.
7. D. Bremner, *Chem. Ber.,* 1986, **22**, 633.

7

HOMOGENEOUS SONOCHEMICAL REACTIONS

It has already been stated[1] that cavitation induces the homolytic cleavage of water and that the primary products are H_2O, H_2O_2 and other high energy intermediates like $HO_2^{\bullet}$, $H^{\bullet}$, $OH^{\bullet}$ and e^- (aq). More complicated mixtures, containing formaldehyde, hydrogen cyanide, imidazole etc. result from the irradiation of N_2, H_2 and CO molecules in water[2].

Following are given some of the interesting homogeneous sonochemical reactions:

7.1 CURTIUS REARRANGEMENT

The cavitational explanation (given above) involved the spontaneous evolution of an exicitied species. A curtius rearrangement is the oldest example. Thus benzoyl azide on sonication in benzene solution undergoes Curtius rearrangement to give phenylisocyanate. Though the rate of formation is much higher than under stirring, the yield is low[3] (Scheme-1).

$$Ph-CO-N_3 \xrightarrow{\text{benzene, RT},))))} Ph-N=C=O + N_2$$

Scheme-1

7.2 SULPHUR EXTRUSION FROM 1, 3, 4-THIADIAZINES

1, 3, 4-Thiadiazines on sonication in a suitable solvent (ethanol)-sulphur extrusion occurs with a low energy cleaning bath as the ultrasound source[4] (Scheme-2).

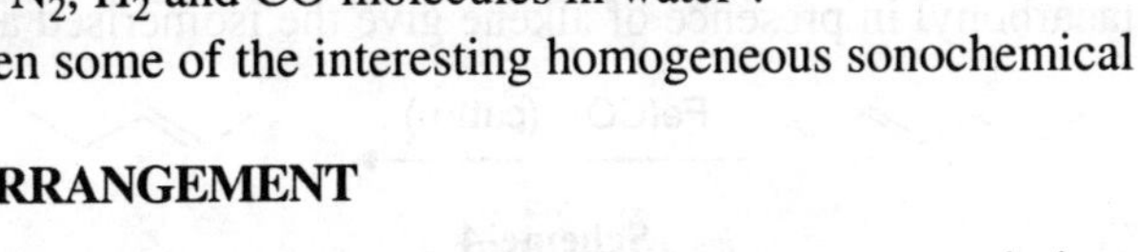

R^1	R^2	R^3	Solvent	Time (h)	Yield (%)
NHAC	Ph	Ph	EtOH	20	87

| SMe | Ph | Ph | Toluene | 14 | 86 |
| NH$_2$ | CO$_2$Et | CH$_3$ | EtOH | 8 | 85 |

Scheme-2

7.3 ISOMERISATION OF MALEIC ACID TO FUMARIC ACID

This is the most interesting example under homogeneous sonochemical reaction. The isomerisation takes place in presence of bromine or alkyl bromide[5]. The mechanism involves sonolysis of bromine molecule to the bromine radical (Scheme-3).

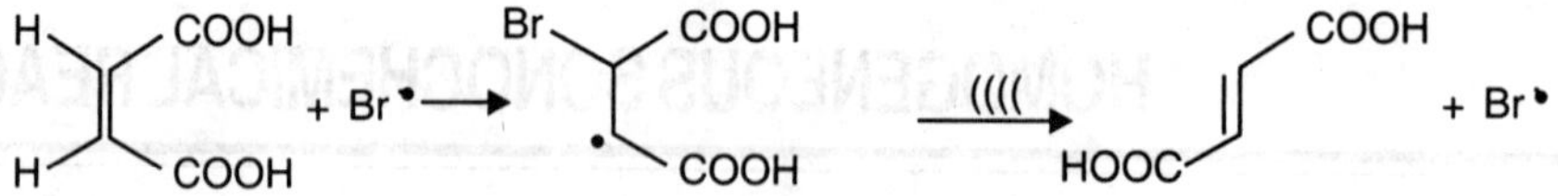

Scheme-3

7.4 ORGANOMETALLIC REACTIONS

7.4.1 Isomerisation of Alkenes

Sonolysis of iron pentacarbonyl in presence of alkene give the isomerised alkene (Scheme-4)[6].

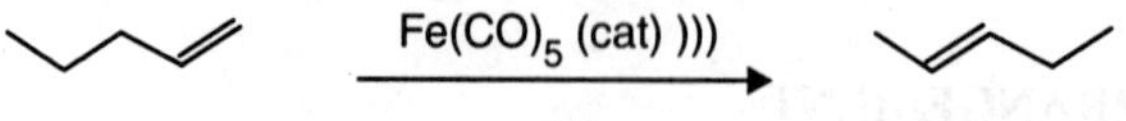

Scheme-4

It is believed that ironpentacarbonyl on sonication gives Fe$_3$(CO)$_{12}$; the yield varies depending on the solvent. Heptane gives the best result

$$Fe(CO)_5 \xrightarrow[\text{heptane}]{))))} Fe_3(CO)_{12} \quad 82\%$$

The rate of isomerisation of alkene is 10^5 higher than in the reaction performed on thermal or photlytic pathway. Similar observations have been made from various metal carbonyl complexes[7].

7.4.2 Annulation

The reaction of cydopentene with appropriate organometallic complex (1) on sonication[8] in toluene for 3 hr gave the annulated cyclopentene (2) in 49% yield. Under thermal conditions the yield was 42% (Scheme-5).

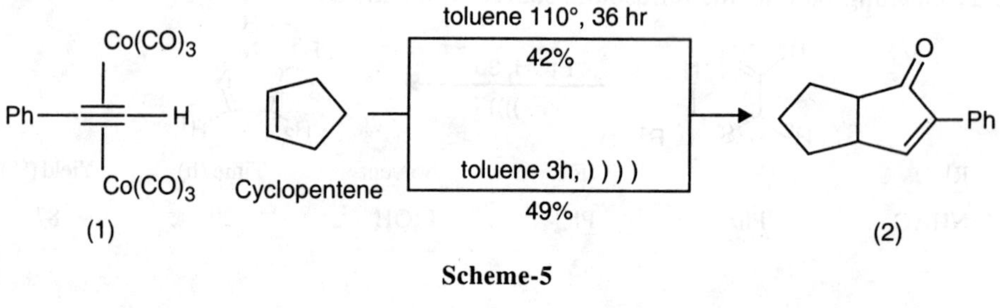

Scheme-5

Using chiral ligand in the organometallic complex (1) leads to the expected cyclopentenone (2) on sonication[9] (Scheme-6).

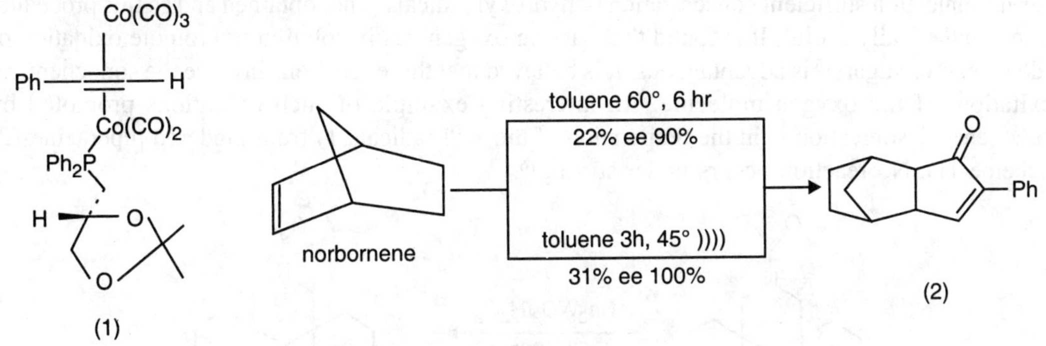

Scheme-6

The snonchemical ligand displacement can also be involved in as alkynone hydration reaction[10] (Scheme-7).

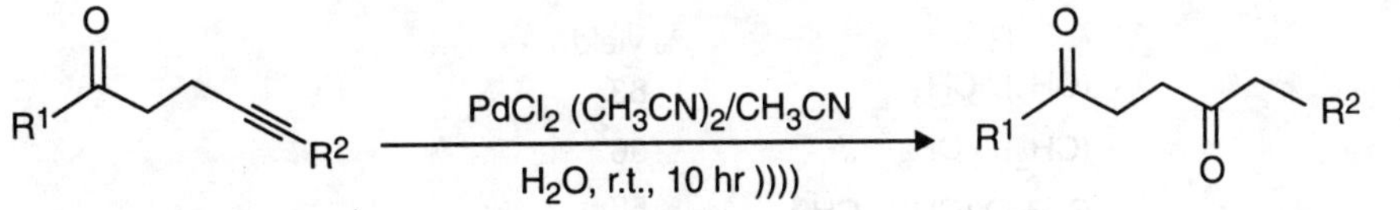

Scheme-7

7.4.3 Grignard Reagents

Magnesium on sonication can be activated. This activated magnesium finds applications[11, 12, 13] in the synthesis of Grignard reagents without the use of activators in ether in common laboratory cleaning bath (Scheme-8).

$$R - X + Mg \xrightarrow{\text{ether} \atop))))} RMgX \ (90\%)$$

Scheme-8

A number of Grignard reagents have been prepared[13] from olefins, e.g., chlorodiene as shown below (Scheme-9).

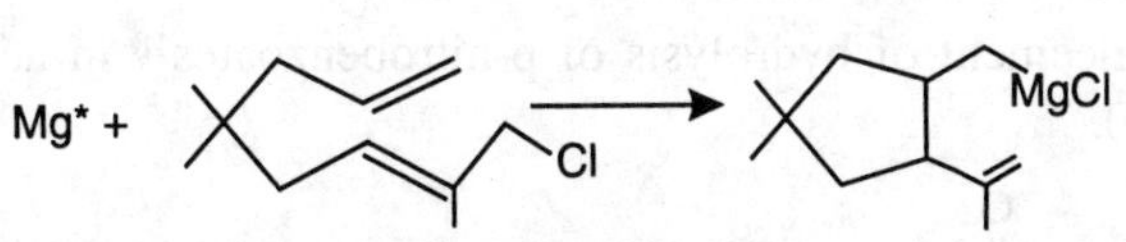

Scheme-9

Similarly, n-propyl, n-butyl and phenyl lithium are prepared[11, 12] in > 90% yield by the reaction of appropriate bromide with lithium wire; sec- and tert-Alkyl bromides require longer period of sonication (Scheme-10).

$$R - X + Li \xrightarrow{))))} R - Li \ (90\%)$$

Scheme-10

7.4 ■ Alternate Energy Processes in Chemical Synthesis

7.5 OXIDATIONS

The oxidizing properties of ultrasound by the generation of hydroxyl radicals has been studied. During sonication sufficient concentration of hydroxyl radicals is not obtained and so this procedure is not synthetically useful. It is found that passing oxygen gas in solution to promote oxidation of aldehydes[14] or sugars[15] is advantageous. It is believed that these reactions involve the sonochemical 'exitation' of the oxygen molecule. An interesting example of such oxidations promoted by homogeneous sonication is in the preparation of nitroxyl radicals(1) from hindered piperazines(2) (Scheme-11). No reaction occurs under stirring[16].

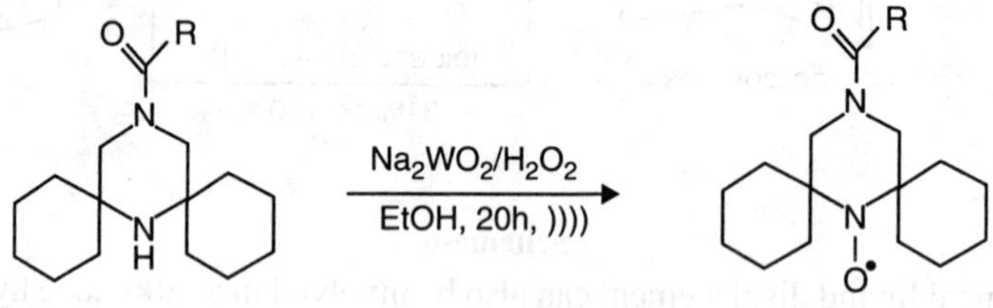

R	% yield
$(CH_2)_4 CH_3$	63
$(CH_2)_{16} CH_3$	86
$C_6H_4O (CH_2)_7 CH_{3-4}$	58

Scheme-11

7.6 SOLVOLYSIS AND HYDROLYSIS

Solvolysis of text butyl chloride in aqueous ethanol takes place[17] (Scheme-12). Rate enhancement of ~2 were observed using a clearing bath.

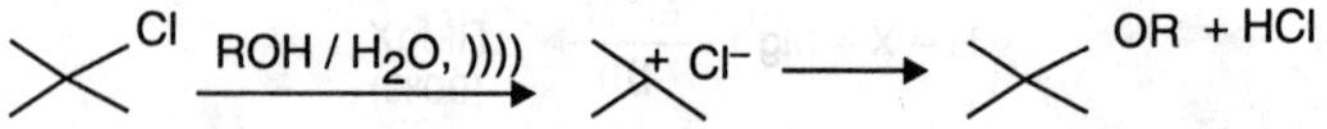

Scheme-12

It is believed that local electrical phenomena raise the ground state energy of the tert-butyl chloride, making the bond breaking easier.

There is rate enhancement of hydrolysis of p-nitrobenzoates[18] in acetonitrile (at pH8) on sonication (Scheme-13).

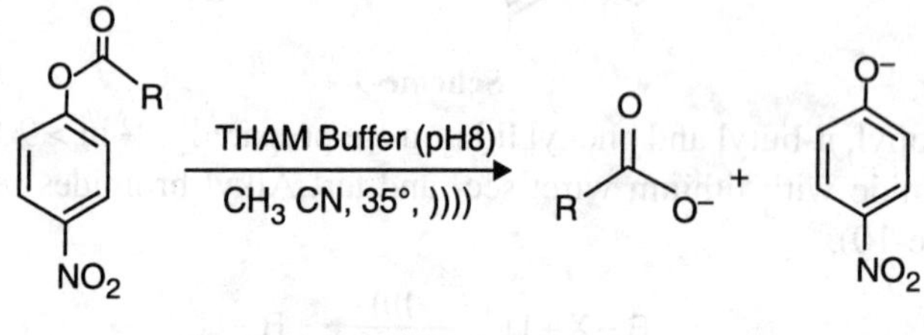

$R = CH_3, CH_2CH_3, i\,C_3H_7, t\text{-}C_4H_9$

Scheme-13

7.7 ADDITION REACTIONS

7.7.1 Diels-Alder Reaction

Diels-Alder reaction is facilitated by sonication. Thus, the addition of dimethylacetylene dicarboxylate to furan derivatives in water at 22–45°C gives quantitative yield of the adduct (Scheme-14).

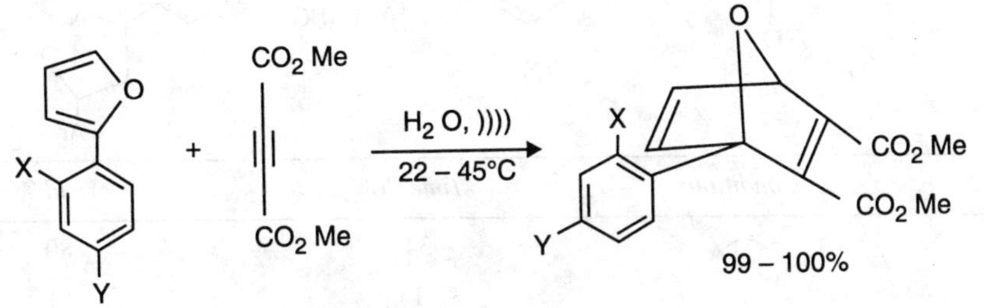

Scheme-14

The Diels-Alder cycloaddition of various dienes (mostly belonging to 1-vinyl cyclohexenes) with o-quinone proceeds very well[19] under ultrasound conditions to give the expected adducts in 59% yield (Scheme-15) compared to 30% under normal reaction conditions. Better results are obtained by soniciation of the neat mixture, and the presence of solvent is detrimental.

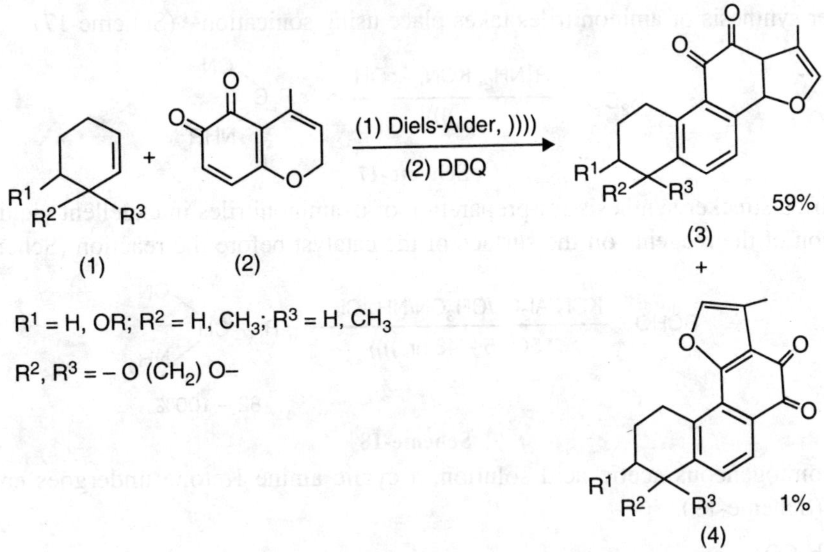

R^1 = H, OR; R^2 = H, CH_3; R^3 = H, CH_3

R^2, R^3 = – O (CH_2) O–

Diel Alder reaction of (1) and (2) (R^1 = H, R^2 = H, R^3 = H)

Solvent	Conditions	Time (h)	Yield	
			(3)	(4)
benzene	retlux	2	30	15
benzene	11 K bar	2	58	9
none	))))	2	59	7

Scheme-15

7.6 ■ Alternate Energy Processes in Chemical Synthesis

7.7.2 1, 3-Dipolar Cycloaddition

Sonication influences 1, 3-dipolar cydoadditions. Thus nitrones (1) add to olefins (2) in excellent yields[20], in times shorter by a factor of 20–40 with comparison to the silent reaction (Scheme-16).

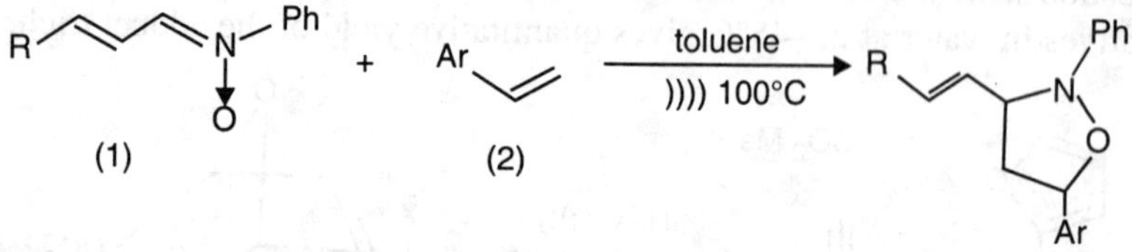

Ar	Conditions	Time (h)	Yield %
Ph	⌒	34	80
Ph	))))	1	81
4-Cl C$_6$H$_4$	⌒	24	75
4-Cl C$_6$H$_4$	))))	1	75

Scheme-16

7.7.3 Strecker Reaction

The Strecker synthesis of aminonitriles takes place using sonication[21] (Scheme-17).

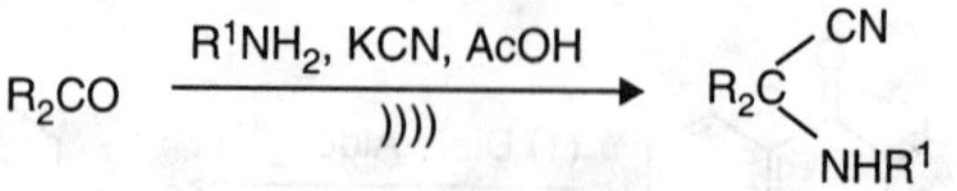

Scheme-17

A modified strecker synthesis for preparation of α-aminonitriles in excellent yield consists of the adsorption of the reagents on the surface of the catalyst before the reaction (Scheme-18).

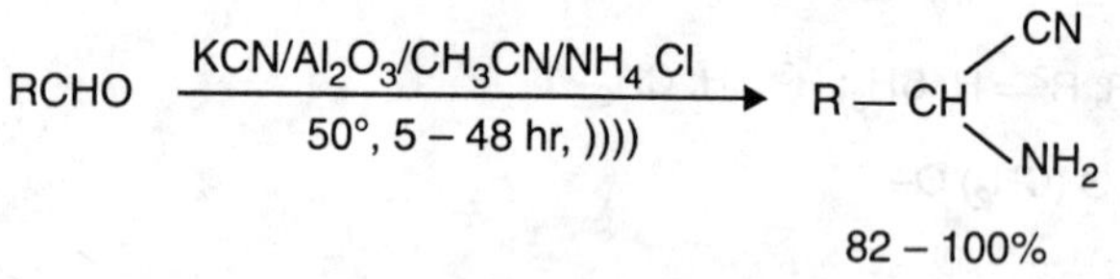

Scheme-18

In an homiogeneous acetic acid solution, a cyclic amine Ketone undergoes easy strecker synthesis[21] (Scheme-19).

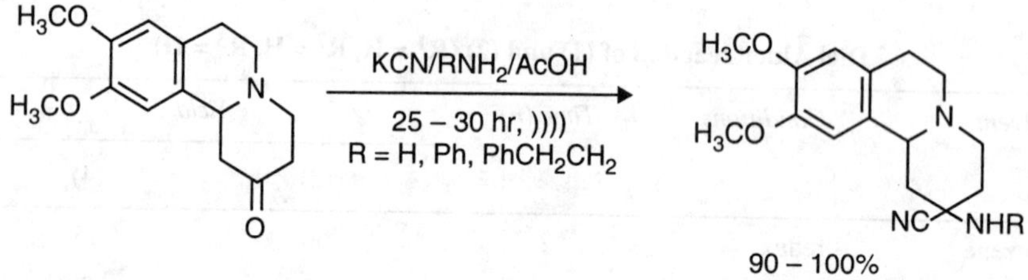

Scheme-19

Basic catalysed addition of thiourea to ferrocenyl chalkones (1) give excellent yields of the product (2), where run under sonication. The thermal process gives mixtures which cannot be separated due to instability of the compounds[22] (Scheme-20):

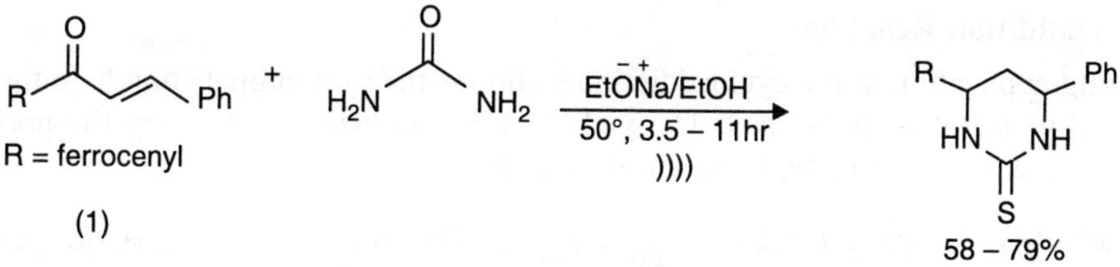

Scheme-20

The addition[23] of magnessium or titanium phenolate (1) to aldehyde(2) by sonication using dichloromethane as the solvent. In this case, the stereoselectivity is sensitive to sonication (Scheme-21).

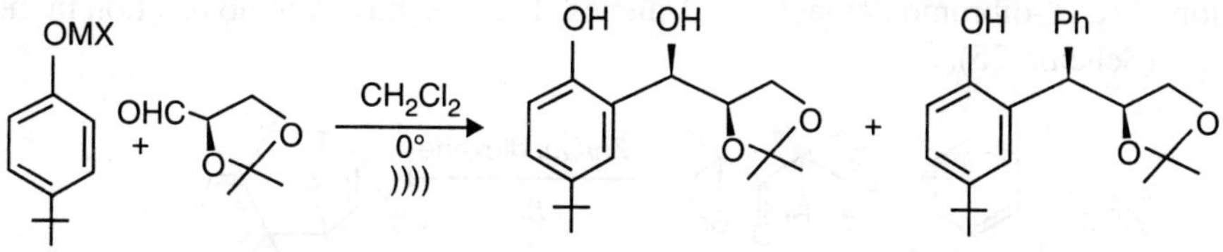

Mx	Conditions	Yield (%)	
		(3)	(4)
Mg Br	↷	27	5
Mg Br	))))	64	6
Ti (OPr-i)$_3$	))))	8	68

Scheme-21

7.7.4 Electrophilic Addition

Besides the nucleophilic addition reaction discussed, the only example of an electrophilic addition is found[24] with the intramolecular aminomercuration of a carbopenam precursor (Scheme-22).

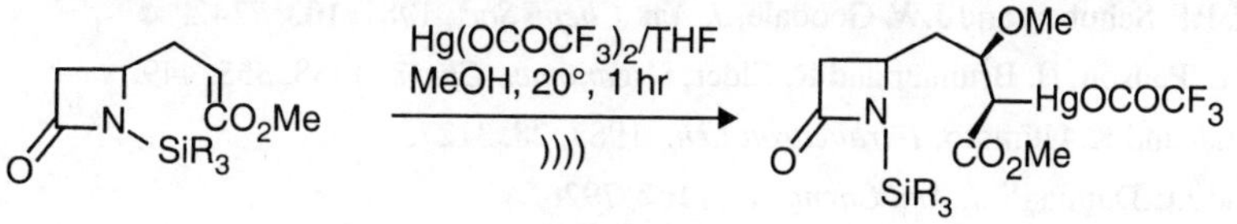

Scheme-22

7.7.5 [2 + 2] Cycloaddition Reactions

The 2 + 2 cycloaddition of dichloroketene to alkenes is also improved by ultrasound[25] (Scheme-23).

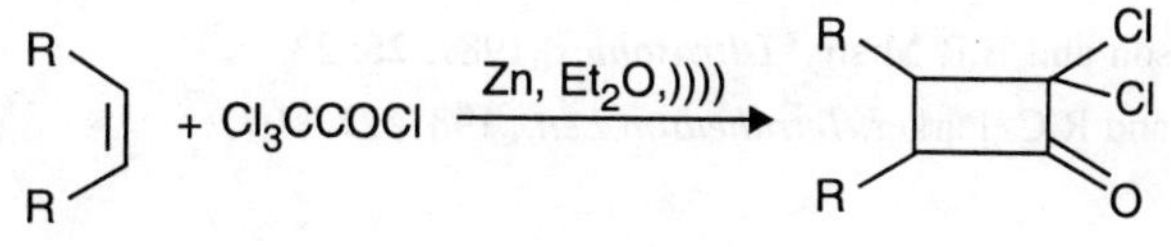

Scheme-23

7.8 ■ Alternate Energy Processes in Chemical Synthesis

7.7.6 Cycloaddition Reactions

An interesting application of the cycloaddition reaction is the cyclopropanation[26]. In this reaction, sonochemically activated Zn is used. The yield is 91% compared to 51% by the normal route. Ultrasonic source is a cleaning bath (50 KHz) (Scheme-24).

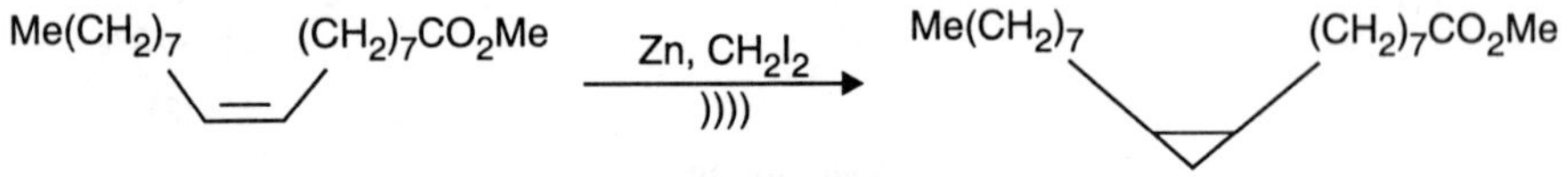

Scheme-24

The method has several advantages and can also be scaled up[27].

Highly hindered bicyclo [3, 2, 1] oct-6-en-3-ones are easily obtained[28] by zinc promoted cycloaddition of α, α'-dibromo ketones to 1,3-dienes. There is, however, no reaction in the absence of ultrasound (Scheme-25).

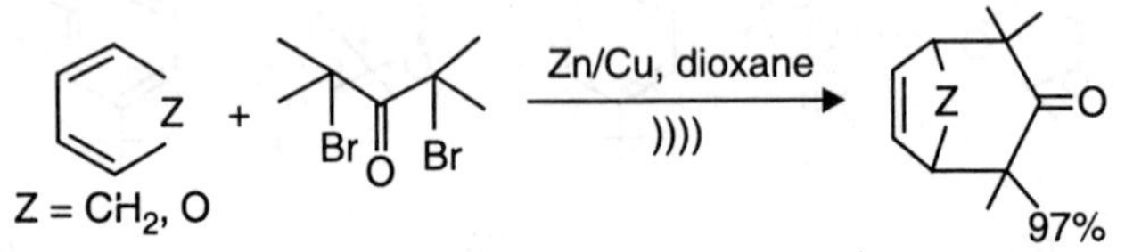

Z = CH_2, O

Scheme-25

NOTES AND REFERENCES

1. C. Einhorn, J. Einhorn and J.L. Luche, *Synthesis*, 1989, 787-813.
2. I.E. Elpiner, *Priroda*, 1968, **11**, 64; CA, 1969, **71**, 77481.
3. C.W. Porter and L. Young, *J. Am. Chem. Soc.*, 1938, **60**, 1497.
4. W.D. Pfeiffer, E. Bulka and R. Miethechan, *Z. Chem.*, 1987, **27**, 296.
5. I.E. Elpiner, A.V. Sokolskaya and M.A. Margulis, *Nature* (Londonn), 1965, **208**, 945.
6. K.S. Suslick, D.A. Hammerton and R.E. Clini, *J. Am. Chem. Soc.*, 1986, **108**, 5641.
7. K.S. Suslick, J.W. Goodale, P.F. Schubert and H.H. Wang, *J. Am. Chem. Soc.*, 1983, **105**, 5781; K.S. Suslick and R.E. Johnson, *J. Am. Chem. Soc.*, 1984, **105**, 6856.
8. K.S. Suslick, P.F. Schubert and J.W. Goodale, *J. Am. Chem. Soc.*, 1981, **103**, 7342.
9. P. Bladon, P.L. Pauson, H. Brunner and R. Elder, *Organomet. Chem.*, 1988, **355**, 449.
10. K. Imi, K. Imai and K. Utimoto, *Tetrahedron Lett.*, 1987, **28**, 3127.
11. J.L.Luche and J.C.Damiano, *J. Am. Chem. Soc.*, **102**, 7926.
12. J.D. Sprich and G.S. Lewandos, *Inorg. Chem-Acta.*, 1982, **76**, 1241.
13. W. Oppolzer and A. Nakao, *Tetrahedron Lett.*, 1986, **27**, 5471.
14. V.L. Starchevskii, E.N. Mokryi and M.A. Margulis, *Russian J. Phys. Chem.*, 1984, **58**, 845.
15. C. Cook and Y. Cho, *Yakhak Hoeji*, 1969, **13**, 16; C.A., 1970, **73**, 45709.
16. V. Kaliska, S. Toma and J. Lesko, *J. Coll. Czech. Chem. Commun.*, 1987, **52**, 2266.
17. J.P. Lorimer, T.M. Mason and B.P. Mistry, *Ultrasonics*, 1987, **25**, 23.
18. D.S. Kristol, H. Klotz and R.C. Parker, *Tetrahedron Lett.*, 1981, **22**, 907.

19. J. Lee and J.K. Sayder, *J. Am. Chem.Soc.*, 1989, **111**, 1522.
20. D.R. Borthakur and J.S. Sandhu, *J. Chem. Soc-Chem. Commun.*, 1988, 1444.
21. J.C. Menendez, G.G. Trigo and M.M. Sollhuber, *Tetrahedron Lett.,* 1986, **27**, 3285.
22. S. Toma, M. Putala and M. Salisova, *Coll. Czech. Chem. Commun.,* 1987, **52**, 395.
23. G. Casiraghi, M. Cornia and G. Rassu, *J. Org. Chem.,* 1998, **53**, 4919.
24. F. Dumas and J. d'Angelo, *Tetrahedron Lett.*, 1986, **27**, 3725.
25. G. Mehta and H.S.P. Rao, *Synth. Commun.*, 1985, **15**, 991.
26. O. Repic and S. Vogt, *Tetrahedron Lett.*, 1982, **23**, 2729.
27. O. Repic, P.G. Lee and N. Giger, *Org. Prepn. Proc. Int.* 1984, **16**, 25.
28. N. Joshi and H.M.R. Hottmann, *Tetrahedron Lett.*, 1986, **27**, 687.

8

HETEROGENEOUS LIQUID-LIQUID REACTIONS

8.0 INTRODUCTION

Ultrasound is much more effective in heterogeneous liquid-liquid reactions. This is because ultrasound waves generate extremely fine emulsions, which result in very large interfacial contact areas between the liquids. This results in a dramatic increase in the reactivity between the species dissolved in separate liquids and, therefore react much faster than even the conventional phase-transfer conditions. However, it has been found that a combination of sonication and PTC gives a better overall effect than either of the two techniques.

Following are described some of the important heterogeneous liquid-liquid reactions.

8.1 ESTERIFICATION

The esterification of carboxylic acids is a widely studied reversible reaction[1]. The esterification is generally carried out in the presence of catalysts like sulphuric acid, p-toluenesulphonic acid, tosyl chloride, polyphosphate ester, dicyclohexylcarbodiimide, graphite phosphate. The reaction requires longer time, and yields are generally low. A simple procedure for the esterification of a variety of carboxylic acids with different alcohols at ambient temperature using ultrasound has been reported[2]. (Scheme-1).

$$RCOOH + R'OH \xrightarrow[))))]{H_2SO_4, R.T.} RCOOR'$$

Scheme-1

Low intensity ultrasound (cleaning bath) has also been used[3] for esterification. (Scheme-2).

$$RCO_2H + R'X \xrightarrow[))))]{KOH, PEG} RCO_2R'$$

X = Cl, Br, I

Scheme-2

8.2 ■ Alternate Energy Processes in Chemical Synthesis

Using the reaction of carboxylic acid with alcohol (Scheme-1) following esters have been prepared[2].

RCOOH	ROH	Time (hrs)	Isolated Yields (%)
Benzoic acid	Methanol	3	95
Benzoic acid	Ethanol	3	89
Benzoic acid	1-Propanol	4	91
Benzoic acid	2-Propanol	5	88
Benzoic acid	t-Butanol	5	46
p-Chlorobenzoic acid	Methanol	5	79
m-Chlorobenzoic acid	Methanol	5	88
p-Nitrobenzoic acid	Methanol	4	76
p-Hydroxybenzoic acid	Methanol	4	81
Salicyclic acid	Methanol	7	79
Cinnamic acid	Methanol	7	96
p-Anisic acid	Methanol	5	92
Phenylacetic acid	Methanol	1.5	93
Phenylacetic acid	Ethanol	1.5	93
Phenylacetic acid	1-Propanol	2	90
Phenylacetic acid	2-Propanol	4	87
Phenylacetic acid	1-Butanol	2	86
Phenylacetic acid	t-Butanol	5	57
Stearic acid	Methanol	2.5	90
Stearic acid	Ethanol	4	92
Oleic acid	Methanol	4	91
Palmitic acid	Methanol	2	87
Succinic acid	Methanol	4	93
Oxalic acid	Methanol	4	79
Fumaric acid	Methanol	4	92
Maleic acid	Methanol	3.5	86

8.2 SAPONIFICATION

Ester hydrolysis is frequently effected under agressive conditions. It can, however, be conducted under milder conditions when sonication is used[4]. The rate increase is attributed to the emulsifying effect. Thus, methyl-2, 4-dimethylbenzoate on saponification (20 KHz), gives 2,4-dimethylbenzoic acid in 94% compared to 15% yield by normal process of heating with aqueous alkali (90 min) (Scheme-3).

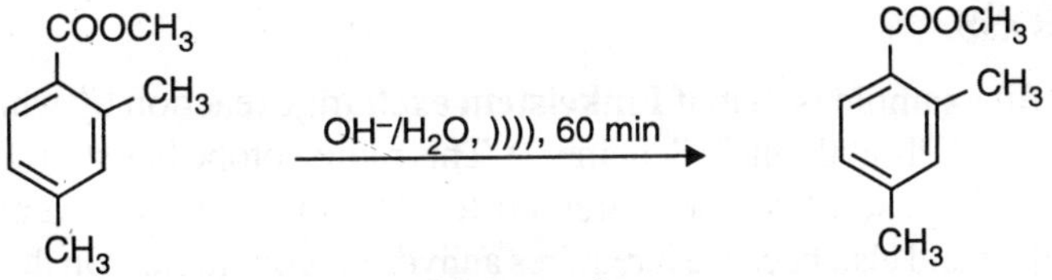

Scheme-3

Ultrasonically (20 KHz) induced rate enhancements for the hydrolysis of a number of 4-nitrophenyl esters of a number of aliphatic carboxylic acids has been reported[4,5].

Saponification of commercially important substances such as glycerides, rape seed oil and wool waxes is greatly accelerated by sonication[3].

8.3 HYDROLYSIS/SOLVOLYSIS

Solvolysis of tert-butyl chloride in aqueous alcohol has been studied[6] using sound waves. There is rate enhancement of ~2 using a cleaning bath. However, use of a probe generator raised the acceleration value to ~20 at 10° (Scheme-4).

Scheme-4

Basic hydrolysis of nitriles to carboxylic acids is also greatly improved by sonication[7] (Scheme-5).

ArCN $\xrightarrow[\text{))))}]{\text{HO}^-/\text{H}_2\text{O}}$ ArCOOH

Scheme-5

Using this procedure, benzoic acid is obtained[6] in 33% yield compared to 15% yield by the usual procedure from the corresponding nitrile in 10 hr. Similarly, β-naphthoic acid is obtained in 90% yield compared to 63% yield by the usual procedure in 6 hrs.

An interesting hydrolysis catalysed by ultrasound involves the synthesis of a Milbemycin D fragment (1)[8] (Scheme-6). In this double hydrolysis followed by acetalization, the role of sonic waves is unclear, but is essential.

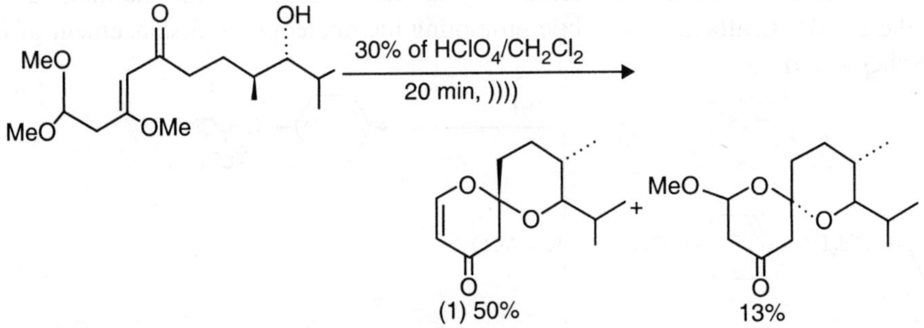

Scheme-6

8.4 ■ Alternate Energy Processes in Chemical Synthesis

8.4 SUBSTITUTIONS

An interestingly useful example is that of **Finkelstein exchange** reaction by which an ω-bramofatty acid is transformed to its ^{123}I-analogue[9] (Scheme-7). This radioisotope has short life and is expensive. The usual method makes use of temperatures up to 180°; This is inconvenient due to thermal instability of the product and also because it requires anhydrous conditions. On the contrary, sonication in a cleaning bath leads to a fast reaction (20 min) with quantitative yield.

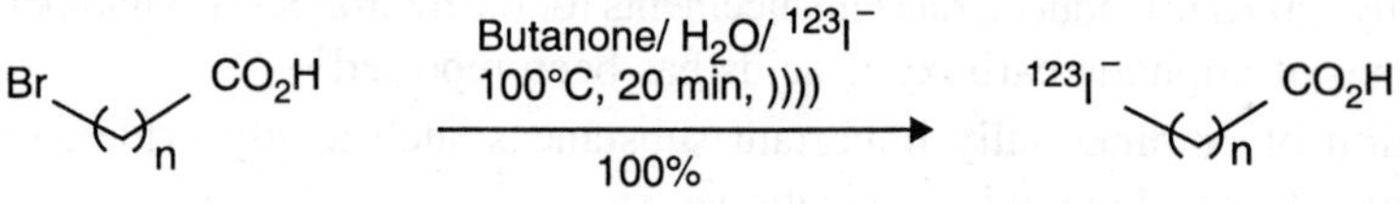

$$Br-(CH_2)_n-CO_2H \xrightarrow[100\%]{\text{Butanone/ } H_2O/\ ^{123}I^-\\ 100°C,\ 20\ min,\ ))))} ^{123}I-(CH_2)_n-CO_2H$$

Scheme-7

The Friedel-Crafts acylation of aromatics is facilitated[10] by ultrasound (Scheme-8).

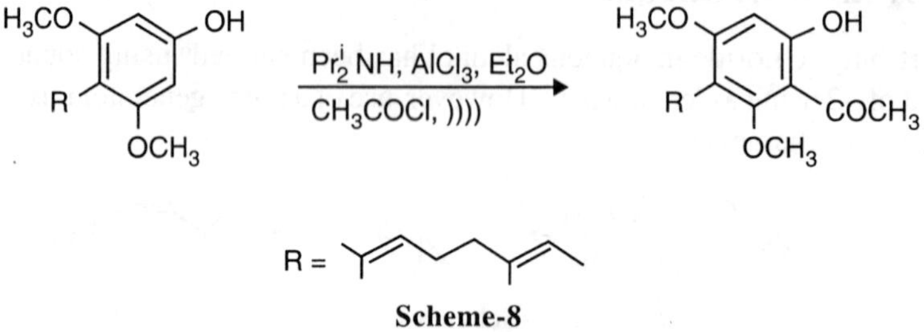

[Structure: 3,4-dimethoxy substituted arene with R group → acylated product with COCH$_3$]

Reagents: Pr_2^i NH, AlCl$_3$, Et$_2$O, CH$_3$COCl,))))

R = [branched alkenyl chain]

Scheme-8

An application[11] of the Friedel-Crafts reaction is used in the carbon-carbon bond formation. Excellent yield (75%) is obtained in sonochemical reaction (Scheme-9).

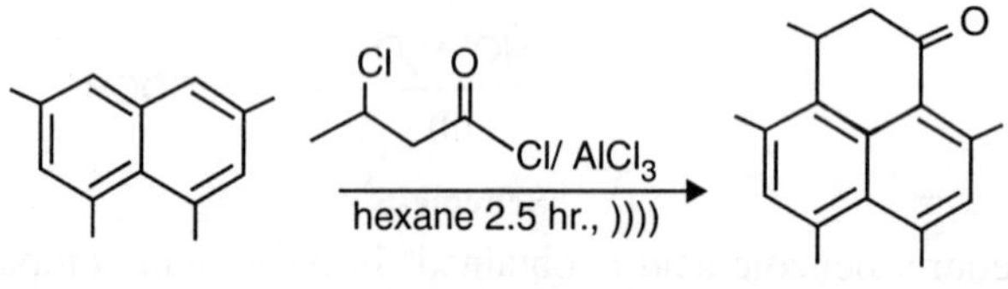

[Structure: naphthalene + ClCOCH$_2$CH$_2$Cl / AlCl$_3$, hexane 2.5 hr.,)))) → tricyclic ketone product]

Scheme-9

Halides can be converted into cyanides. Thus, the reaction of benzyl bromide in toluene with potassium cyanide is catalysed by alumina on sonication to give[12] the substitution product, viz., benzyl cyanide in 76% yield. In this case, the formation of the Friedel-Craft acylation product is not observed. Without the use of ultrasound 83% of the Friedel-Craft product is obtained. A probable explanation[2] involves ultrasonic dispersion of potassium cyanide on the alumina surface which decreases the Friedel-Crafts activity while promoting the nucleophilic displacement of CN$^-$ on its surface (Scheme-10).

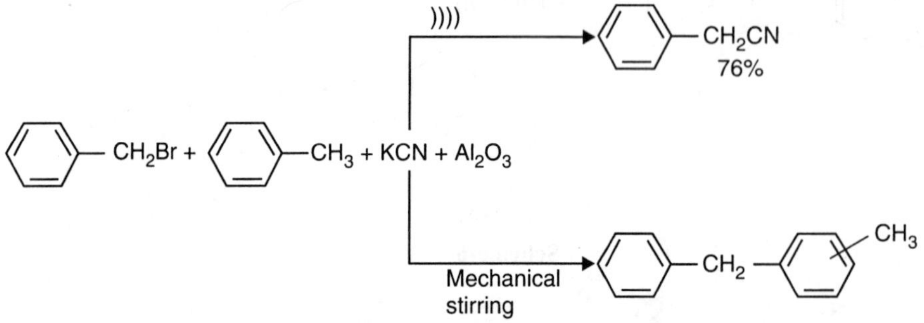

[Scheme showing: PhCH$_2$Br + PhCH$_3$ + KCN + Al$_2$O$_3$ → (with)))) sonication) PhCH$_2$CN 76%; → (with Mechanical stirring) PhCH$_2$–C$_6$H$_4$–CH$_3$]

Scheme-10

Aromatic acyl cyanides are obtained from the corresponding acyl chloride by treatment with potassium cyanide in acetonitrile at 50° and subjecting it to sonication even in the absence of phase transfer condition[13]. In the absence of ultrasound, the reaction is facilitated by the presence of traces of water but proceeds slowly (Scheme-11).

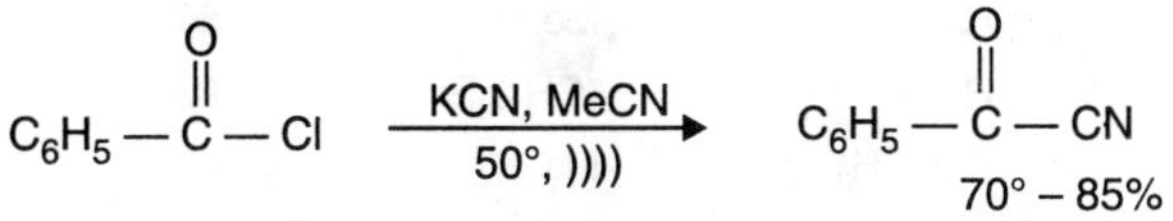

$$C_6H_5-\underset{O}{\overset{O}{\|}}C-Cl \xrightarrow[50°,))))]{KCN, MeCN} C_6H_5-\underset{O}{\overset{O}{\|}}C-CN$$

70° – 85%

R = H, 2 — CH_3, 3 — CH_3, 4 — CH_3, 4—OCH_3)

Scheme-11

The synthesis of azides from primary alkyl halides and aqueous sodium azide is also facilitated[14] by ultrasound (Scheme-12).

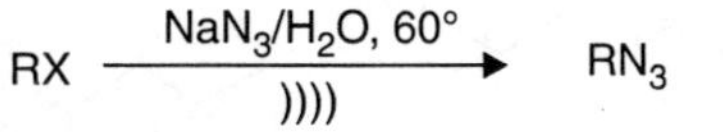

$$RX \xrightarrow[))))]{NaN_3/H_2O, 60°} RN_3$$

R = propargyl or allyl; yield 60 – 90%
R = Alkyl; yield 20%
R = allyl; yield 86%

Scheme-12

The alkyl bromides on sonication with KSCN in the presence of a quaternary ammonium salt gives the corresponding sulphocyanide in 62% yield compared to 43% yield under usual procedures (Scheme-13).

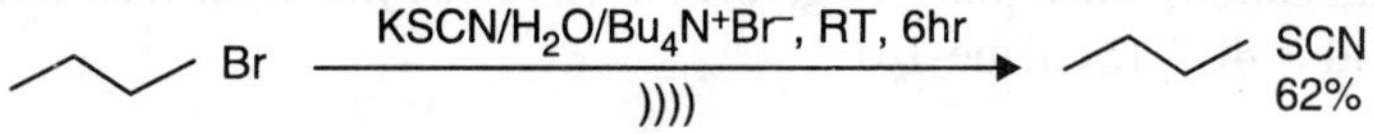

$$\sim\!\!\sim\!\!Br \xrightarrow[))))]{KSCN/H_2O/Bu_4N^+Br^-, RT, 6hr} \sim\!\!\sim\!\!SCN$$

62%

Scheme-13

Following the same principle, allytic and propargylic halides undergo an easy substitution with aqueous sodium azide[14] (Scheme-14).

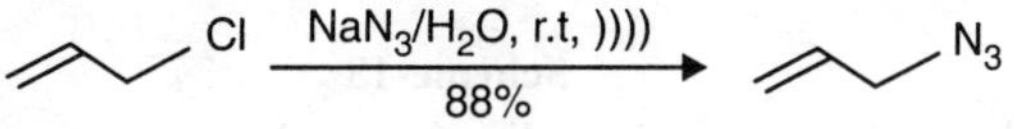

$$\sim\!\!\!\diagup\!\!\!Cl \xrightarrow[88\%]{NaN_3/H_2O, r.t,))))} \sim\!\!\!\diagup\!\!\!N_3$$

Scheme-14

Alkylation of thiocarbonic acid salts (1) has also been achieved using sonication[15] (Scheme-15).

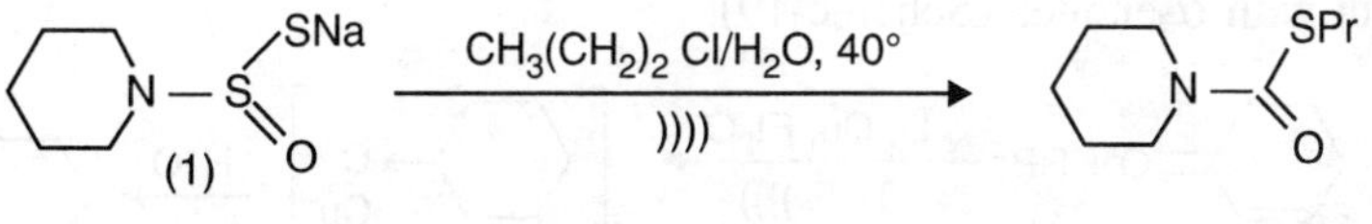

Piperidine-N-C(=O)-SNa (1) $\xrightarrow[))))]{CH_3(CH_2)_2\,Cl/H_2O, 40°}$ Piperidine-N-C(=O)-SPr

Scheme-15

The above illustration (Scheme-13) particularly useful due to thermal sensitivity of the starting material.

8.5 ADDITIONS

Sodium azide adds on to several alkanoate esters (1). A satisfactory selectivity is obtained under sonication[16] (Scheme-16).

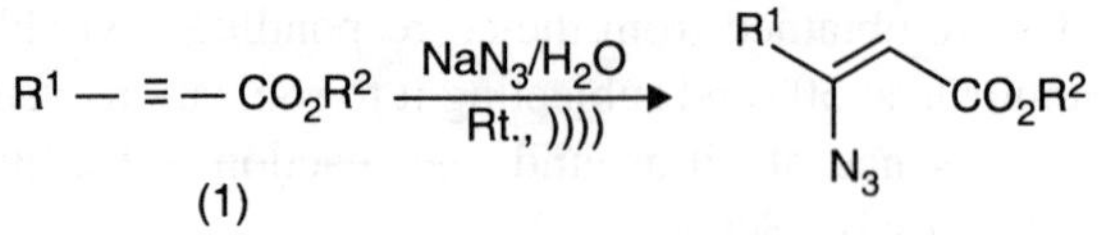

8.6 ■ Alternate Energy Processes in Chemical Synthesis

$$R^1-\equiv-CO_2R^2 \xrightarrow{NaN_3/H_2O}{Rt.,))))} \text{(alkene with } R^1, CO_2R^2, N_3\text{)}$$

(1)

R^1	R^2	Yields (%)
H	Me	86
H	$(CH_2)_2Br$	83
CH_2CH_3	Me	62

Scheme-16

A hydroformylation has been developed for the synthesis of aldehydes from olefins. Thus, 1-hexene and water containing a soluble rhodium catalyse on sonication under a stream of carbon monoxide and hydrogen. A mixture of heptanal (1) 2-methylhexanal (2) is obtained with a rate 2–3 times higher than without ultrasound (Scheme-17)[17, 18].

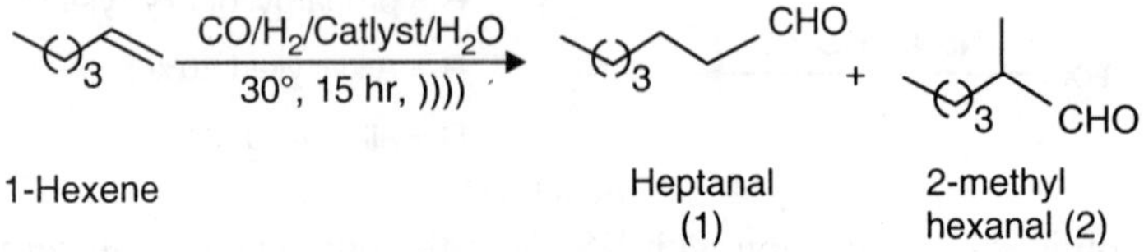

$$\text{1-Hexene} \xrightarrow{CO/H_2/Catlyst/H_2O}{30°, 15\ hr,))))} \text{Heptanal (1)} + \text{2-methyl hexanal (2)}$$

Scheme-17

Hydrocyanation of an aromatic aldehyde with sodium cyanide and benzene sulphonyl chloride in toluene/H_2O (sonication) takes place to give 94% yield of cyanohydrin sulfonate ester. Without sonication, the yield is 40% (Scheme-18)[19].

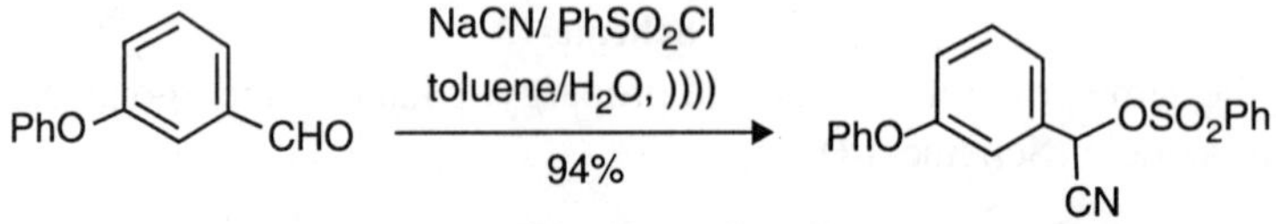

$$PhO\text{-}Ar\text{-}CHO \xrightarrow{NaCN/PhSO_2Cl}{toluene/H_2O,)))), 94\%} PhO\text{-}Ar\text{-}CH(OSO_2Ph)(CN)$$

Scheme-18

1,4-Addition to α, β-unsaturated carbonyl compounds is traditionally[20] carried out by organo copper reagents. However, significant improvements in yields, rates and ease of experimental technique is observed[21] when the organo copper compounds are generated by sonication (bath 50 KHz) of copper (I) compounds, organic halides, and lithium sand in diethyl ether-THF at 0°, are allowed to react with α-enones (Scheme-19).

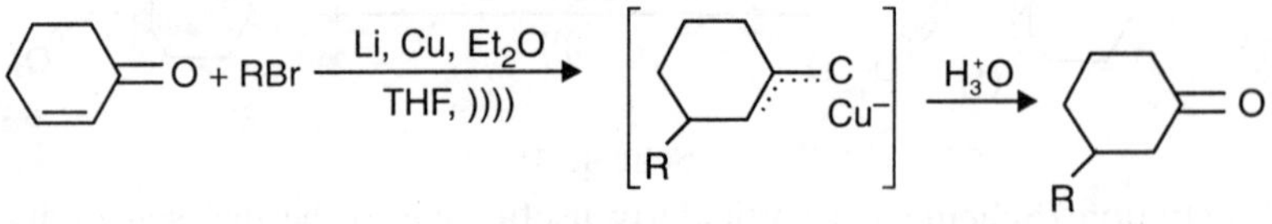

$$\text{cyclohexenone} + RBr \xrightarrow{Li, Cu, Et_2O}{THF,))))} [\text{intermediate with R, Cu}^-] \xrightarrow{H_3^+O} \text{3-R-cyclohexanone}$$

(R = n – Bu 89%)

Scheme-19

Similarly, arylzinc compounds are generated ultrasonically[22] (cleaning bath); these are also excellent reagents for 1,4-addition to α-enones and α-enals in presence of catalytic amount of Ni $(acac)_2$ (Scheme-20).

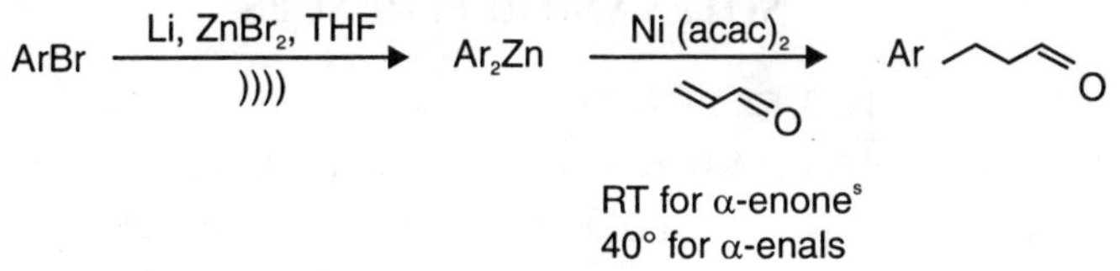

$$\text{ArBr} \xrightarrow[\text{))))}]{\text{Li, ZnBr}_2\text{, THF}} \text{Ar}_2\text{Zn} \xrightarrow[\sim\!\!\!\sim\!\!\text{O}]{\text{Ni (acac)}_2} \text{Ar}\!\sim\!\!\sim\!\!\text{O}$$

RT for α-enone^s
40° for α-enals

Scheme-20

The allylic bromide adds regioselectively[23] in the presence of zinc-silver couple to various aldehydes and ketones except benzophenone). Sonication increases the yield by about 50% and in one-third of time in comparison to a refluxed reaction. However, using magnesium instead of Zn-Ag reverses the regioselectivity (Scheme-21).

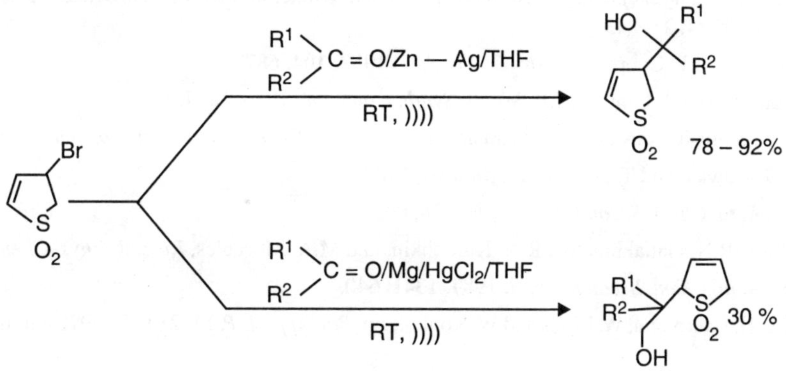

Scheme-21

Trifluoromethylation is possible[24] by the sonification of ketones with perfluoroalkyl halides (Scheme-22).

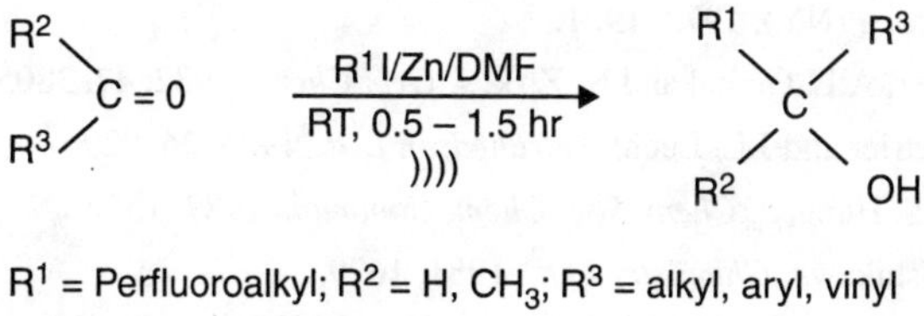

$$\begin{array}{c} R^2 \\ R^3 \end{array}\!\!C=O \xrightarrow[\text{RT, 0.5 - 1.5 hr}]{\text{R}^1\text{I/Zn/DMF}} \begin{array}{c} R^1 \quad R^3 \\ \diagdown\!\!\diagup \\ C \\ \diagup\!\!\diagdown \\ R^2 \quad OH \end{array}$$
))))

R^1 = Perfluoroalkyl; R^2 = H, CH_3; R^3 = alkyl, aryl, vinyl

Scheme-22

Methylenation of carbonyl group requires complex reagents. It can now be accomplished easily by the Simmons-Smith reagent (Zn/CH_2I_2) with sonication[25] (Scheme-23).

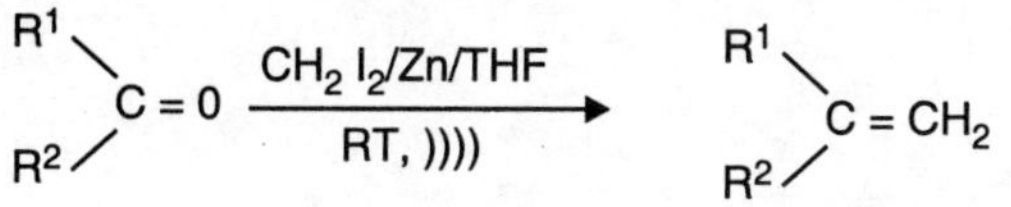

$$\begin{array}{c} R^1 \\ R^2 \end{array}\!\!C=O \xrightarrow[\text{RT,))))}]{\text{CH}_2\text{I}_2\text{/Zn/THF}} \begin{array}{c} R^1 \\ R^2 \end{array}\!\!C=CH_2$$

Scheme-23

The reaction is generally applicable to aldehydes, and not to ketones, with benzaldehyde (R^1 = Ph; R^2 = H), the yield is 70% in 20 min.

8.8 ■ Alternate Energy Processes in Chemical Synthesis

NOTES AND REFERENCES

1. N.O.V. Sonntag, *Chem. Rev.*, 1953, **52**, 2937.
2. J.M. Khurana, P.K. Sahoo and G.C. Maikap, *Synth. Commun.*, 1990, 2267.
3. R.S. Davidson, A. Safdar, J.D. Spencer and D.W. Lewis, *Ultrasonics* 1987, **25**, 35.
4. S. Moon, L. Duclin and J.V.Croney, *Tetrahedron Lett.*, 1979, 391.
5. D.S. Kristol, H. Klotz and R.C. Parker, *Tetrahedron Lett.*, 1981, **22**, 907.
6. J.P. Lorimer, T.M. Mason and B.P. Mistry, *Ultrasonics*, 1987, **25**, 23.
7. J. Elguero, P. Goya, J. Lissavetzky and A.M. Valdemillos, *C.R. Acad Sci.* Paris, 1984, **298**, 877.
8. M.T. Crimmins, W.G. Hollis and D.M. Bankatis-Davis., *Tetrahedron Lett.*, 1987, 28, 3651.
9. J. Mertens, W. Vanryckeghem, A. Bossuyt, P. Vanden Winkel and R. Vanden Driessche, *J. Label Comp and Radiophurm*, 1984, 21, 843.
10. D.M. Trost and B.P. Coppola, *J. Am. Chem. Soc.*, 1982, **104**, 6879.
11. P. Boadjout, W.H. Ohrbom and J.B. Woell, *Synth. Commun.*, 1986, **16**, 401.
12. T. Ando, S. Sumi, T. Kawate, J. Ichihara and T. Haafusa, *J. Chem. Soc. Chem. Commun.*, 1984, 439.
13. T. Ando, J. Kawate and T. Hanafusa, *Synthesis*, 1983, 637.
14. H. Priebe, *Acta. Chem. Second. Ser B.* 1984, **38**, 895.
15. R.B. Valitov, R.N. Galiakhmetov, R.N. Kurochkin, and M.A. Morgules, *Russ. J. Phys. Chem.*, 1985, **59**, 1782.
16. H. Priebe, *Acta. Chem. Second. Ser. B*, 1987, **1341**, 640.
17. B. Cornils, H. Bahrmann, W. Lipps and W. Konkol, *Eu. Pat. Appl. E.P.* 173219 (ClCO7C45/50); C.A., 1986, 105, 81168.
18. B. Cornils, H. Bahrmann, W. Lipps and W. Konkol, *German Patent* 3511428 (Cl. CO6C47/02) (1987); C.A., 1988, **108**, 21493.
19. S.Kawada and T. Kurokawa, *Japanese patent* JP62, 106, 07387, 106, 073 (ClCO7Cl143/68) (1987), C.A., 1988, *108*, 21493.
20. G.H. Posner, *Org. React.* (NY), 1972, **19**, 1.
21. J.L. Lunche, C. Petrier, A.L. Gemal and N. Zirk, *J. Org. Chem.*, 1982, **47**, 3805.
22. J.C.S. Barboza, C. Petrier and J.L. Luchl, *Tetrahedron Lett.*, 1985, **26**, 829.
23. H. Tso, T. Chou and S. Hung, *J. Chem. Soc. Chem. Commun.*, 1987, 1552.
24. T. Kitazume and N. Ishikawa, *Chemistry Lett.*, 1981, 1679.
25. J. Yamashita, Y. Inou, T. Kando and H. Hashimoto, *Bull. Chem. Soc.* Japan, 1984, **57**, 2335.

9

HETEROGENEOUS SOLID-LIQUID REACTIONS

9.0 INTRODUCTION

As already stated heterogeneous solid-liquid reactions are of two types. In Type I, solid serves as one of the reagents and is consumed during the reaction. In Type II reactions, the cavitational erosion is the major effect which is observed when ultrosonic waves propagate towards, or in the vicinity of a solid.

Following are given some of the interesting heterogeneous solid-liquid reactions.

9.1 ALKYLATIONS

9.1.1 N-Alkylation

Secondary amines can be N-alkylated under sonication in the presence of a phase transfer reagent, polyethylene glycol monomethyl ether was necessary. Thus, benzopyrrole on methylation with methyl iodide in toluene in the presence of potassium hydroxide and PTC gives the corresponding N-methyl ether in 65% yield. However, under normal conditions (20°, 5 hr) the yield is 60% (Scheme-1).

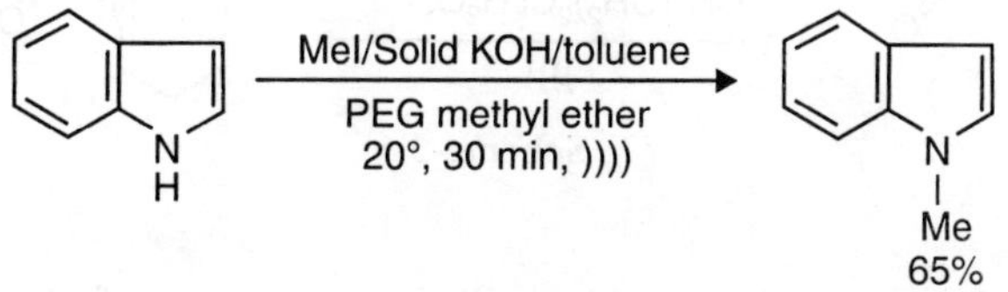

MeI/Solid KOH/toluene
PEG methyl ether
20°, 30 min,))))

65%

Scheme-1

Similarly, diphenylamine on sonication with benzyl bromide gives[1] 98% yield of the corresponding N-benzylated product in 1 hr, compared to 70% in 48 hr under normal conditions (reflux) (Scheme-2).

9.2 ■ Alternate Energy Processes in Chemical Synthesis

$$Ph_2NH \xrightarrow[\substack{PEG\ methyl\ ether \\ 20°,\ 1\ hr,\ ))))}]{PhCH_2Br/solid\ KOH/toluene} Ph_2NCH_2Ph\ \ 98\%$$

Scheme-2

In the absence of PTC, the reaction does not take place under sonication. This indicates that the increase in reactivity is not simply a matter of interfacial contact area.

N-Alkylation of crown compounds can be effected[2] conveniently under sonication (Scheme-3).

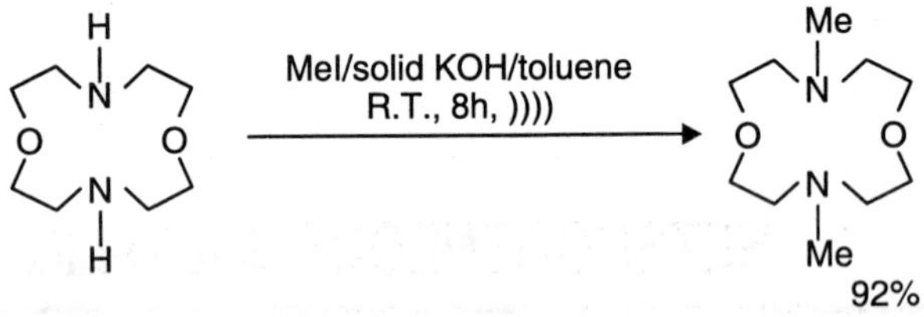

Scheme-3

9.1.2 C-Alkylation

C-Alkylation of isoquinoline derivatives has been effected[3] by sonication using PTC as catalyst (Scheme-4).

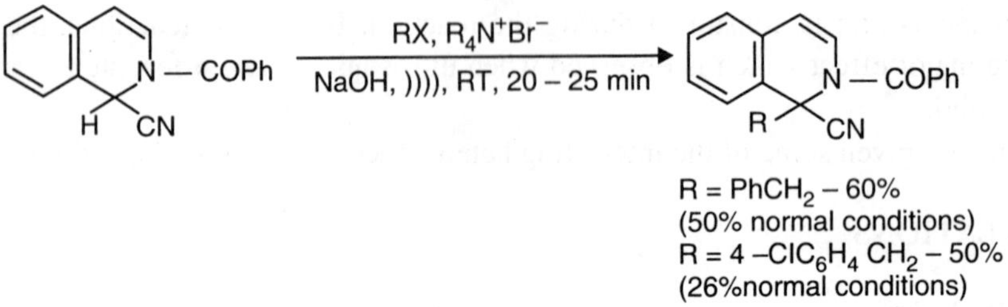

R = PhCH$_2$ – 60%
(50% normal conditions)
R = 4 –ClC$_6$H$_4$ CH$_2$ – 50%
(26% normal conditions)

Scheme-4

Alkylation of thiocarbamic acid salts has also been carried out under sonication.[4] This procedure is found to be advantageous because of the thermal sensitivity of the starting material. Without sonication the reaction is very much slow (Scheme-5).

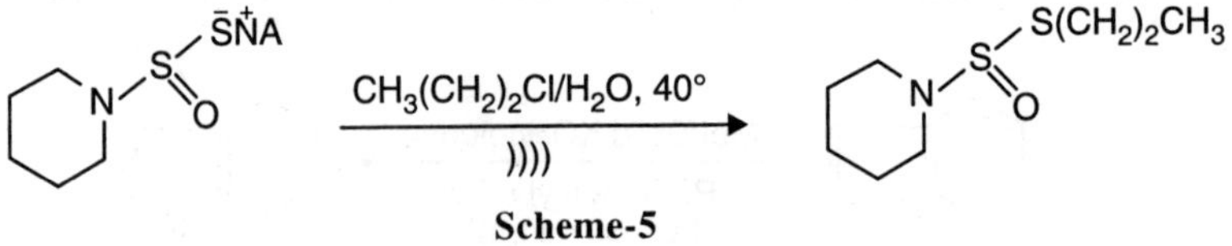

Scheme-5

9.1.3 O-Alkylation

Primary alcohols on sonication[5] with benzyl bromide in the presence of silver oxide gives 72% yield of the O-benzylated product. Under normal conditions without sonication the reaction yield is low with poor reproducibility (Scheme-6).

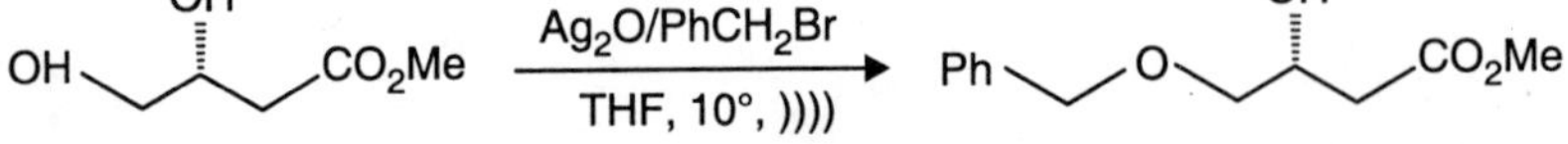

Scheme-6

Similarly, ether formation of a phenol group in 5-hydroxychromone is quantitative when sonicated with a probe generator[6] (Scheme-7).

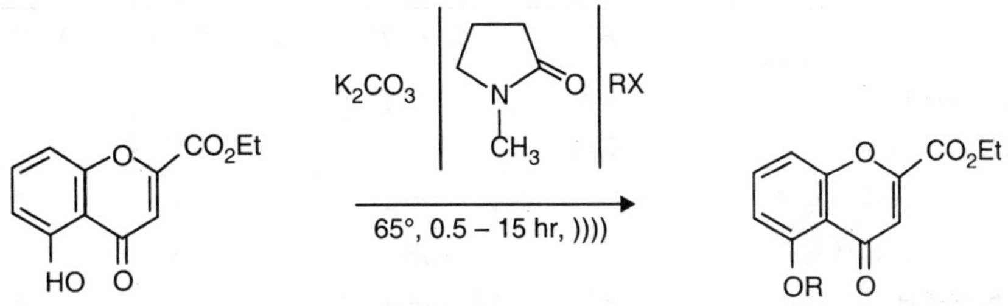

Scheme-7

Low intensity ultrasound (cleaning bath) has beneficial effects in the synthesis of ethers. Thus, ethyl phenyl ether is obtained[7] in 80% yield on sonication compared to 44% in stirred control reaction (Scheme-8).

$$C_2H_5OH + C_6H_5X \xrightarrow[\text{))))}]{\text{KOH, PEG}} C_2H_5OC_6H_5 \quad 80\%$$

Scheme-8

In the above O-alkylation, potassium carbonate is broken into small particles in N-methylpyrrolidone solvent which liberates high energy by cavitation. The same effect explains the high yield[8] of diethyl carbonate by the sonochemical modification of the reaction of ethyl bromide with potassium carbonate.

9.1.4 S-Alkylation

Arylthiols, aralkylthiols and alkanethiols can be alkylated[9] in presence of K_2CO_3/DMF at ambient temperature (Scheme-9).

$$RSH + R'X \xrightarrow[\text{))))}]{K_2CO_3/DMF} RSR'$$

Scheme-9

The reaction is accelerated[9] under sonication.

9.2 OXIDATIONS

In the case of oxidations, the usual inorganic reagents require water as a cosolvent, which can make the reaction less efficient and also make the work-up procedure difficult. Solid potassium permanganate can be used to oxidize alcohols in benzene or hexane suspension under sonication[10] (Scheme-10).

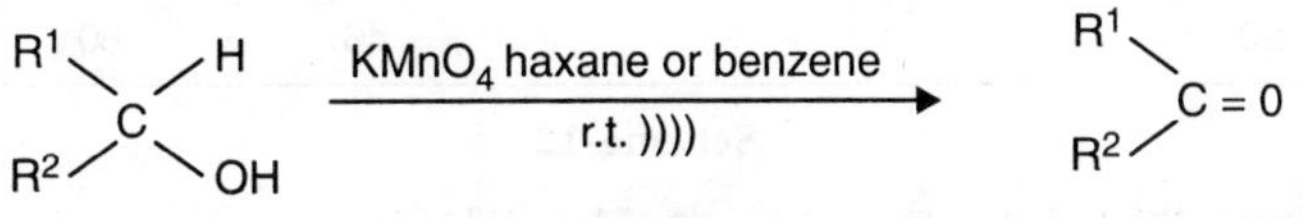

Scheme-10

It has been shown that in case of oxidation of octan-2-ol ($R^1 = C_6H_{13}$; $R^2 = H$) the corresponding ketone is obtained in 92.8% yield (5 hr) compared to 2% yield by mechanical stirring. Similarly cyclohexanol gives 53% yield of cyclohexanone by oxidation under sonication (5 hr) compared to

9.4 ■ Alternate Energy Processes in Chemical Synthesis

4% yield under usual conditions. In a similar way, oxidation of cinnamyl aldehyde under sonification gives 82% of cinnamic acid (3 hr) compared to 4% under usual conditions.

R′	R²	Conditions	Time (hr)	Yield (%)
$CH_3(CH_2)_5$	CH_3	⌒	5	2
$CH_3(CH_2)_5$	CH_3	))))	5	92
$-(CH_2)_5-$		⌒	5	4
$-(CH_2)_5-$		))))	5	53
PhCH=CH	H	⌒	3	4
PhCH=CH	H	))))	3	82

It has been shown[11] that crystalline manganese dioxide, which is of low reactivity, is activated upon sonification. Thus, sonication of cinnamyl alcohol, geraniol or 1-phenylethanol with MnO_2 gives the corresponding aldehydes. In these oxidations pentane, hexane or octane is used as solvent — best results are obtained by using a less volatile medium (octane) (Scheme-11).

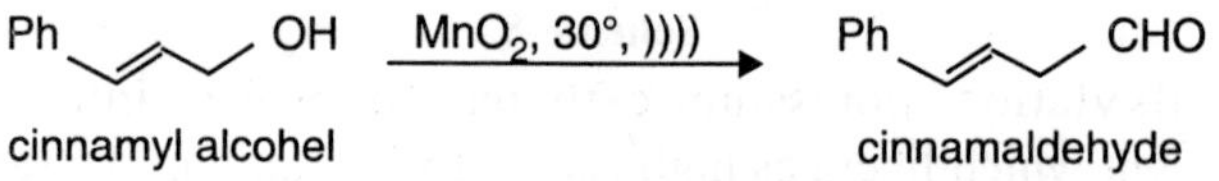

$$\text{Ph}\diagdown\diagup\text{OH} \xrightarrow{MnO_2,\ 30°,\ ))))} \text{Ph}\diagdown\diagup\text{CHO}$$

cinnamyl alcohel   cinnamaldehyde

Time (min)	Yield (%) of cinnamaldehyde		
	pentane	hexane	octane
10	23	28	46
15	32	35	63
20	44	51	73

Scheme-11

The role of solvent in MnO_2 oxidation is also clear from the following example (Scheme-12).

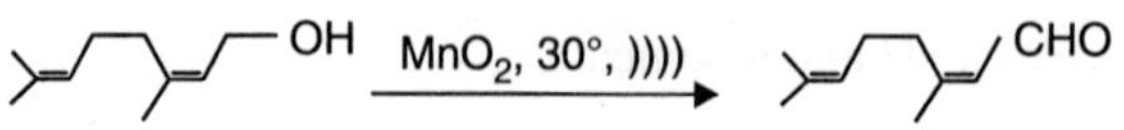

$$\diagup\diagdown\diagup\diagdown\text{OH} \xrightarrow{MnO_2,\ 30°,\ ))))} \diagup\diagdown\diagup\diagdown\text{CHO}$$

Time (min)	Yield (%) of the aldehyde in solvent	
	pentane	hexane
60	19	55
75	34	67
90	48	80

Scheme-12

A number of benzylic halides have been oxidised[12] with aqueous sodium hypochlorite[13] at ambient temperature on ultrasonic irradiation. The oxidations are believed to be proceeding *via* benzylic hypohalides (Scheme-13).

$$\text{ArRCHX} + \text{NaOCl} \xrightarrow[))))]{\text{CH}_3\text{CN, RT}} \text{ArRCHOCl}$$

$$\text{ArRCHOCl} + \text{NaOCl} \xrightarrow[))))]{\text{CH}_3\text{CN, RT}} \text{ArRCO}$$

Scheme-13

The above method has also been used for the oxidation of mesylates, tosylates, amines, oximes and alcohols.

9.3 REDUCTIONS

Sonication plays an important role[14] in the reactivity of platinum, palladium and rhodium black. Thus, formic acid and palladium on carbon are an efficient couple for the hydrogenation[15] of a wide range of alkenes at room temperature in the presence of low intensity ultrasonic fields (cleaning bath, 50 KHz) (Scheme-14).

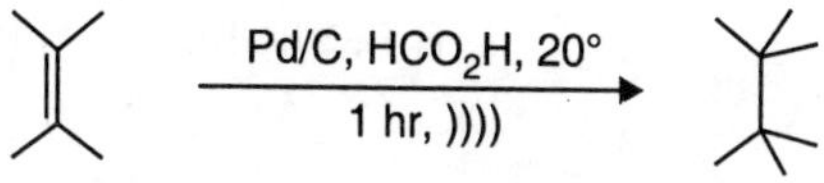

Scheme-14

Similarly, hydrazine-palladium on carbon couple is also useful for the hydrogenation of alkenes in ethanol at room temperature using an ultrasonic bath[16]. A commercially useful example of a sonochemically enhanced catalytic reaction is the ultrasonic hydrogenation of soyabean oil[17]. This has considerable advantage over the currently used batch methods which require much longer reaction times.

Sonication has also been used[17] for the hydrogenolysis of benzyl ethers with H_2/Pd – C in methanol in the presence of acetic acid. Introductin of deuterium with Raney nickel in sugar derivatives[18] and p-bromoacetophenone[19] has been achieved (Scheme-15).

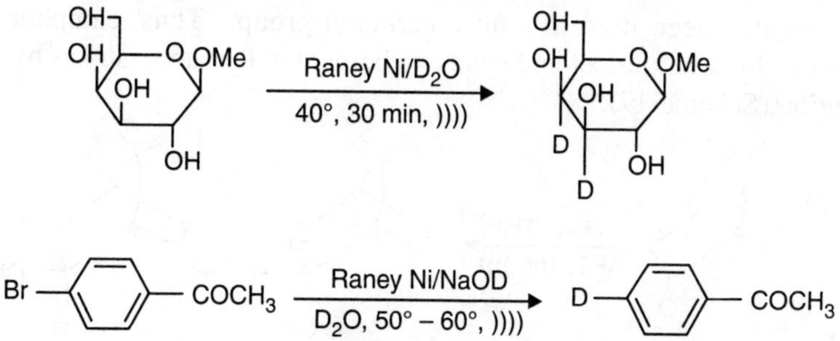

Scheme-15

In the last case, a mixture of mono, di- and tri-deuterated compounds is obtained without sonication.

Sonication also increases the activation of nickel powder[20] which, in turn, has been used for reduction of alkenes.

Catalytically, active nickel can be obtained by sonochemical reduction of its salts such as chloride with zinc powder. It has been shown that under these conditions the excess of metallic zinc is activated and reduces the water present in the medium producing hydrogen gas[21]. In this way,

9.6 ◼ Alternate Energy Processes in Chemical Synthesis

not only the catalyst but also the reagent is produced *in situ* with maximal efficiency and safety. This process has been used for the reduction of carbon-carbon double bonds in α, β-unsaturated carbonyl compounds which is reduced much faster than the carbonyl group. The variation in the conditions, especially the pH permits[22] the selective reduction of C = C in preference to C = 0 (Scheme-16).

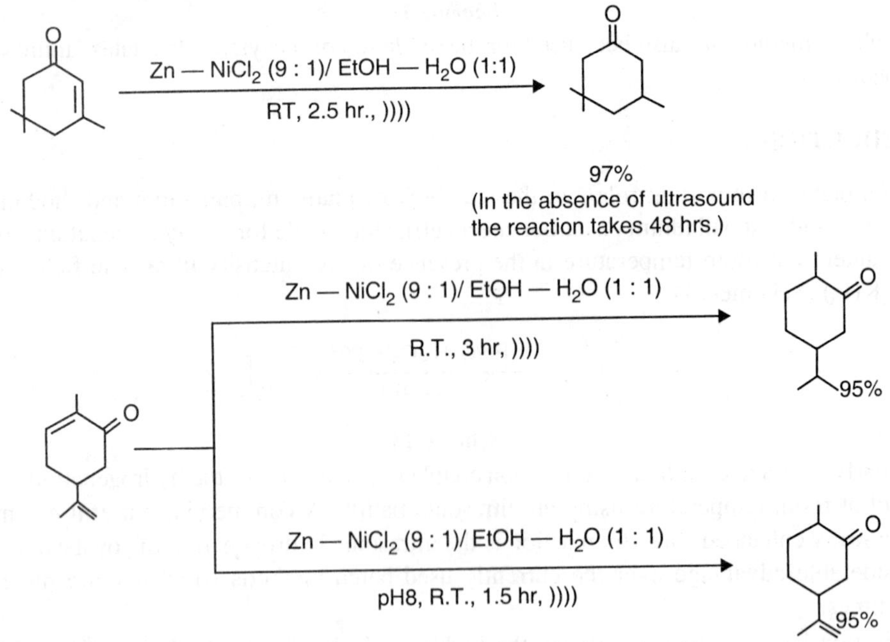

97%
(In the absence of ultrasound the reaction takes 48 hrs.)

95%

95%

Scheme-16

9.3.1 Reduction of Carbonyl Group

Sonication[23] has also been used to reduce carbonyl groups. Thus, camphor on sonication in tetrahydrofuran yields a mixture of endo and exo borneol in the same ratio as by using the metal in ammonia solution (Scheme-17).

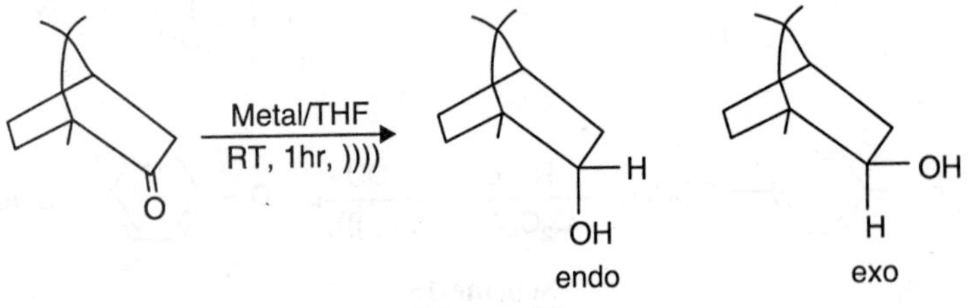

Scheme-17

The endo product is obtained in 73, 68 or 42% by the use of Li, Na or K as the metal in the above reduction.

Carbonyl groups with less negative redox potential such as quinones or α-diketones can be reduced on sonication with zinc in presence of trimethyl-chlorosilane[24]. In this case, tetrahydrofuran is found to be superior than diethyl ether.

The Clemmensen reduction can also be carried out by sonication in improved yields.[25]

9.3.2 Reduction of Aromatic Rings

Reduction of aromatic rings has been investigated only in a very limited case. For example, N-protected indole on sonication in the presence of lithium and trimethylchlorosilane undergoes a clean reduction to give the dihydrocompound (Scheme-18).

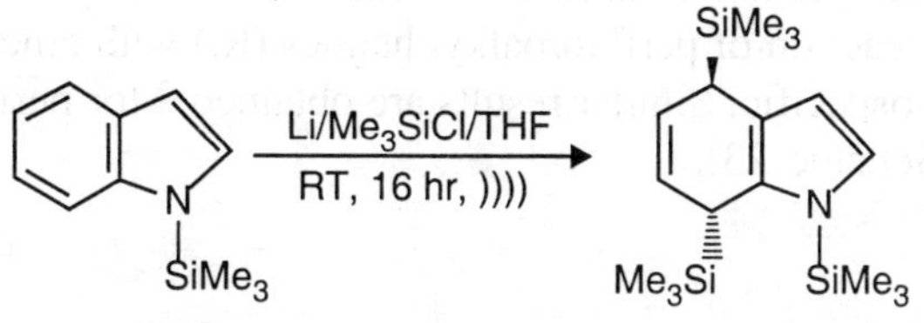

Scheme-18

Extension of this process to other aromatic rings, followed by benzoquinone oxidation of the dihydro intermediate, provides a satisfactory method for ring silation[26, 27].

Aromatic halides, which are known to be reluctant towards reduction by hydrides, undergo a substitution by hydride with excellent yields in dimethoxyethane (DME) solution[28] (Scheme-19).

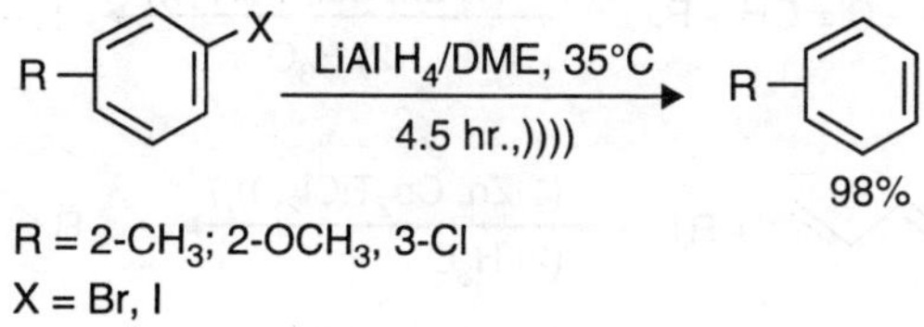

$R = 2-CH_3; 2-OCH_3, 3-Cl$
$X = Br, I$

Scheme-19

9.4 HYDROBORATION

Hydroboration is considerably enhanced[29] by low intensity ultrasound, especially in heterogeneous systems. Thus, tricyclohexylborane is obtained in 1 hr by the reaction of cyclohexene with $BH_3 \cdot SMe_2$; this traditionally requires 24 hrs at 25° (Scheme-20).

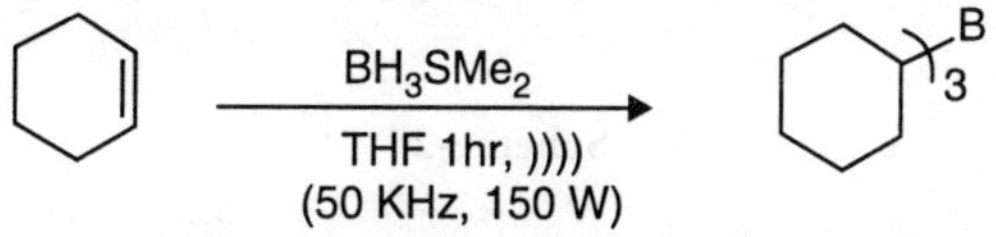

Scheme-20

Thus, sonication is shown to provide an interesting alternative and to provide excellent yields in short periods. The products obtained are stable under ultrasonic irradiation, even those with boron-halogen bond. Even chiral reagent can be obtained conveniently from α-pinene and 9-Borobicyclo [3.3.1] nonane (9-BBN) (Scheme-21).

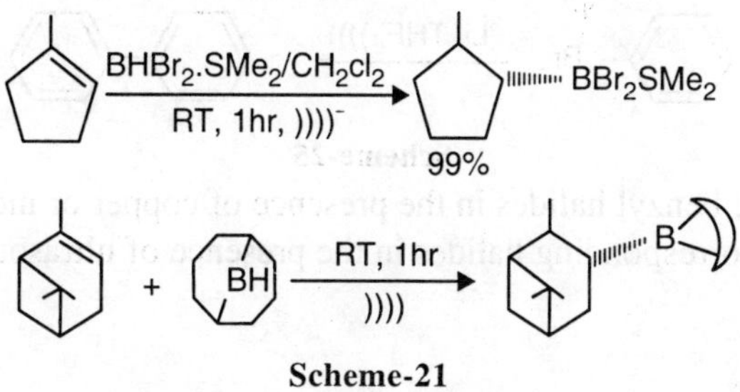

Scheme-21

In the above two cases the normal preparations take 5 hrs.

9.5 HYDROSILATION AND HYDROALKYLATION

Hydrosilation of alkenes can be achieved[30, 31] in presence of platinum-carbon catalyst in an ultrasonic cleaning bath at much lower temperature 30° than is used for without sonication (Scheme-22). Similarly, hydroperfluoroalkylation of alkynes is achieved sonically with perfluoroalkylcuprates (generated *in situ* by the reaction of perfluoroalkyl halides (R_f) with zinc and copper (I) iodide) in THF. The reaction is regiospecific. Similar results are obtained[32] for perfluoroalkylation of dienes catalysed by Cp_2TiCl_2 (Scheme-23).

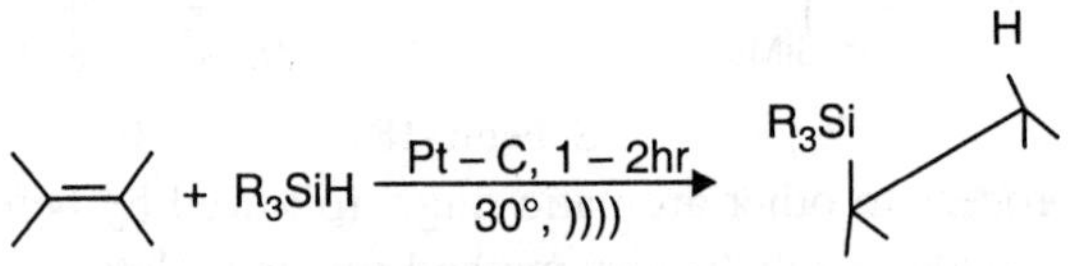

Scheme-22

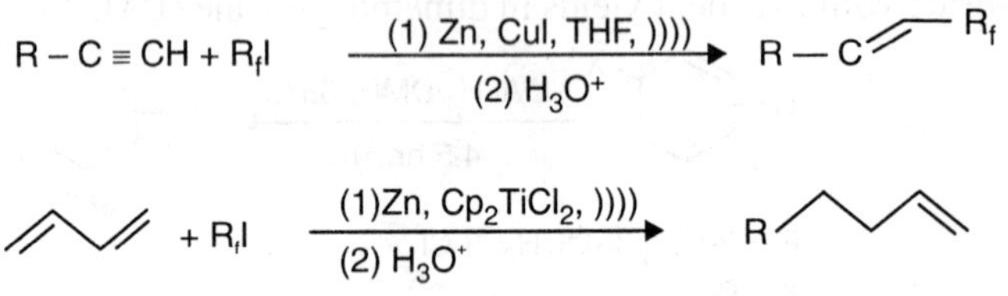

Scheme-23

An interesting application of hydroalkylation is the preparation of dienes[33] by the reaction of sonochemically generated highly functionalized allyl zinc intermediates with alkynes (Scheme-24).

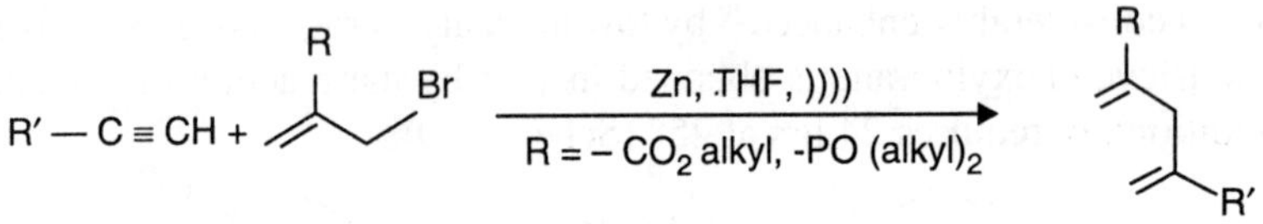

Scheme-24

9.6 COUPLING REACTIONS

Sonication is effective in promoting[31, 34] the homocoupling of organometallic intermediates obtained by the reaction of alkyl, aryl or vinyl halides with lithium wire in THF (ultrasonic bath, 117 W, 50 KHz). No reaction takes place in the absence of ultrasound (Scheme-25).

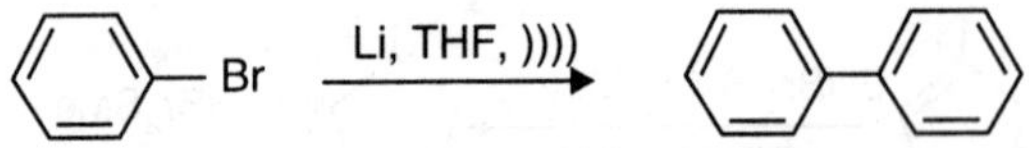

Scheme-25

Similarly, coupling of benzyl halides in the presence of copper or nickel powder generated by lithium reduction of the corresponding halides in the presence of ultrasound gives high yields[35] of dibenzyl (Scheme-26).

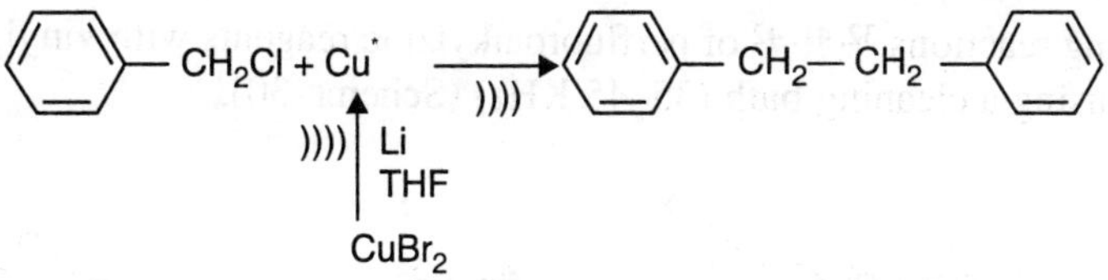
CH2Cl + Cu*
))))
)))) Li
THF
CuBr2
CH2— CH2

Scheme-26

In the classical Ullmann coupling, copper has to be reacted at high temperature. On sonication, this requirement is modified as activation occurs by reducing the size of the metal powder[36]. Breakage of the particles brings in contact with the reactive solutions a fresh surface, the reactivity of which is not hindered by the usual oxide layer. Thus, sonication[37] of o-iodonitrobenzene with copper powder in the presence of DMF at 60° gives 70% yield of the product in 95 min (Scheme-27).

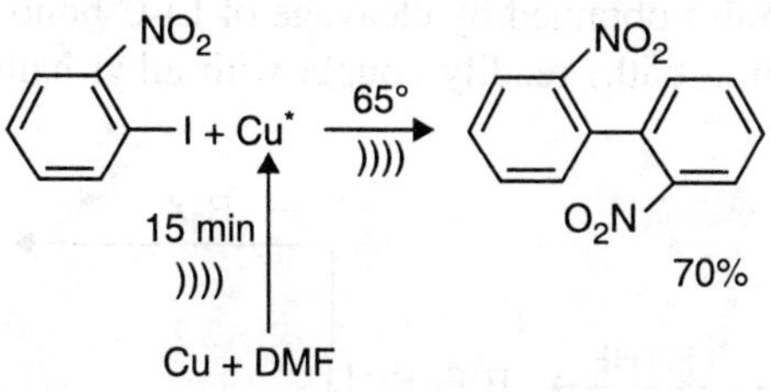
NO2
I + Cu*
65°
))))
15 min
))))
Cu + DMF
NO2
O2N
70%

Scheme-27

Sonication of aryl sulphonates in presence of *in situ* generated nickel(0) complex is an interesting Ullmann type coupling[38]. This method works best for triflates (R = CF_3). However, for tosylates (R = 4 – $CH_3C_6H_4$) the yields are low (Scheme-28).

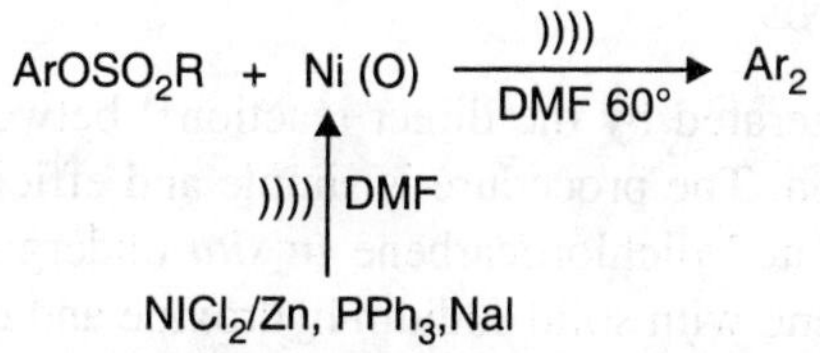
ArOSO2R + Ni (O)
))))
DMF 60°
Ar2
)))) DMF
NICl2/Zn, PPh3,NaI

Scheme-28

This process is proved to be of interest in the case of the formation of silicon-silicon bonds. Various chlorosilanes can be coupled by sonication in the presence of lithium[39]. the highly hindered tetramesityldisilane is prepared in good yield from dichlorodimesitylsilane using lithium under sonication[40] (Scheme-29).

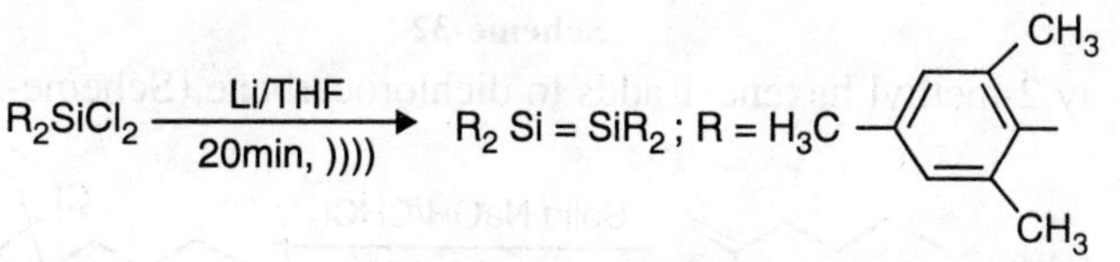
R2SiCl2
Li/THF
20min,))))
R2 Si = SiR2 ; R = H3C
CH3
CH3

Scheme-29

9.10 ■ Alternate Energy Processes in Chemical Synthesis

Cross coupling reactions.[32, 41, 42] of perfluoroalkylzinc reagents with vinyl, allyl or aryl halides can be achieved using a cleaning bath (35–45 KHz) (Scheme-30).

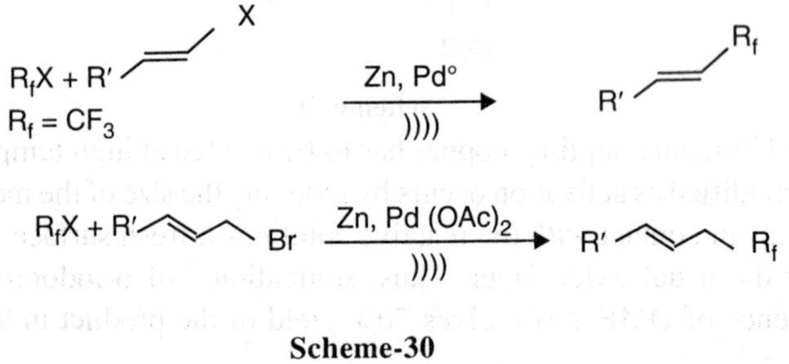

$$R_f X + R'\text{-CH=CH-X} \xrightarrow{\text{Zn, Pd}^\circ,))))} R'\text{-CH=CH-}R_f$$
$$R_f = CF_3$$

$$R_f X + R'\text{-CH}_2\text{-CH=CH-Br} \xrightarrow{\text{Zn, Pd(OAc)}_2,))))} R'\text{-CH}_2\text{-CH=CH-}R_f$$

Scheme-30

Lithium dialkylphosphides obtained by cleavage of Li-C bond in THF in the presence of low intensity ultrasound (cleaning bath) readily couple with alkyl halides to give[43, 44] phosphonanes (Scheme-31).

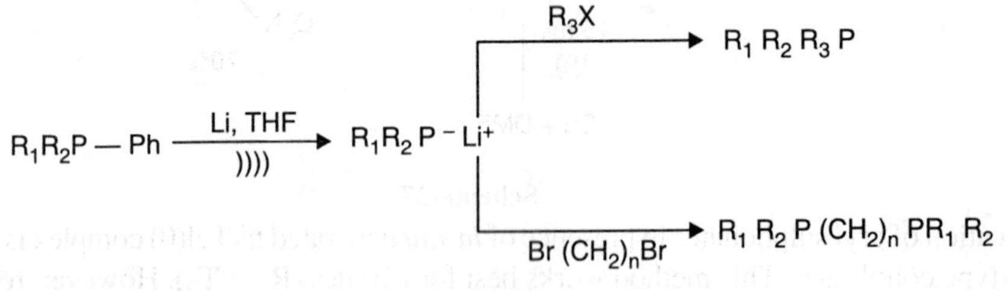

$$R_1R_2P\text{—Ph} \xrightarrow{\text{Li, THF},))))} R_1R_2P^-Li^+ \begin{cases} \xrightarrow{R_3X} R_1 R_2 R_3 P \\ \xrightarrow{Br(CH_2)_nBr} R_1 R_2 P(CH_2)_n PR_1 R_2 \end{cases}$$

Scheme-31

9.7 DICHLOROCARBENE

Dichlorocarbene can be generated by the direct reaction[45] between powdered sodium hydroxide and chloroform by sonication. The procedure is simple and efficient, and avoids the use of phase transfer catalyst. The generated dichlorocarbene *in situ* undergoes addition reaction to alkenes. Thus, the sonication of styrene with solid sodium hydroxide and chloroform by stirring gives 96% yield of the adduct in 1 hr. However, the yield is reduced to 38% (20 hr stirring only) and 31% (16 hr stirring only) (Scheme-32).

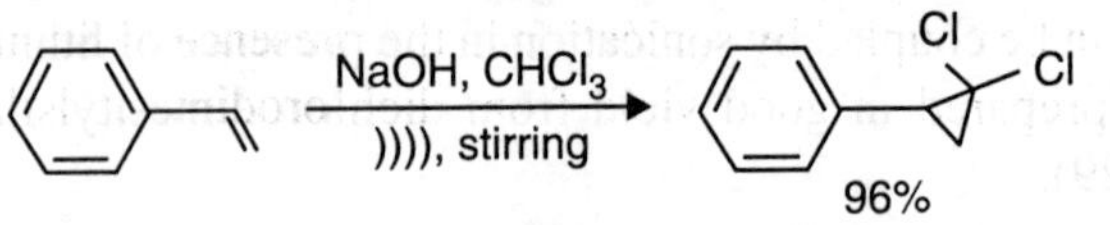

$$Ph\text{-CH=CH}_2 \xrightarrow{\text{NaOH, CHCl}_3,)))), \text{stirring}} Ph\text{-cyclopropane-}CCl_2 \quad 96\%$$

Scheme-32

In a similar way 2-methyl hexene-1 adds to dichlorocarbene (Scheme-33).

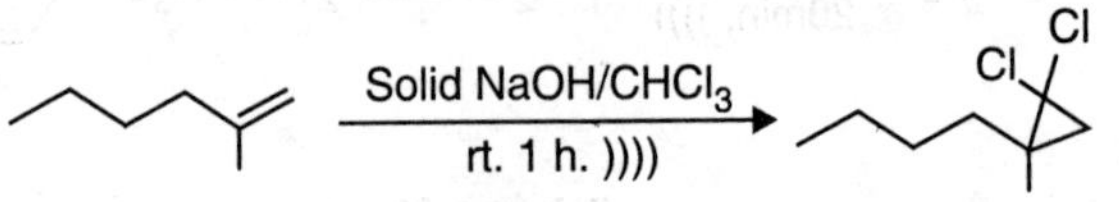

$$\text{CH}_2=\text{C(CH}_3)\text{CH}_2\text{CH}_2\text{CH}_2\text{CH}_3 \xrightarrow{\text{Solid NaOH/CHCl}_3, \text{rt. 1 h. })))} \text{dichlorocyclopropane adduct}$$

Scheme-33

9.8 SOME ULTRASONICALLY INDUCED ORGANIC REACTIONS

9.8.1 Bouveault Reaction

Organolithium reagents are obtained by sonication of aryl halides with lithium with low ultrasonic intensity. Their application in the Bouveault reaction has resulted in higher yields of product aldehydes than the traditional methods[46] (Scheme-34).

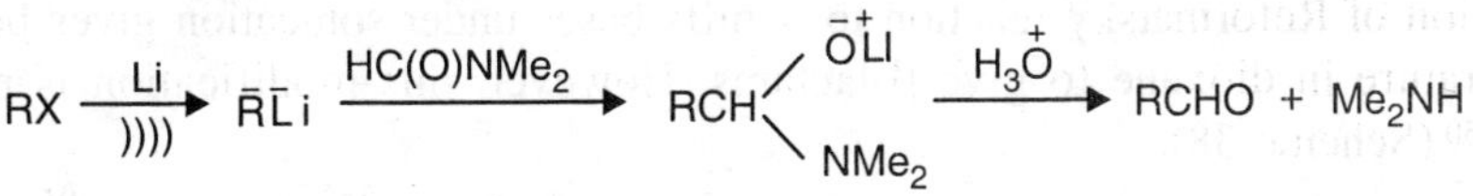

$$RX \xrightarrow{Li,))))} R\bar{L}i \xrightarrow{HC(O)NMe_2} RCH\begin{matrix}\bar{O}\overset{+}{L}i \\ NMe_2\end{matrix} \xrightarrow{H_3\overset{+}{O}} RCHO + Me_2NH$$

Scheme-34

In non-ultrasonic Bouveault reaction, which suffers from numerous side reactions, the method is improved when DMF is replaced by more elaborate and expensive formamide, $Me_2NCH_2CH_2N(Me)CHO$.

A simplification of this method is by sonication of any aryl halide and amide with excess lithium for 15 min followed by dropwise addition of 1-bromobutane, sonification for 30 min more gives the o-substituted aldehydes (Scheme-35).

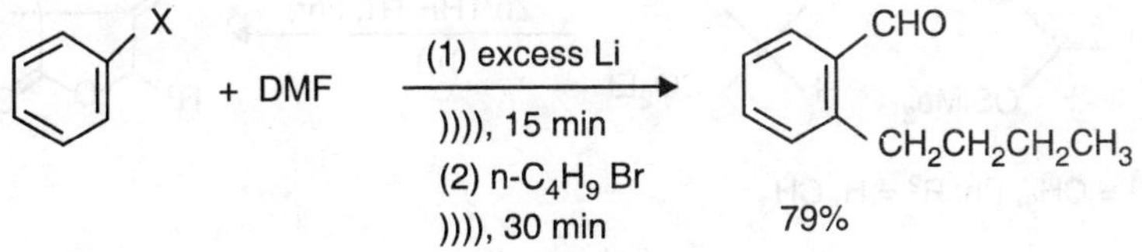

Ar-X + DMF → (1) excess Li,)))), 15 min (2) n-C_4H_9Br,)))), 30 min → o-(CHO)C_6H_4-$CH_2CH_2CH_2CH_3$ 79%

Scheme-35

Use of iodomethane in place of n-butyl bromide in the above reaction gives o-tolualdehyde[47].

9.8.2 Cannizzarro Reaction

The Cannizzarro Reaction under heterogeneous conditions catalyzed by barium hydroxide is considerably accelerated by low intensity ultrasound (cleaning bath). The yields are 100% after 10 min, whereas no reaction is observed during this period without the use of ultrasound[48] (Scheme-36).

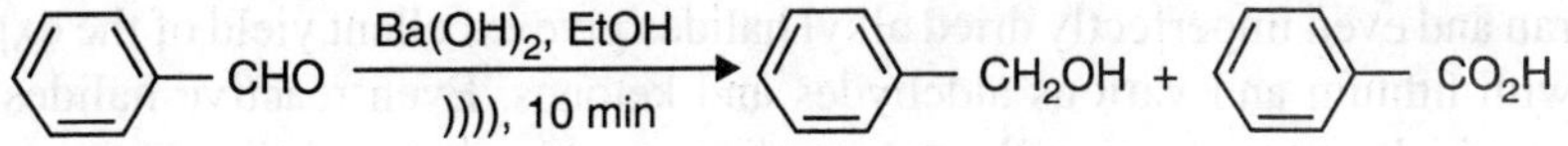

PhCHO $\xrightarrow{Ba(OH)_2, EtOH,)))), 10\text{ min}}$ PhCH$_2$OH + PhCO$_2$H

Scheme-36

9.8.3 Strecker Synthesis

(See Section 7.7.3).

9.8.4 The Reformatsky Reaction

The Reformatsky reaction under sonication gives excellent yields with respect to more traditional methods, e.g., those employing activated zinc or trimethylborate as a consolvent[49]. In the sonication procedure it is necessary to activate the metal with iodine and to run the reaction in dioxane. Under optimal conditions quantitative yields of β-hydroxyester is obtained (Scheme-37).

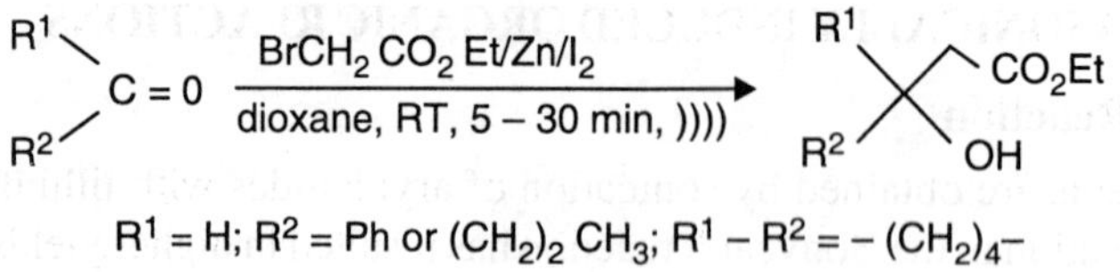

9.12 ■ *Alternate Energy Processes in Chemical Synthesis*

$$R^1R^2C=O \xrightarrow[\text{dioxane, RT, 5-30 min, })))]{BrCH_2CO_2Et/Zn/I_2} R^1R^2C(CO_2Et)(OH)$$

$R^1 = H; R^2 = Ph$ or $(CH_2)_2CH_3$; $R^1 - R^2 = -(CH_2)_4-$

Scheme-37

Application of Reformatsky reaction to Schiffs bases under sonication gives better results at room temperature in dioxane to give β-lactams. However, this modification is not of general applicability[50] (Scheme-38).

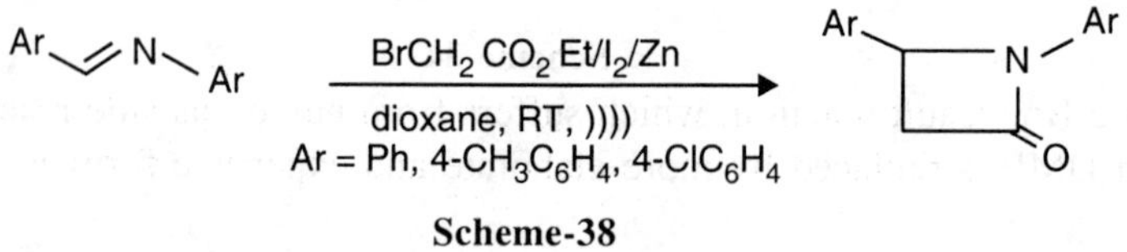

$$Ar-CH=N-Ar \xrightarrow[\text{dioxane, RT, })))]{BrCH_2CO_2Et/I_2/Zn} \text{β-lactam}$$

$Ar = Ph, 4-CH_3C_6H_4, 4-ClC_6H_4$

Scheme-38

The Reformatsky reaction with nitriles leads to the formation of imines which readily hydrolyse to the ketones. This has made available keto-γ-butyrolactones in fair yields[51] (Scheme-39).

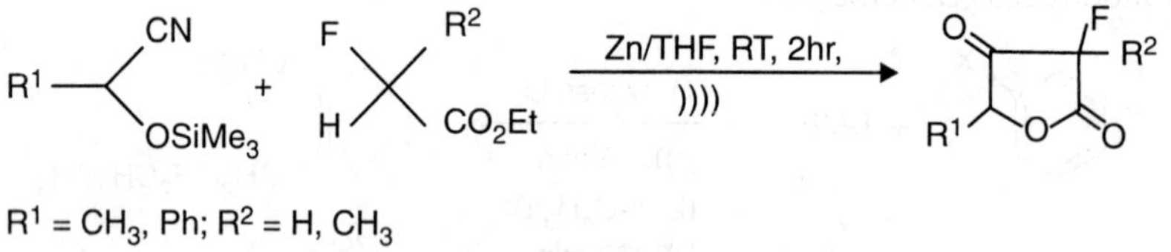

$$R^1-C(CN)(OSiMe_3) + F-CH(R^2)-CO_2Et \xrightarrow[))))]{Zn/THF, RT, 2hr,} \text{keto-γ-butyrolactone}$$

$R^1 = CH_3, Ph; R^2 = H, CH_3$

Scheme-39

9.8.5 The Barbier Reaction of Carbonyl Compounds

The reaction of a ketone with an organometallic reagent (obtained *in situ*) to give the corresponding alcohol is known as the Barbier reaction. It is to be noted that the same products are also obtained by using a ketone and a Grignard reagent, which is obtained separately. Thus, the Barbier reaction offers time-saving and also the possibility of utilising organometallics which are unstable. The main disadvantage is that in the normal Barbier reaction only reactive alkyl halides can be used. This difficulty is overcome if the reaction is carried out by sonication[52]. The reaction is carried out in tetrahydrofuran and even imperfectly dried alkyl halides give excellent yield of the expected alcohol by reaction with lithium and various aldehydes and ketones. Even reactive halides, e.g., allyl or benzyl halides which generally give Wurtz coupling can also be used. Some examples are given below (Scheme-40).

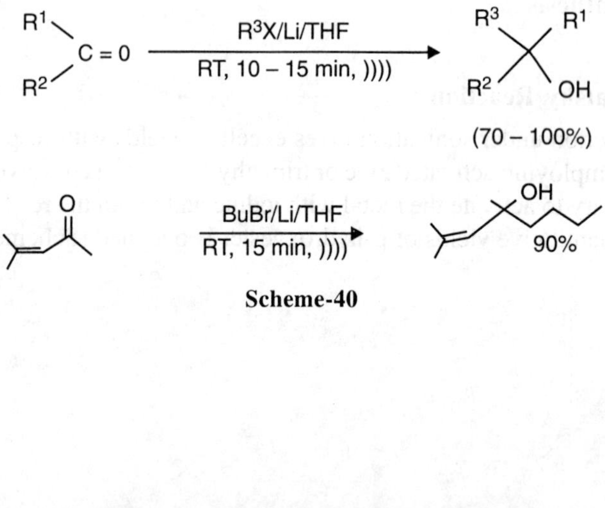

$$R^1R^2C=O \xrightarrow[\text{RT, 10-15 min, })))]{R^3X/Li/THF} R^3R^2R^1C-OH$$

(70 – 100%)

$$\text{MeC(O)CH=CMe_2} \xrightarrow[\text{RT, 15 min, })))]{BuBr/Li/THF} \text{alcohol, 90\%}$$

Scheme-40

A synthetic intermediate in the synthesis of sesquiterpene is obtained by a sonochemical Barbier reaction[53] (Scheme-41).

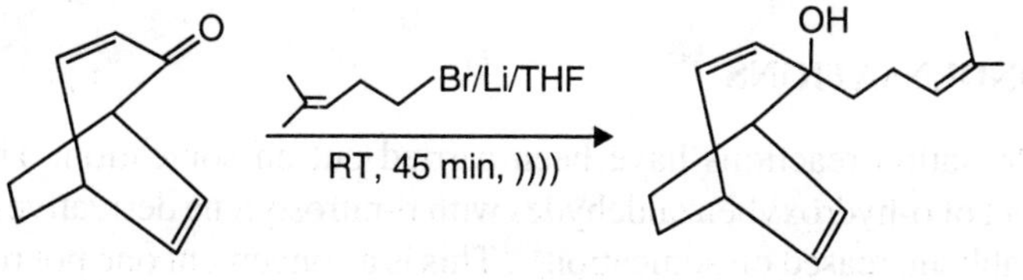

Scheme-41

A total synthesis of pentalenic acid has been achieved by the reaction of dimethyl-cyclopentenone with 5-bromo-1-hexene and lithium with sonication[54] (Scheme-42).

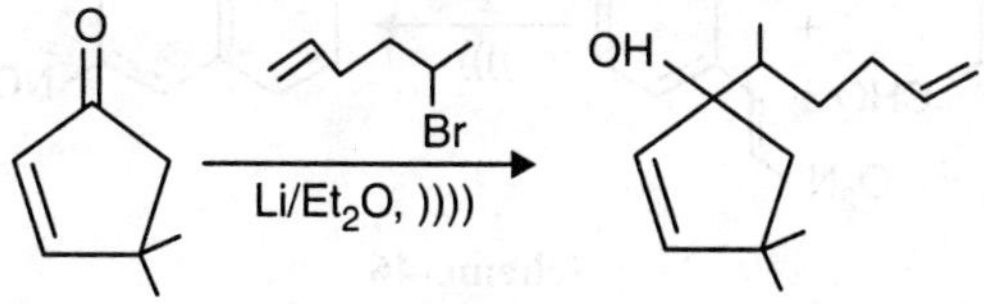

Scheme-42

Benzylhalides can also be used in Barbier reaction. It is to be noted that there is no Wurtz type coupling[55, 56] (Scheme-43).

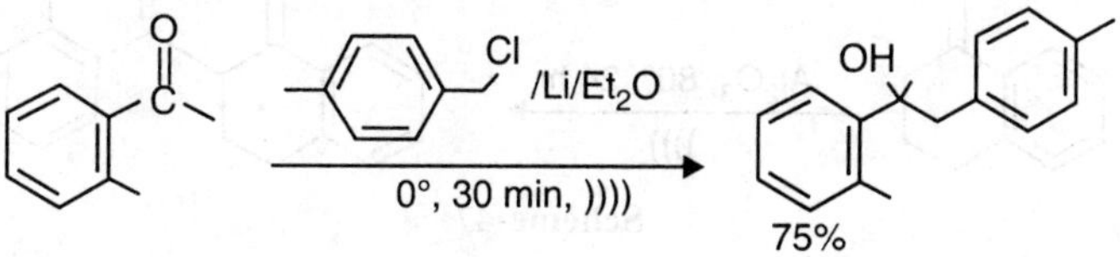

Scheme-43

Chloromethyllithium is an unstable reagent. It is obtained *in situ* by the reaction of chloroidomethane by sonication[57] with lithium in tetrahaydrofuran. This reagent reacts with ketones to give epoxides in high yields. The aldehydes, however, give exposides or chlorohydrins (Scheme-44).

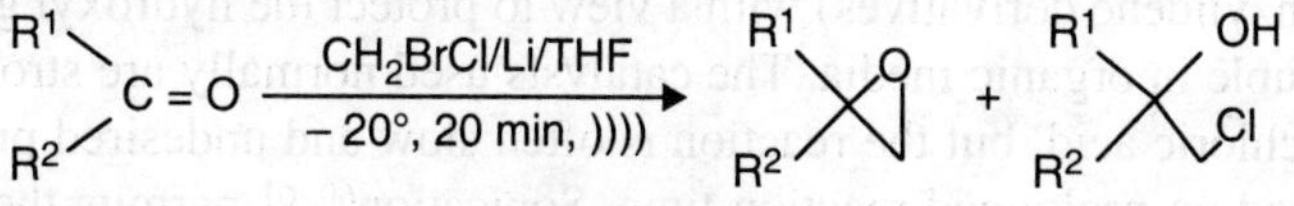

Scheme-44

With benzaldehyde the chlorohydrin is obtained in 90% yield. However, with phenyl amyl ketone (R^1 = Ph; R^2 = n – $(CH_2)_5$) the epoxide is obtained in 91% yield.

9.8.6 Dieckmann Cyclisation

On sonication potassium is easily transformed to a silvery blue suspension in toluene. The ultrasonically dispersed potassium effects Dieckmann and Thorpe-Ziegler cyclisations under favourable conditions[58] (Scheme-45).

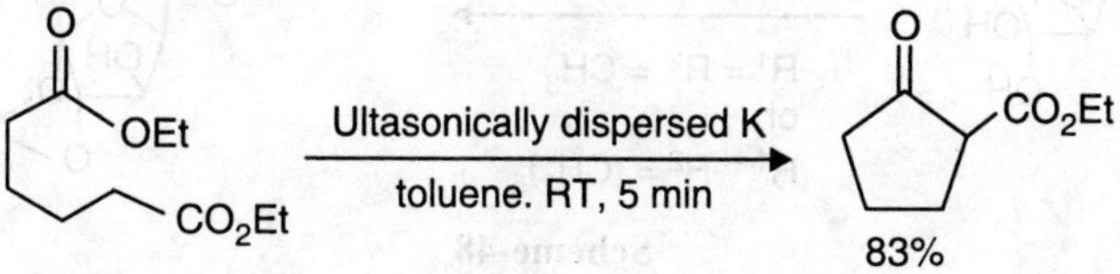

Scheme-45

9.14 ■ *Alternate Energy Processes in Chemical Synthesis*

Similarly, ultrasonically dispersed sodium is also obtained in toluene and can be used in various reactions.

9.9 CYCLOCONDENSATIONS

A number of condensation reactions have been carried out on sonication. Thus, the yield of the condensation product of o-hydroxybenzaldehydes with β-nitrostyrene derivatives using basic alumina catalyst is considerably increased on sonication[59]. This is a convenient one pot route for the synthesis of 3-nitro-2H-chromenes (Scheme-46).

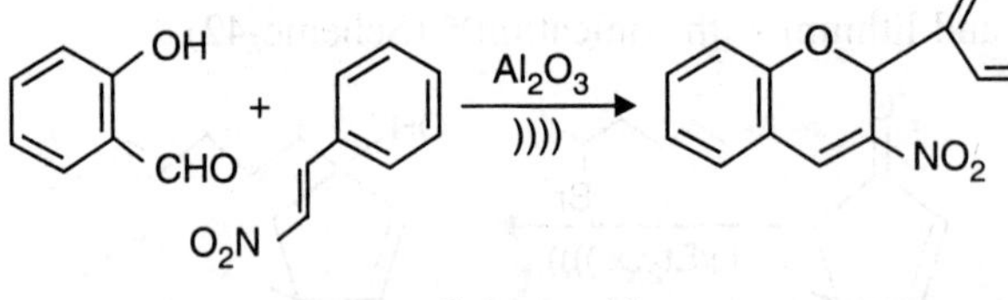

Scheme-46

the rate of aldol dimerisation of ketones catalyzed by basic alumina is also increased by sonication[60] (Scheme-47).

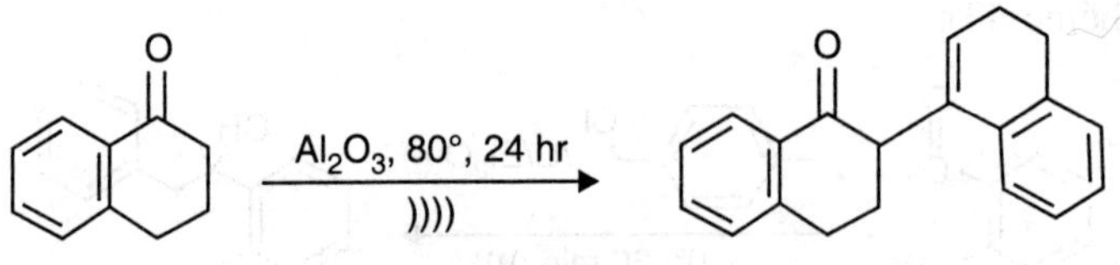

Scheme-47

9.10 CARBOHYDRATES-FORMATION OF ACETALS AND BENZYLIDENE DERIVATIVES OF ALKYLGLYCOPYRANOSIDES

Many syntheses in carbohydrate series start with the formation of acetals (isopropylidene, cyclohexlidene, benzylidene derivatives) with a view to protect the hydroxyl groups and rendering the compounds soluble in organic media. The catalysts used normally are strong protic acids, e.g., sulphuric acid, perchloric acid, but the reaction is often slow and undesired products including tar formation are formed on prolonged reaction time. Sonication[60, 61] permits the reaction time to be considerably shortened and better yields. Thus, the two vicinal hydroxyl groups in glucose can be protected by treatment with acetone or cyclopentanone on sonication in the presence of concentrated sulphuric acid at room temperature in 60–75% yields compared to 5–10% yield without sonication (Scheme-48).

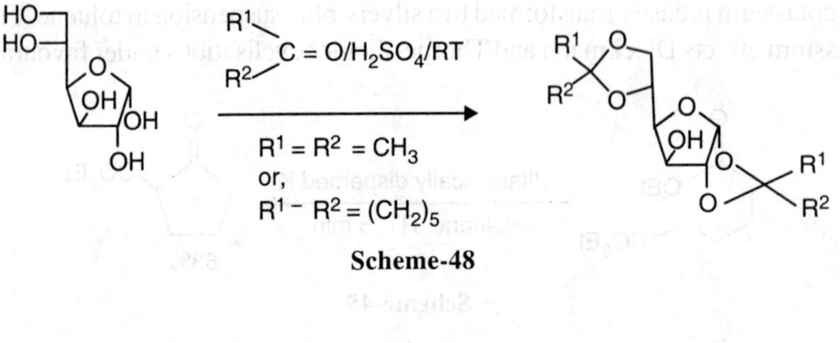

$R^1 = R^2 = CH_3$
or,
$R^1 - R^2 = (CH_2)_5$

Scheme-48

Similarly, the benzylidene derivatives of alkyl glycopyranoside are achieved[62] on sonication. Initially, benzaldehyde is made to react with fused zinc chloride under sonication to give the complex catalyst. Subsequent reaction with methylglycopyranoside and sonication, the benzylidene derivative is obtained (Scheme-49).

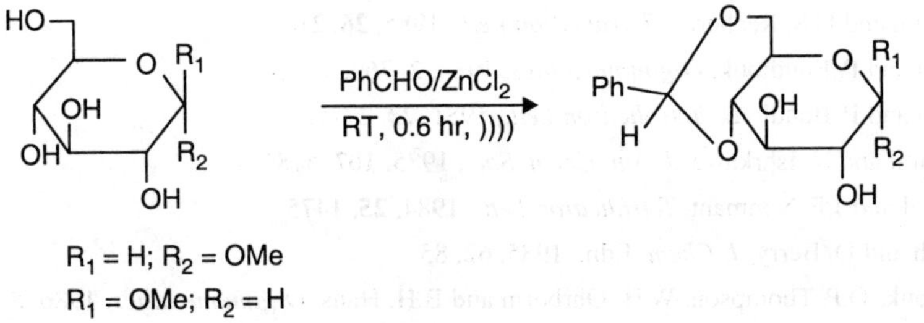

$R_1 = H; R_2 = OMe$
$R_1 = OMe; R_2 = H$

Scheme-49

NOTES AND REFERENCES

1. R.S. Davidson, A.M. Palel, A. Safdar and D. Thornthwaite, *Tetrahedron Lett.*, 1983, **24**, 5907.
2. J. Jurczak and R. Ostaszewki, *Tetrahedron Lett.*, 1988, **29**, 959.
3. J. Ezgerra and J. Alvorez-Builla, *J. Chem. Soc. Chem. Commun.*, 1984, 54.
4. P.B. Valitov, R.N. Galiakhmetov, A.K. Kurochkin and M.A. Margulis, *Russian J. Phys. Chem.*, 1985, **59**, 1782.
5. R.D. Walkup and R.T. Cunningham, *Tetrahedron Lett.*, 1987, **28**, 4019.
6. T.J. Mason, J.P. Lorimer, A.T. Turner and A.R. Harris, *J. Chem. Res.(S)*, 1988, 80.
7. R.S. Davidson, A. Safdar, J.D. Spencer and D.W. Lewis, *Ultrasonics*, 1987, **25**, 35.
8. T. Fujinami, S. Sato and S. Sakai, *Chemistry Lett.*, 1981, 749.
9. J.M. Khurana and P.K. Sahoo, *Syn. Commun.*, 1992, 1691.
10. J. Yamawaki, S. Sumi, T. Ando and T. Hanafusa, *Chemistry Lett.*, 1983, 379.
11. T. Kimura, M. Fujita and T. Ando, *Chemistry Lett.*, 1988, 137.
12. J.M. Khurana, P.K. Sahoo, S.S. Titus and G.C. Maikap, *Synth. Commun.*, 1990, 1357.
13. Sekera and Marvel, *J. Am. Chem. Soc.*, 1933, **55**, 345; *A Text Book of Practical Organic Chemistry*, A.I. Vogel, IV Edn., Longman, N.Y., 1907, p. 605.
14. A.W. Maltsev, *Russ. J. Phys. Chem.*, 1976, **50**, 995.
15. J. Jurczak and R. Ostaszewki, *Tetrahedron Lett.*, 1988, **57**, 2335.
16. D.H. Shin and B.H. Han, *Bull. Korean. Chem. Soc.*, 1985, **6**, 247.
17. K.J. Moulton, S.Koritala and E.N. Frankel, *J. Am. Oil Chem. Soc.*, 1983, **60**, 1257.
18. E.A. Cioffi and J.H. Prestegard, *Tetrahedron Lett.*, 1986, **27**, 415.
19. M. Jashiro, M. Nokayama, Y. Nakamura, A. Aoki, K. Takigawa, I. Meada, I. Tago and M. Yoshida, *Jap. Pat.* JP 61.53, 28886, 53, 228 (1986); CA, 1986, 105, 133517.
20. K.S. Suslick and D.J. Casadonte, *J. Am. Chem. Soc.*, 1987, **109**, 3459.
21. C. Petrier and J.L. Luchi, *Tetrahedron Lett.*, 1989, **28**, 2347.
22. C. Petrier and J.L. Luchi, *Tetrahedron Lett.*, 1989, **28**, 2351.
23. J.W. Huffman, W. Liao and R.H. Wallace, *Tetrahedron Lett.*, 1987, **28**, 3315.
24. P. Boudjouk and J. So, *Synthetic Commun.*, 1986, **16**, 775.

9.16 ■ Alternate Energy Processes in Chemical Synthesis

25. W.P. Reeves, J.A. Murry, D.W. Willoughby and W.J. Friedrick, *Synthetic Commun.*, 1988, **18**, 1961.
26. A.G.M. Barrett, I.A. Dauzonne, I.A. Oneil and A.J. Renaud, *J. Org. Chem.*, 1984, **49**, 4409.
27. A.G.M. Barrett and I.A. O'Neil, *J. Org. Chem.*, 1988, **53**, 1815.
28. B.H. Han and P. Boudjouk, *Tetrahedron Lett.*, 1982, **23**, 1643.
29. H.C. Broun and U.S. Racheria, *Tetrahedron Lett.*, 1985, **26**, 2187.
30. B.H. Han and P. Boudjouk, *Organometallics*, 1983, **2**, 769.
31. B.H. Han and P. Boudjouk, *Tetrahedron Lett.*, 1981, **22**, 2757.
32. T. Kitazumi and N. Ishrkawa, *J. Am. Chem. Soc.*, 1985, **107**, 5186.
33. P. Knochel and J.F. Nonmant, *Tetrahedron Lett.*, 1984, **25**, 1475.
34. T.D. Lash and D. Berry, *J. Chem. Edn.*, 1985, **62**, 85.
35. P. Boudjouk, D.P. Thompson, W.H. Ohrborm and B.H. Hans, *Organometallics*, 1986, **5**, 1257.
36. J. Lindly, T.J.Mason and J.P. Lorimer, *Ultrasonics*, 1987, **25**, 45.
37. L. Lindley, J.P. Lorimer and T.J. Mason, *Ultrasonics*, 1986, **24**, 292.
38. T. Yamashita, Y. Inoue, T. Kondo and H. Hashimoto, *Chem. Lett.*, 1986, 407.
39. P. Boudjouk and B. Hans, *Tetrahedron Lett.*, 1981, **22**, 3813.
40. P. Boudjouk, B. Hans and K.R. Anderson, *J. Amer. Chem. Soc.*, 1982, **104**, 4992.
41. T. Kitazume and N. Ishikawa, *Chemistry Lett.*, 1981, 1679.
42. N. Ishikawa and T. Kitazume, *European Patent*, 0082 252 Al, 1982.
43. T.S. Chou, J.J. Ying and C-H. Tsao, *J. Chem. Res(S)*, 1985, 18.
44. T.S. Chou, C-H. Tsao and S.C. Hung, *J. Org. Chem.*, 1985, **50**, 4329.
45. S.L. Regen and A. Singh, *J. Org. Chem.*, 1982, **47**, 3361.
46. C. Petrier, A.L. Gemal and J.L. Luche, *Tetrahedron Lett.*, 1982, **23**, 3361.
47. J.L. Luche, *Ultrasonics*, 1987, **25**, 40.
48. A. Fuentes and J.V. Sinisterra, *Tetrahedron Lett.*, 1986, **27** 2967.
49. B. Han and P.J. Boudjouk, *J. Org. Chem.*, 1982, **47**, 5050.
50. A.K. Bose, K. Gupta and M.S. Manhas, *J. Chem. Soc. Chem. Commun.*, 1984, **86**.
51. T. Kitazume, *Synthesis*, 1986, 855.
52. J.L. Luche and J.C. Damiano, *J. Am. Chem. Soc.*, **102**, 7926.
53. T. Uyehara, J. Yamada, K. Ogata and T. Kato, *Bull. Chem. Soc.* , Japan, 1985, **58**, 211.
54. M. Ihara, M. Katogi, F. Fukumoto and T. Kametani, *J. Chem. Soc. Chem. Commun.*, 1987, 721.
55. I.C. Burkow, K.L. Sydnes and D.C.N. Ubeda, *Acta. Chem. Scand. Ser B.*, 1989, **B41**, 235.
56. S.B. Singh and G.R. Pettit, *Syunth. Commun.*, 1987, **17**, 877.
57. C. Einhorn, C. Allavena and J.L. Luche, *J. Chem. Soc. Chem. Commun.*, 1988, 333.
58. J.L. Luche, C. Petrier and C. Duputy, *Tetrahedron Lett.*, 1985, **26**, 753.
59. R.S. Varma and G.W. Kabalka, *Heterocycles*, 1985, **23**, 189.
60. B.C. Barot, D.W. Sullins and E. Eisenbraum, *Synth. Commun.*, 1984, **14**, 397.
61. S. Akiya and S. Okui, *J. Pharm. Soc.* Japan, 1951, **71**, 182.
62. C. Chittendin and J.L. Luche, *Carbohydrate Res.*, 1986, **155**, 258.

10

MISCELLANEOUS APPLICATIONS

Following are given some miscellaneous applications of ultrasound.

10.1 POTASSIUM SUPEROXIDE

It is used in the preparation[1] of anion of acetonitrile. The formed anion condenses with various aldehydes (Scheme-1).

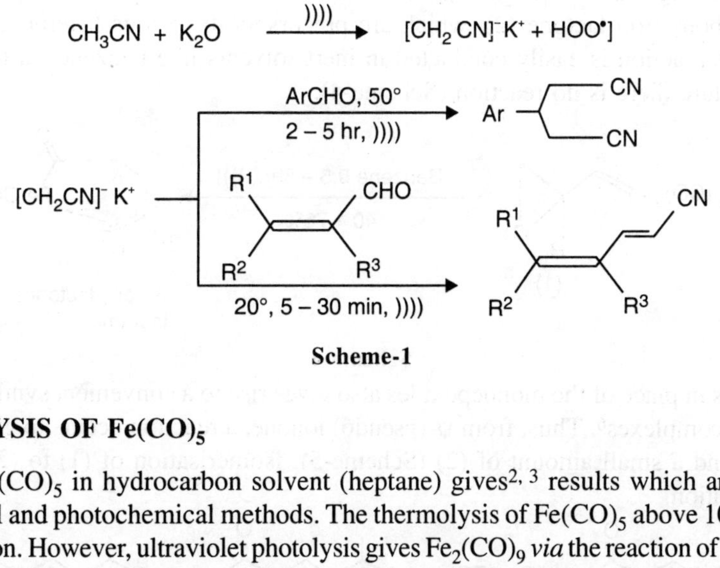

$$CH_3CN + K_2O \xrightarrow{))))} [CH_2CN]^- K^+ + HOO^\bullet$$

$[CH_2CN]^- K^+$
— ArCHO, 50°, 2 – 5 hr,)))) → Ar−CH(CN)−CH_2−CN (Ar with two CN groups)
— $R^1R^2C=CR^3$−CHO, 20°, 5 – 30 min,)))) → $R^1R^2C=CR^3$−CH=CH−CN

Scheme-1

10.2 SONOLYSIS OF Fe(CO)₅

Sonolysis of $Fe(CO)_5$ in hydrocarbon solvent (heptane) gives[2,3] results which are different both from the thermal and photochemical methods. The thermolysis of $Fe(CO)_5$ above 100° gives mainly finely divided iron. However, ultraviolet photolysis gives $Fe_2(CO)_9$ via the reaction of the intermediate, $Fe(CO)_4$ with $Fe(CO)_5$. On the other hand, sonolysis yields $Fe_3(CO)_{12}$. Use of heptane as solvent gives > 82% yield on sonication. The formation of $Fe_3(CO)_{12}$ during sonication is considered to arise from the coordinatively unsaturated species $Fe(CO)_3$ (Scheme-2).

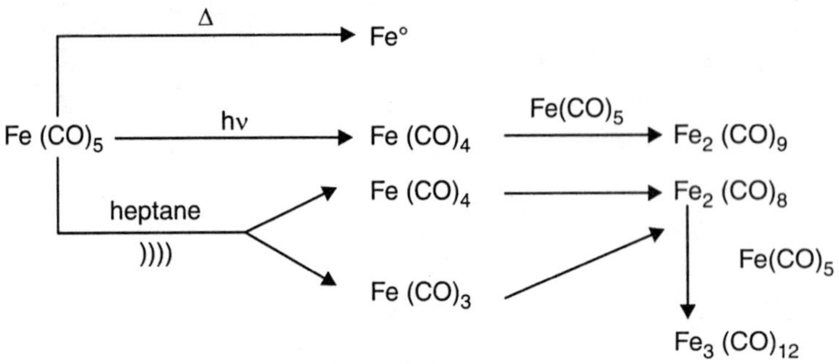

10.2 ■ Alternate Energy Processes in Chemical Synthesis

Scheme-2

$Fe_2(CO)_9$ has been used under sonication for a number of synthesis[4].

Thus, diiron nonacarbonyl [$Fe_2(CO)_9$] on sonication at –20°C in hexane suspension with anthracene readily forms the compound (1) (Scheme-3). It is not possible to obtain (1) by other chemical methods[5].

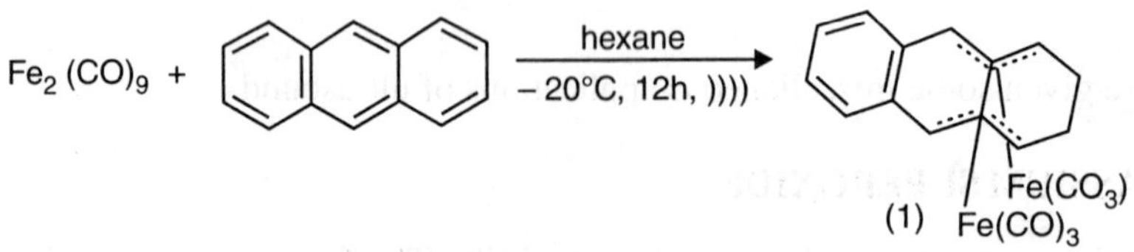

Scheme-3

The same starting material [$Fe_2(CO)_9$] on reaction with Vinyl epoxides (1) under sonication gave allyl tricarbonyliron lactone (2), which are precursors to various heterocyclic compounds[6] (Scheme-4). The reaction is easily conducted in inert solvents like benzene. In this solvent non-sonocical procedure there is no reaction (Scheme-4).

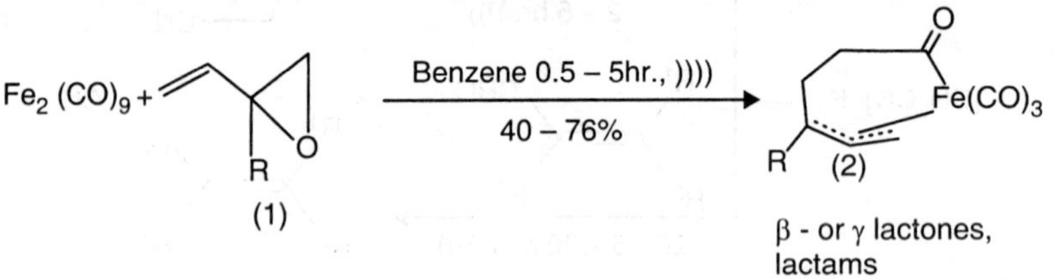

β - or γ lactones, lactams

Scheme-4

Using dienes in place of the monoepoxides also gives rise to a convenient synthesis of n⁴-diene tricarbonyl iron complexes[6]. Thus, from ψ-(pseudo) ionone, a reaction occurs to give 72% yield of a complex (1) and a small amount of (2) (Scheme-5). Isomerisation of (1) to (2) is possible by prolonged sonication.

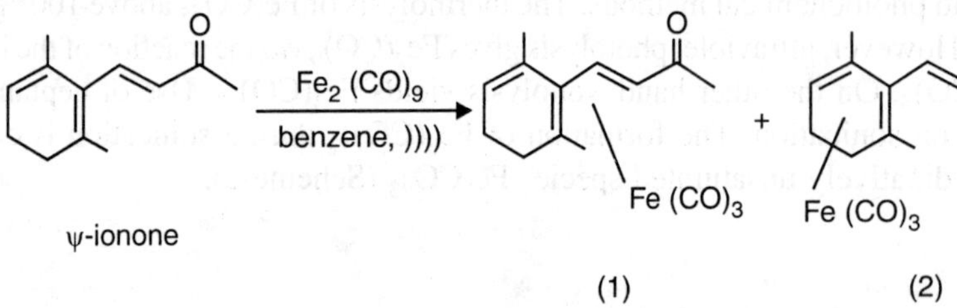

ψ-ionone

Scheme-5

10.3 OXYMERCURATION OF OLEFINS: SYNTHESIS OF α-TERPINOL

Oxymercuration of olefins is a very well known reaction to form carbon-oxygen bonds. Generally it is performed with mercuric acetate or trifluoroacetate. Since other salts of mercury are not easily available. It has now been shown that almost any mercuric salt can be prepared from mercuric oxide and an organic acid under sonication[7]. It is possible that the salt preparation and the oxymercuration step can be carried out in a one pot reaction under sonication and excellent yields are obtained. Thus, limonene on sonication with mercuric acetate in THF–H$_2$O gives 80% yield of α-terpinol (Scheme-6).

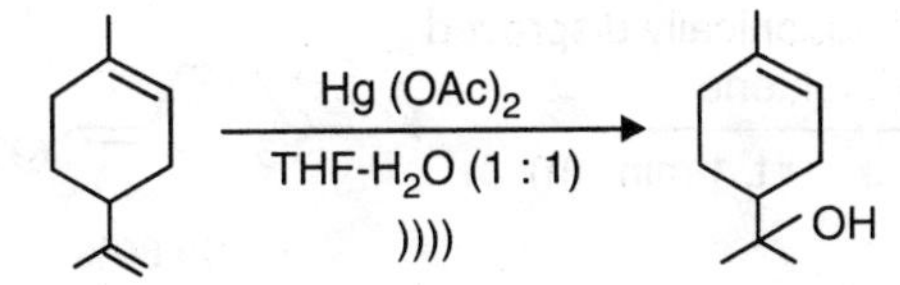

Reagent	Conditions	Time (min.)	Yield (%)
Hg(OAc)$_2$		30	48
HgO/t-BuCO$_2$H	rt))))	7	80
HgO/C$_7$F$_{15}$CO$_2$H	rt))))	5	80

Scheme-6

10.4 ACTIVATION OF NICKEL POWDER

Nickel powder can be activated[8] by sonication in octane medium. The activity of sonicated nickel powder increases by a factor of ~10^3. Using this technique, alkenes can be rapidly transformed to alkanes (Scheme-7).

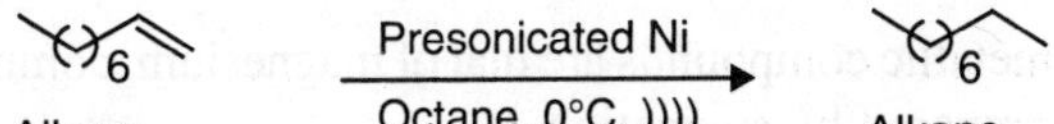

Scheme-7

The use of sonochemically activated nickel powder can offer interesting advantages over the classical Raney Nickel Catalyst. For making activated nickel powder, it should be sonicated in octane for about 1 hour.

10.5 ULTRASONICALLY DISPERSED POTASSIUM

As already stated (Section 9.8.6), on sonication potassium is easily transformed to a silvery blue suspension in toluene and that ultrasonically dispersed potassium is used in Dieckmann and Thorpe-Zieglar cyclisations under favourable conditions.

Sodium needs xylene to be dispersed (solvent effect)[9]. Ultrasonically dispersed sodium has not been used in synthesis. However, ultrasonically dispersed potassium finds a number of synthetic applications.

Ultrasonically dispersed potassium effects reduction of sulphur-carbon bonds in cyclic sulfone (1)[10]. This procedure provides excellent yields of open chain sulfones (2) and (3) after methylation of the intermediate (Scheme-8).

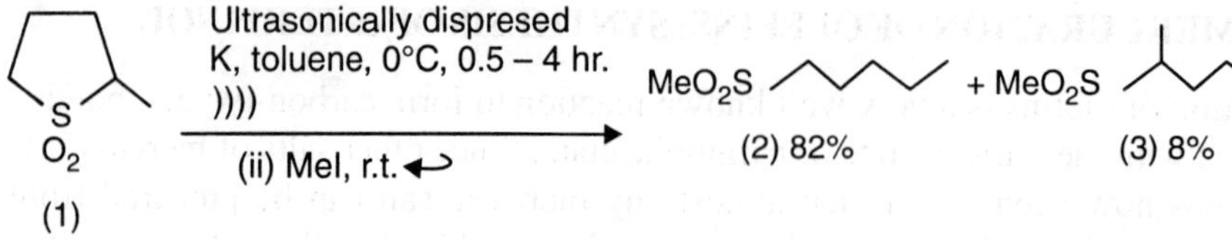

10.4 ■ Alternate Energy Processes in Chemical Synthesis

Scheme-8

In a similar way, the 3-sulfene (1) gives the dienes (2) and (3) with a satisfactory stereoselectivity[11] (Scheme-9).

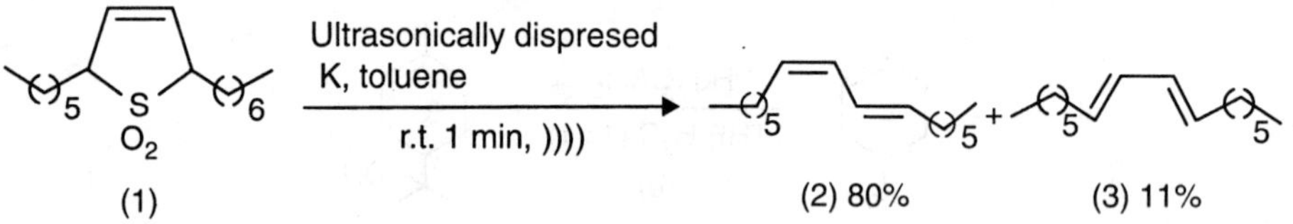

Scheme-9

Cyclopentadienyl iron couple (1) could also be synthesised in 10 min using the above technique (Scheme-10)[12]. The classical method takes 7 days.

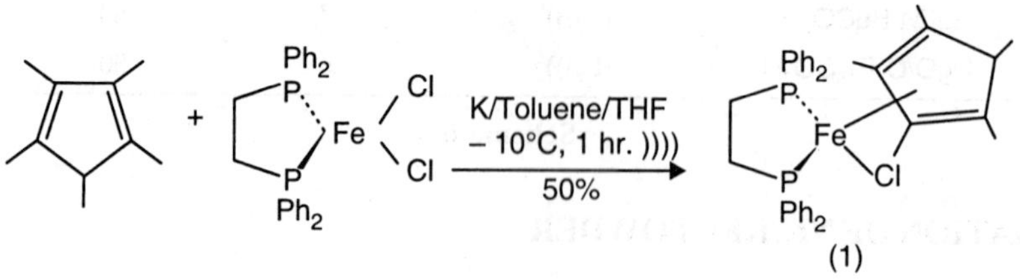

Scheme-10

10.6 ORGANOMETALLIC COMPOUNDS

The most common organometallic compounds are that of magnesium, commonly known as Grignard Reagents. These are best prepared by sonication.

10.6.1 Organolithium Compounds

Organo lithium compounds, viz., isopropenyl lithium was conveniently prepared under sonication[13]. Organo lithium compounds are frequently used in synthesis and make use of primary, secondary or tertiary butyllithium. Since these reagents can be conveniently prepared by sonication, their preparation and *in situ* reaction was successfully conducted in the presence of variety of substrates. One of the more important is diisopropylamine, which by metalation yields highly useful lithium diisopropylamide (LDA). This new ultrasonic method permits a quite direct and rapid preparation of LDA in high yield (92%) in the Solvent needed for further use. In General tetrahydrofuran is used (Scheme-11).

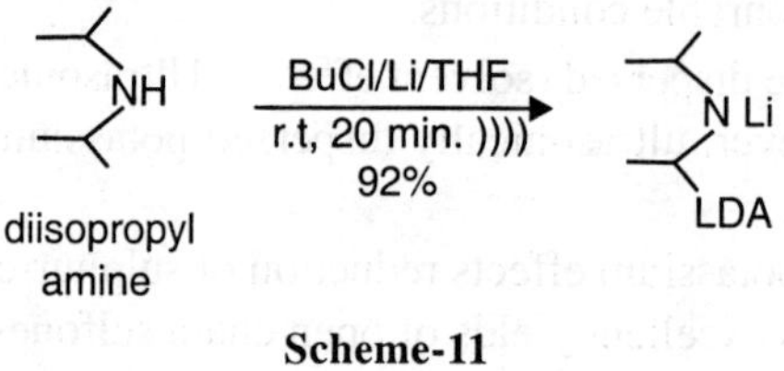

Scheme-11

It has also been shown that the preparation of LDA in a discrete step can be unnecessary, as shown with the generation of isobutyric acid dianion. Some other metalation examples are given in Scheme-12.

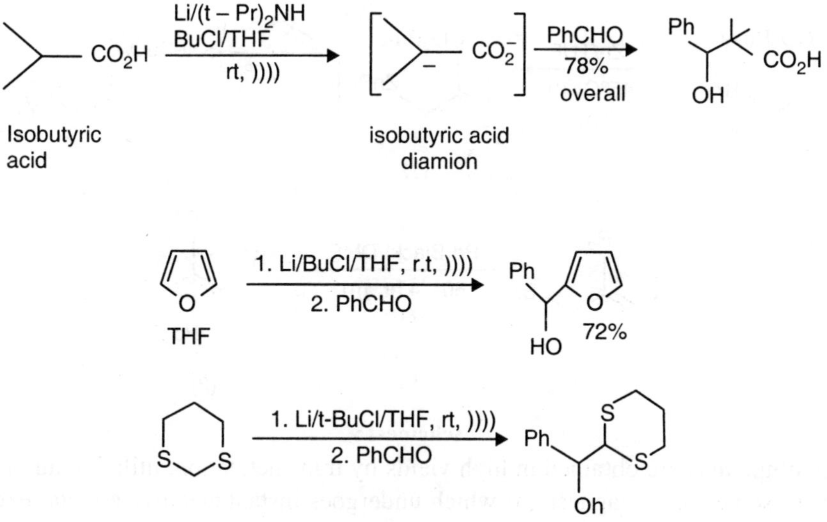

Scheme-12

10.6.2 Organoaluminium Compounds

Organoaluminium compounds were also obtained by sonication. Thus, tris (1-methyl-2-butenyl dialuminium tribromide[11]) was first prepared[14] (Scheme-13).

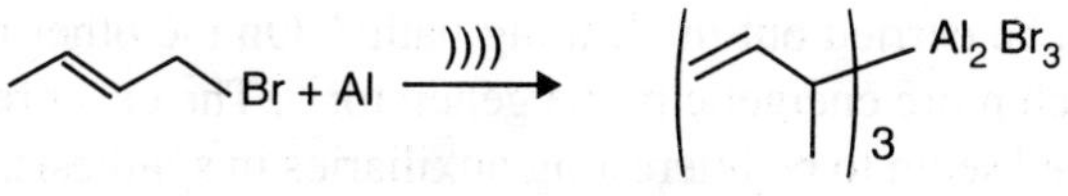

Scheme-13

The method was extended for the preparation of trimethyl dialuminium triiodide (Scheme-14)[15]. Excellent yield (96%) is obtained in 3 hr by sonication at room temperature. The silent process (2 hr room temperature) gave no positive results. The reaction product can be transformed to trimethylaluminium by reaction with triethylaluminium. Sonochemically this process is accomplished in 2.5 hr at room temperature, while at 100° and 6.5 hr are necessary to obtain the same yield by a silent process (Scheme-14).

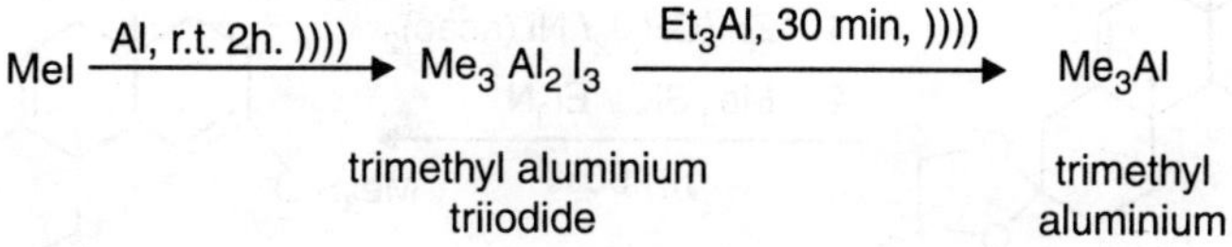

Scheme-14

10.6 ■ Alternate Energy Processes in Chemical Synthesis

10.6.3 Organo Zinc and Palladium Compounds

Allylic derivatives of zince (1)[16] and palladium (2)[17] can also be prepared by a direct sonochemical reaction from the metal (Scheme-15).

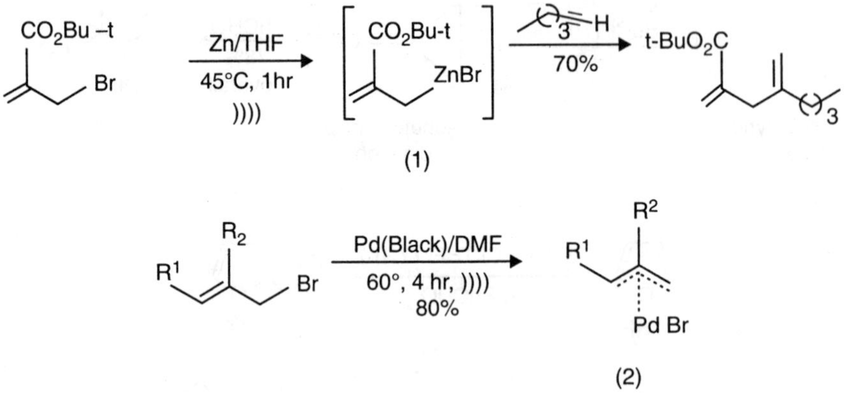

Scheme-15

Organozinc compounds are obtained in high yields by transmetalation, utilizing an organic halide and lithium (in some cases magnesium) which undergoes instantaneously a metal exchange with zinc chloride present *in situ*[18]. Scheme-16 describes the preparation of diaryl zinc using an aryl halide.

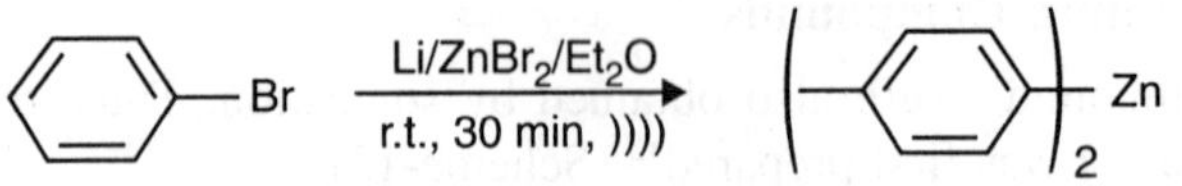

Scheme-16

The above reaction could be carried out in cleaning bath[19]. On the other hand, dialkyl compounds require the use of the much more energetic probe generator[20]. The exact reason for this is not clear. The reagents, thus obtained seem to be interesting auxiliaries in synthesis, since they add smoothly to α-enones, in cases where the more popular copper derivatives are unsuccessful. β-Cuparenoal (1)[21] and polycyclic compounds (2)[22] were prepared in excellent yields (Scheme-17).

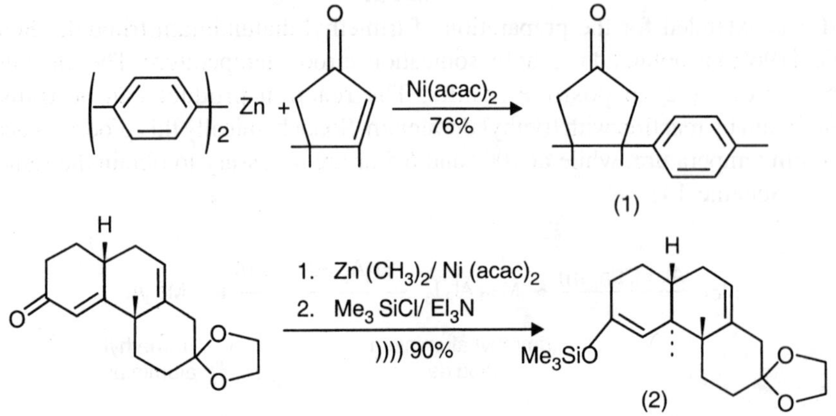

Scheme-17

10.6.4 Synthesis of Trialkyl Boranes

Trialkyl boranes are prepared in high yield by formation of a Grignard reagent followed by *in situ* transmetalation with borontrifluoride etherate[23]. A similar preparation of triethylborane can also be achieved by sonication of bromoethane with aluminium triethyl borate[24] (Scheme-18).

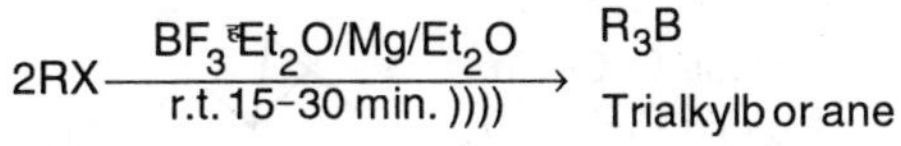

$$2RX \xrightarrow[\text{r.t. 15-30 min. })))]{BF_3 \cdot Et_2O/Mg/Et_2O} R_3B \text{ Trialkylborane}$$

R	Yield
Pr	100
Ph	97
PhCH$_2$	99
Allyl	94

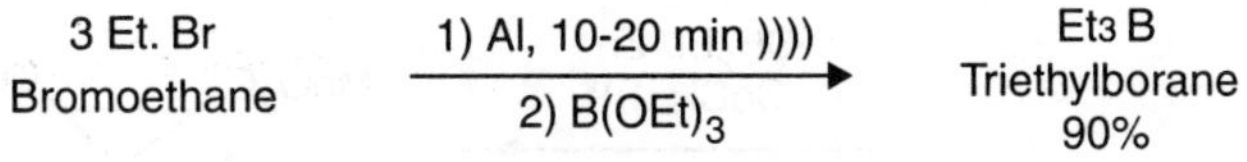

$$3 \text{ Et. Br (Bromoethane)} \xrightarrow[\text{2) B(OEt)}_3]{\text{1) Al, 10-20 min))))}} Et_3B \text{ Triethylborane } 90\%$$

Scheme-18

10.7 SYNTHESIS OF ALDEHYDES FROM HALIDES

Various halides on sonication with lithium, dimethylformamide in THF give 76–80% yield of the corresponding aldehydes[25] (Scheme-19).

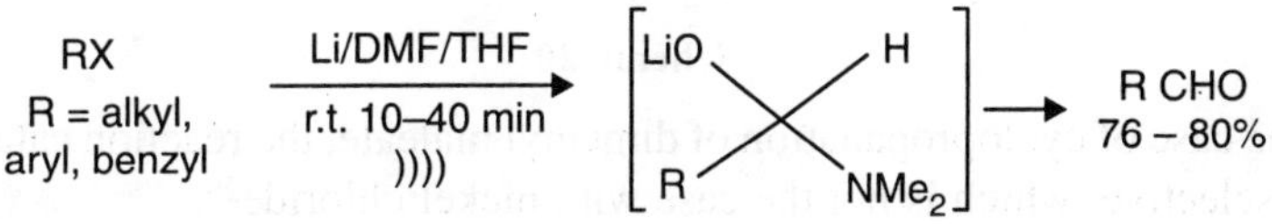

$$RX \text{ (R = alkyl, aryl, benzyl)} \xrightarrow[\text{r.t. 10-40 min))))}]{Li/DMF/THF} \left[\begin{array}{c} LiO \quad H \\ \times \\ R \quad NMe_2 \end{array} \right] \rightarrow R\,CHO \quad 76-80\%$$

Scheme-19

10.8 SONOCHEMICAL METHYLENATION OF ALKENES AND CARBONYL COMPOUNDS

Methylenation of alkenes with CH_2I_2 in presence of zinc is referred to as Simmons Smith Cyclopropanation. One case has already been referred to in Section 7.7.6. It is possible to use low grade zinc ('mossy' zinc), with high yields and reproducibility[26]. It is possible to carryout the reaction on a large scale in specially designed apparatus[27]. This method has been slightly modified to be run with dibromo methane, much cheaper than the diiodo analogue[28]. Some illustrations of sonochemical methylenation using CH_2I_2 and CH_2Br_2 are given in Scheme-20.

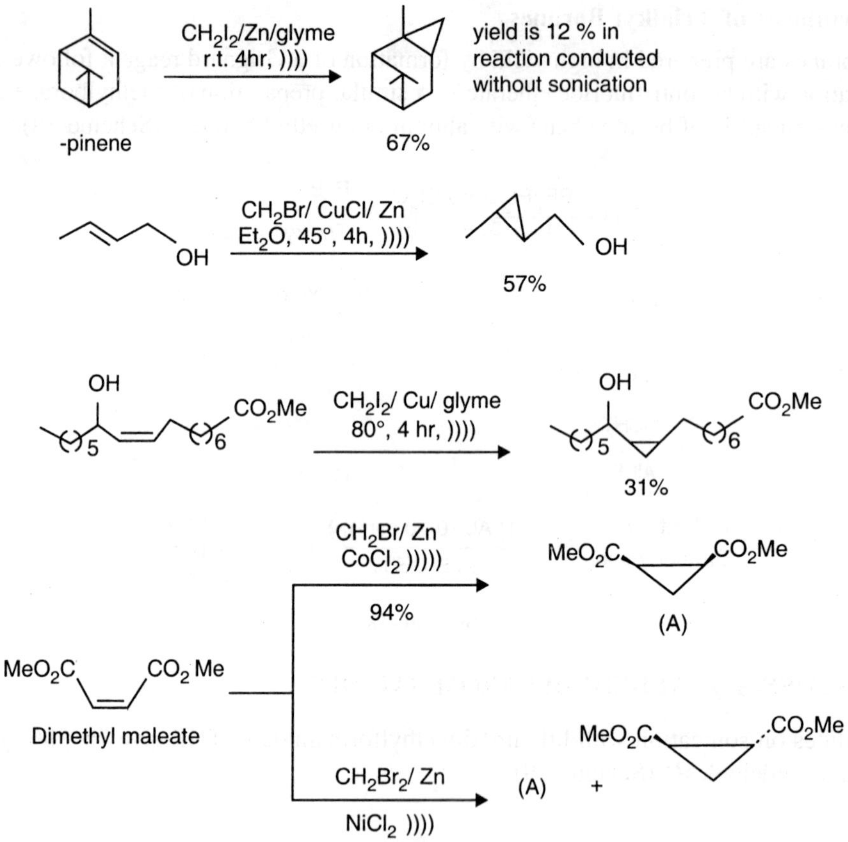

10.8 ■ Alternate Energy Processes in Chemical Synthesis

Scheme-20

As seen, in the case of cyclopropanation of dimethyl maleate, the reaction catalysed with cobalt chloride is stereoselective, which is not the case with nickel chloride[29].

Methylenation of carbonyl group, which frequently requires complex reagents, can be easily accomplished by the Simmons-Smith reagent (Zn/CH_2I_2) with sonication[30] (Scheme-21). This reaction, however, does not seem to be general, as the procedure cannot be applied to Ketones in high yields.

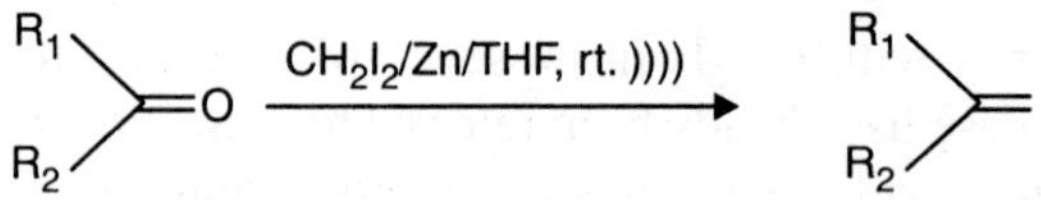

R_1	R_2	Time (min.)	Yield (%)
Ph	H	20	70
I-naphthyl	H	120	54
$(CH_2)_2CH_3$	H	20	71
$-(CH_2)_3-$		60	18

Scheme-21

10.9 SODIUMPHENYLSELENIDE

The useful reagent sodiumphenylselenide is obtained[31] by the reductive cleavage of diphenylselenide with sodium. It is obtained as a colloidal suspension on sonication. Its reactivity with various substrates, viz., halides, tosylates, epoxides etc. is much higher than that of a conventionally prepared salt (Scheme-22).

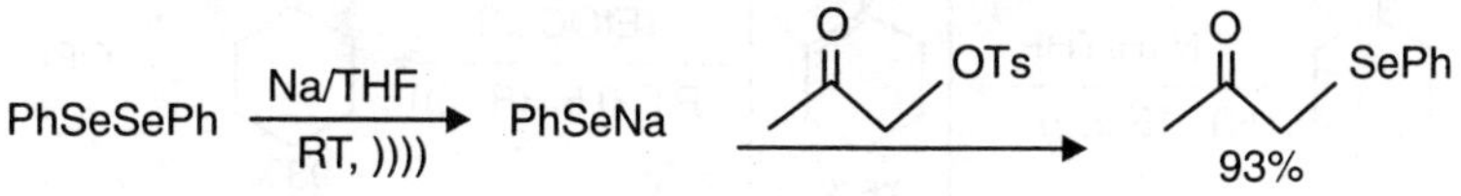

Scheme-22

10.10 ARYLAMIDES

Aryhalides on treatment with alkyl isocyanates in presence of alkali metals and magnesium under sonochemical Barbier conditions give secondary arylamides[32] (Scheme-23).

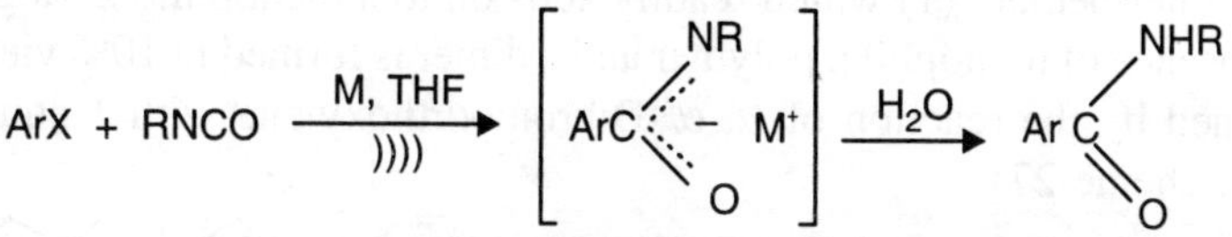

Scheme-23

The intermediates are obtained in good yield when M = Mg and can be ortholithiated when M = Na; the ortho lithiated intermediates react with a wide range of electrophiles to give ortho substituted arylamides (Scheme-24).

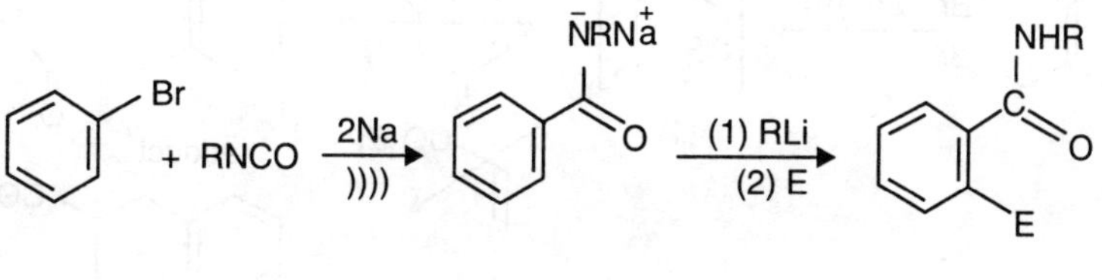

Scheme-24

10.11 SPIROKETONES

The synthesis of spiroketones from cycloalkanones has been effected successfully in excellent yield[33] by sonication with appropriate alkyl halides in presence of potassium *tert*. butoxide in benzene (Scheme-25).

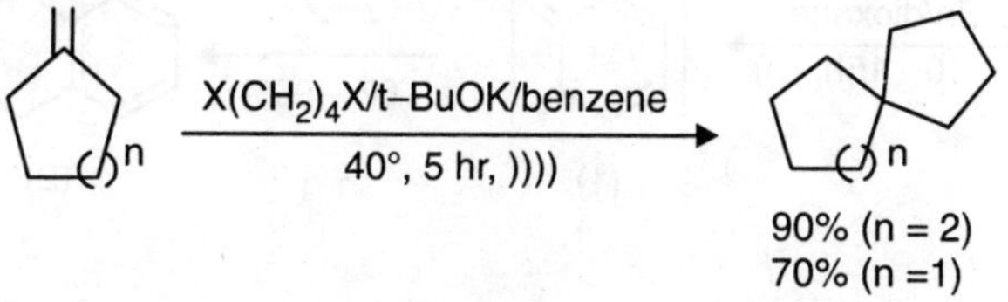

90% (n = 2)
70% (n = 1)

Scheme-25

10.10 ■ Alternate Energy Processes in Chemical Synthesis

10.12 β-KETO-THINOESTERS

Cyclohexanone can easily be converted[34] into enolate by sonication with sodium hydride in tetrahydrofuran in 75% yield compared to 18% under normal reactions. Subsequent sonication of the enolate with trithiodicarbonate gave the β-ketothinoester (Scheme-26).

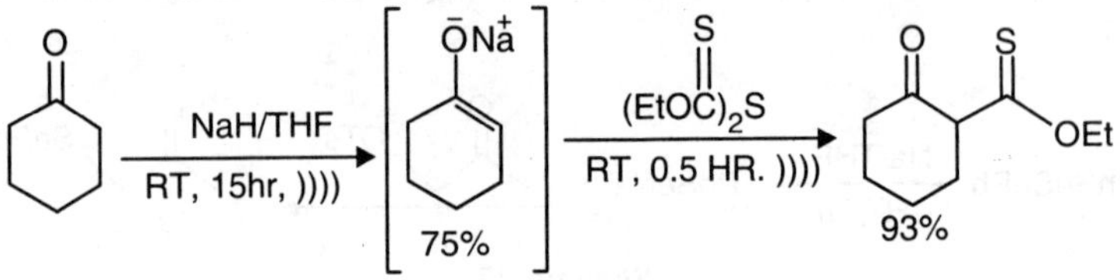

Scheme-26

10.13 DEHALOGENATION

α, α'-Dibromoorthoxylene in dioxane on treatment with zinc in the presence of ultrasonic irradiation gives a o-xylylene intermediate (1) which readily adds on to a dienophile giving high yield of the adducts[35]. In the absence of dienophile, polymer and a dimer is formed in 10% yield. However, only the dimer is obtained by the reaction of α, α'-Dibromoorthoxylene with 1 equivalent lithium in ultrasonic bath[36] (Scheme-27).

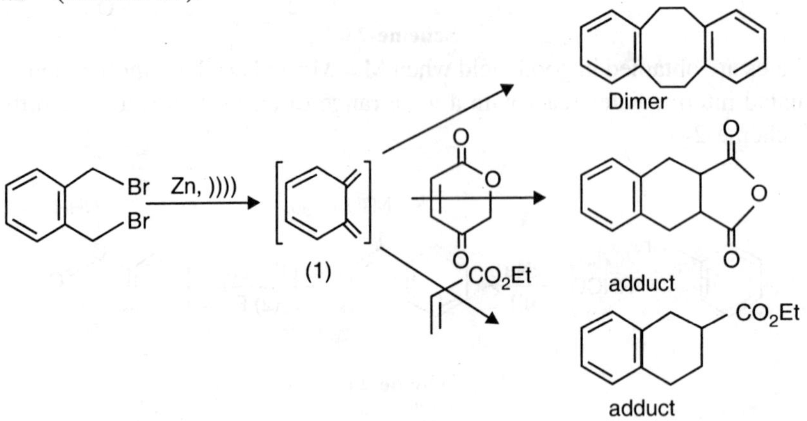

Scheme-27

The o-xylydene intermediate (1) obtained in Scheme-27 can be trapped *in situ* by the dienophile present in the medium. Excellent yields of (2) were obtained (Scheme-28)[37]. It is noteworthy that the course of the reaction is completely different when lithium is used instead of zinc (Scheme-28)[36].

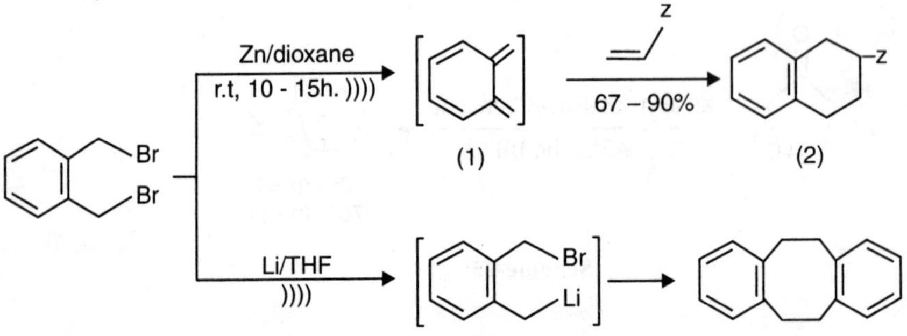

Scheme-28

Miscellaneous Applications ■ 10.11

Gem-Dihalogenocyclopropanes can be dehalogenated on sonication in presence of lithium, sodium or magnesium in tetrahydrofuran (but not pentane). This shows the importance of solvent in sonochemical reactions[38] (Scheme-29).

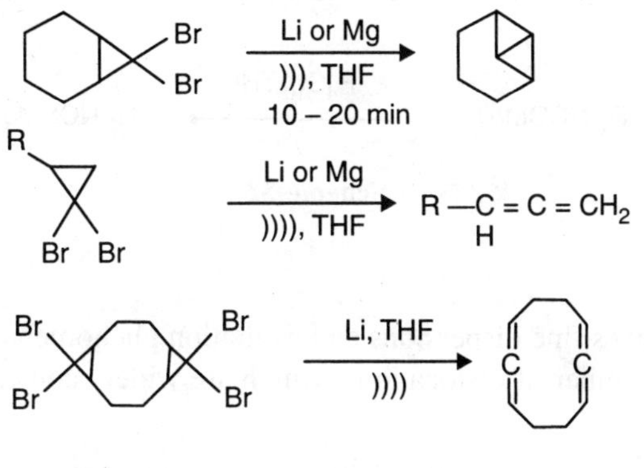

Scheme-29

6-Bromopenicillinate esters can be debrominated[39] with zinc in dioxane under ultrasonic irradiation. This method is efficient, clean and cheap than employing the usual debrominating agents Bu_3SnH or Pd –C/H_2 (Scheme-30).

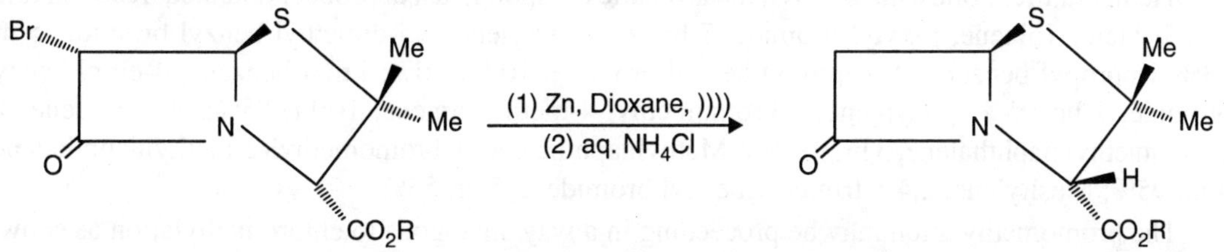

Scheme-30

Reductive dehalogenation of aryl halides with nickel (II) chloride and zinc in aqueous HMPA is facilitated by low intensity ultrasound.[40] Alternatively, lithium aluminium hydride can also be used.[41] In the latter case, the yield is 97% in 5 hr from bromobenzene compared to only 21% in 24 hr under non-ultrasonic conditions (Scheme-31).

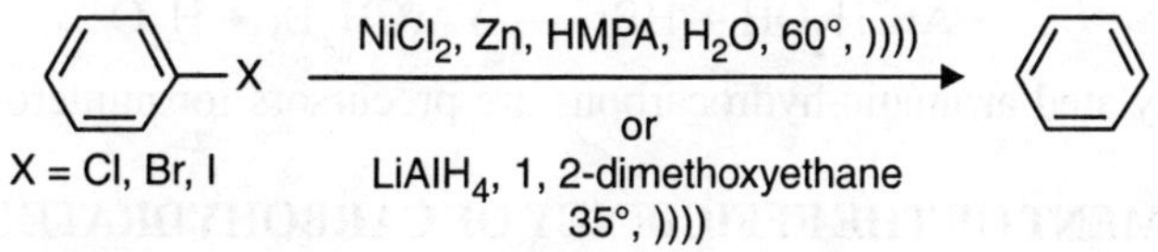

Scheme-31

10.14 THIOAMIDES

The rate of preparation of thioamides by the reaction of the respective amide with P_4S_{10} in dry tetrahydrofuran is enhanced by using ultrasonic irradiation[42]. This method has additional advantage of requiring 1–1.5 equiv. of P_4S_{10} rather than large excess used in the traditional method. (Scheme-32).

$$R_2NCOMe \xrightarrow[))))]{P_4S_{10} \text{ THF}} R_2NCSMe$$

Scheme-32

10.15 CATALYSIS

The ultrasonic field produces fine dispersions and cavitation phenomena give rise to clean surfaces containing an increased number of dislocations, which are widely considered to be the active sites in catalysis[43].

10.16 BROMOMETHYLATION OF AROMATICS

The bromomethylations of aromatics like toluene, ethyl benzene, cumene, p-xylene, p-cymene, mesitylene, naphthalene and 2-methylnaphthalene has been carried out in quantitative yield on sonication[44] with paraformaldehyde, acetic acid and aqueous hydrobromic acid (48%) for 2–4 hr at room temperature. Following is given the aromatic compound used, product obtained, reaction time and % yield. Toluene, p-xylyl bromide, 7 hr, 90%; p-xylene, 2,5-dimethyl benzyl bromide, 5 hr, 93%, isopropyl benzene, 4-isopropyl benzyl bromide, 10 hr, 80%; Ethyl benzene, 4-ethyl benzyl bromide, 7 hr, 95%; p-Cymene, 3-bromomethyl-4-methylcumene, 10 hr, 85%; Naphthalene, 1-bromomethyl naphthalene, 3 hr, 92%; 2-Methylnaphthalene, 1-bromomethyl-2-methylnaphthalene, 4 hr, 95%; Mesitylene, 2,4,5-trimethylbenzyl bromide, 2.5 hr, 55%.

The bromomethylation may be proceeding in a way analogous to chloromethylation as shown below:

$$(CH_2=O)_n \xrightarrow{H^+} [CH_2 = \overset{+}{O}H \rightleftharpoons \overset{+}{C}H_2 - OH]$$

$$ArH + \overset{+}{C}H_2OH \longrightarrow ArCH_2OH + H^+$$

$$ArCH_2OH + HBr \longrightarrow ArCH_2Br + H_2O$$

The bromomethylated aromatic hydrocarbons are precursors for numerous organic synthesis.

10.17 ENHANCEMENT OF THE EFFICIENCY OF CARBOHYDRATE SYNTHESIS

Due to biological relevance of carbohydrates,[45–47] the synthesis of oligosaccharides or carbohydrate derivatives has been pursued[48, 49]. The synthesis of carbohydrate derivatives is a challenging task due to the complexity and diversity of carbohydrates found in nature. It is found[50] that sonication-mediated reactions are compatible with traditionally employed conditions. The reaction time is considerably reduced giving excellent yields in most cases. Following are given five types of commonly employed reactions in carbohydrate synthesis.

10.17.1 Manuplication of Hydroxyl Groups

The procedures adopted for manuplication of hydroxyl group include, acetylation, protection/deprotection and acyl group migration.

Acetylation of carbohydrate usually take several hours. A typical example is the acetylation of trehalose derivative (1), the reaction is difficult to reach completion, probably due to the steric hindrance. However, using sonication, the acetylation of all four hydroxyl groups of (1) could be achieved in few minutes (Scheme-33).

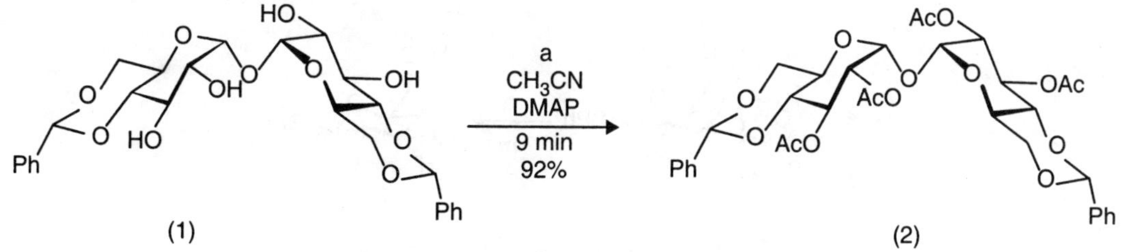

Scheme-33

In a similar way deprotection of various protecting group of the hydroxyl group such as acetyl, benzylidene and trityl groups can be completed[50] in a few minutes (Scheme-34).

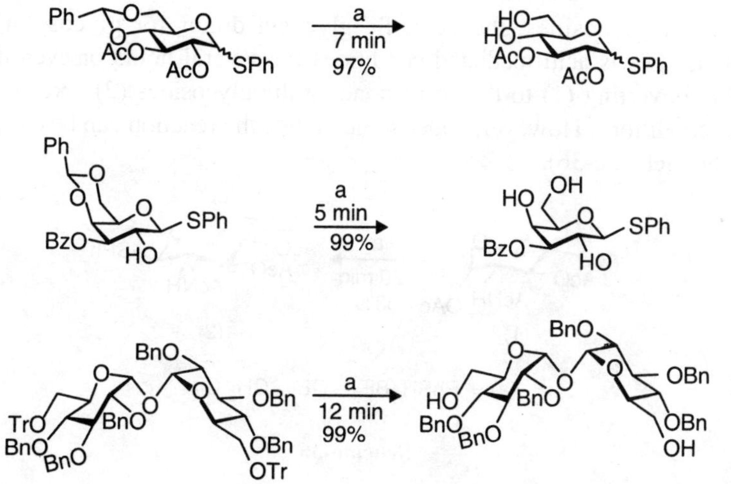

Scheme-34

A general protocol for regiospecific migration of acetyl groups has been reported[51]. This reaction generally takes days to complete. However, under sonication[50], the reaction time can be shortened to several minutes (Scheme-35).

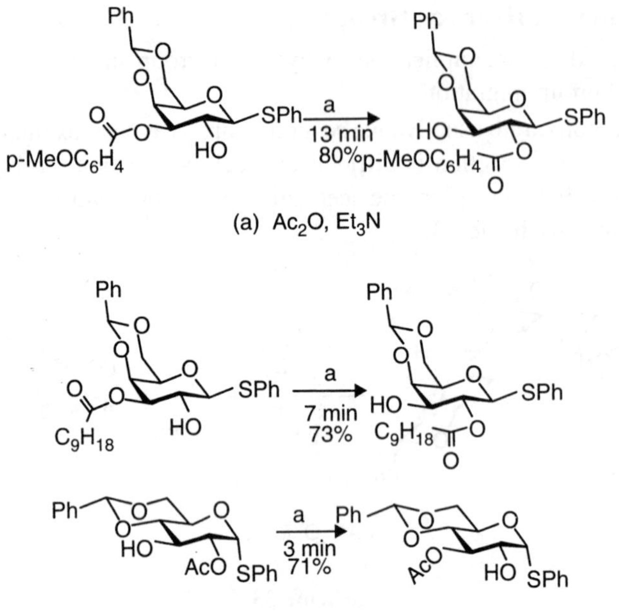

10.14 ■ *Alternate Energy Processes in Chemical Synthesis*

[Scheme showing glycoside reactions: first reaction 13 min, 80%; second reaction 7 min, 73%; third reaction 3 min, 71%]

(a) Ac$_2$O, Et$_3$N

(a) Ac$_2$O, Et$_3$N

Scheme-35

10.17.2 Synthesis of Thioglycosides

The synthesis of thioglycosides, the versatile glycosyl donor for glycosylation is generally accomplished under Lewis acid mediated conditions for several hours or even days[52]. Thus, the reaction time for converting (1) to the corresponding thioglycosides (2) takes as long as a week under traditional conditions. However, under sonication[50], the reaction can be completed in several minutes upto 2 hr (Scheme-36).

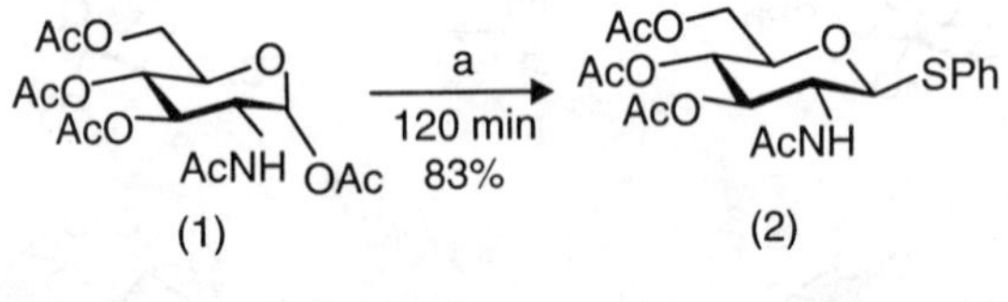

[Scheme showing conversion of (1) to (2): 120 min, 83%]

(a) = PhSH, BF$_3$–OEt$_2$, CH$_2$Cl$_2$

Scheme-36

10.17.3 Synthesis of Azidoglycosides

Azidoglycosides, commonly used for the synthesis of various N-glycosides are normally synthesised using TMSN$_3$ in presence of Lewis and Catalyst for several hours[53]. By using sonication[50], azidoglycosides can be prepared from the corresponding glycosyl bromides in a few minutes via a substitution reaction even without the presence of Ag(I) (Scheme-37). It should be noted that some of the azido compounds of small molecular weights are known to be explosive and should be handled carefully.

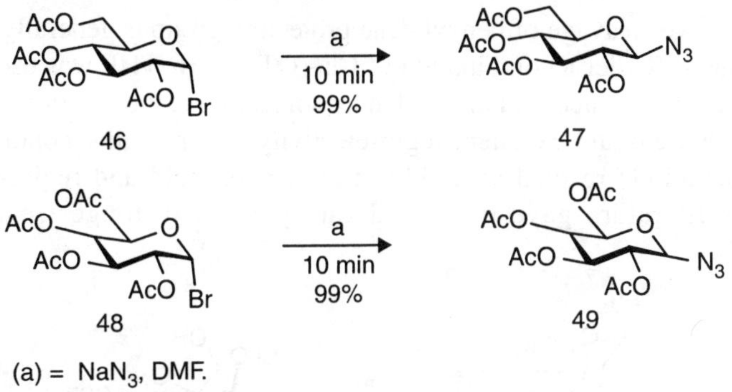

Scheme-37

10.17.4 Dipolar Cycloaddition

1,3-Dipolar cycloaddition (Scheme-38) is useful for incorporating carbohydrate components or for the preparation of glycomimetics. Using sonication[50], the reaction time can be shortened from days to less than an hour with excellent regioselectivity giving the formation of 1,4-disubstituted 1,2,3-triazoles. Some examples are given in Scheme-38.

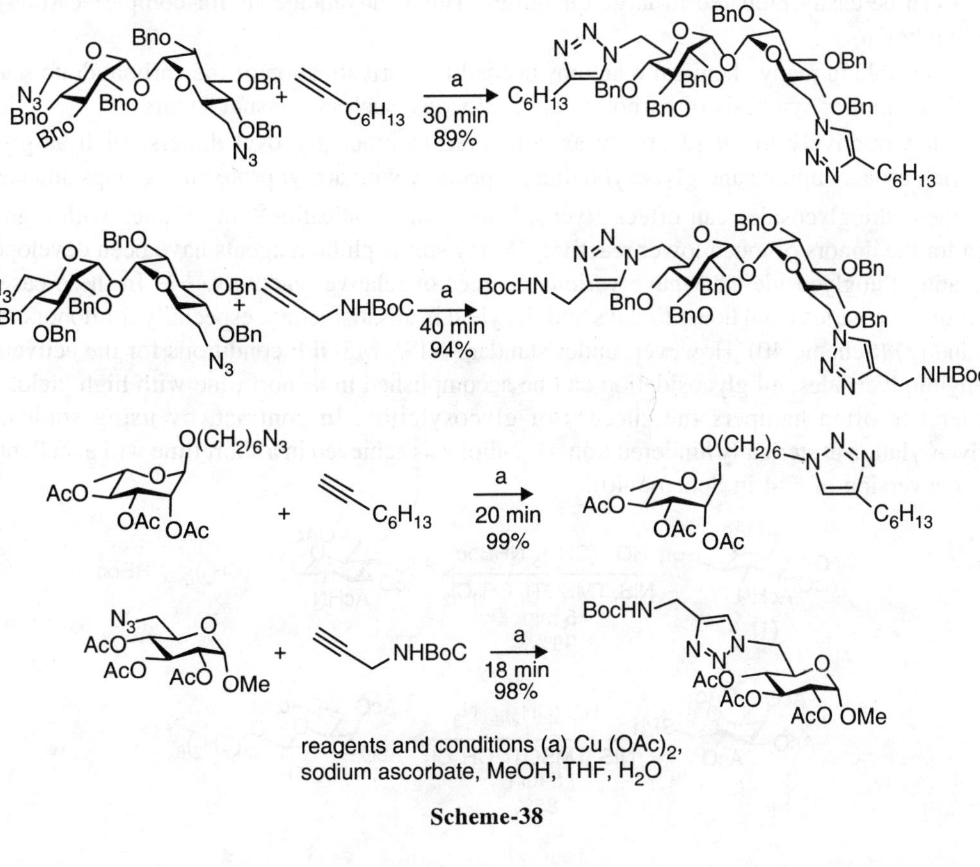

reagents and conditions (a) Cu (OAc)$_2$, sodium ascorbate, MeOH, THF, H$_2$O

Scheme-38

10.16 ■ *Alternate Energy Processes in Chemical Synthesis*

10.17.5 Reductive Cleavage of Benzylidene Protecting Group

Regioselective reductive cleavage of benzylidene protecting group is generally used[54] for complex carbohydrate synthesis. Earlier a combination of Cu–(OTf)$_2$/BH$_3$·THF was used for regioselective reduction of benzylidene in glucopyranoside. Under sonication[50], the reaction can be accomplished in minutes with high yield and excellent regioselectivity under similar conditions. As an typical example the compound (1) reacted smoothly in high rate, yield and regioselectivity while the counterpart in the literature gave only moderate yield with longer reaction time (20 hr) (Scheme-39).

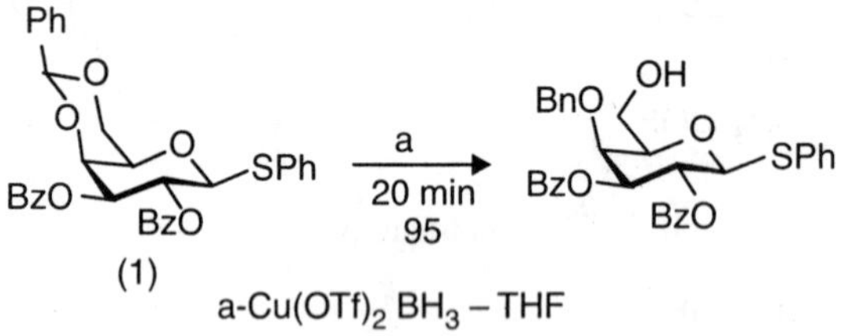

Scheme-39

10.17.6 Glycosylation with Phenylthioglycosides

Phenylthioglycosides are good reagents for glycosylation due to following reasons.

— It can be easily prepared in large quantities. This is advantageous for complex carbohydrate synthesis.
— It is stable in many chemical reactions needed for derivatization of the carbohydrate scaffold. This simplies synthesis of carbohydrate analogous, such as unusual sugars.
— It has relatively lower reactivity as compared to other glycosyl doners, such as glycosyl trichloroacetimidate and glycosyl halide, especially with acetyl protecting groups attacked.

Phenylthioglycosides can effect glycosidation under sonication[50] in minutes with high yield even for the donors of much lower reactivity. Many sulfur-philic reagents have been developed for activating thioglycosides that have various degrees of relative reactivity.[52, 55] In these cases, the reaction time varies from hours to days and the yield is unsatisfactory, especially for donors such as (1) and (2) (Scheme-40). However, under standard NIS/TMSOFF conditions for the activation of phenylthiglycosides, all glycosidation can be accomplished in a short time with high yield. Steric hinderance often hampers the success of glycosylation. In contrast, by using sonication a diglycosylation at sterically hindered trans-1-2-diol was achieved in a short time will excellent yield (see conversion of 3–4 in Scheme-40).

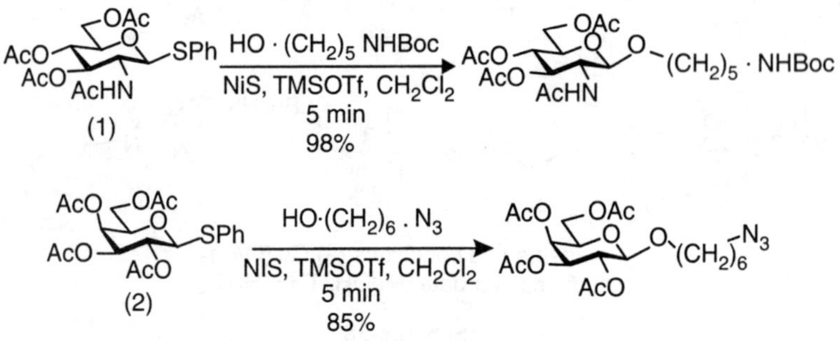

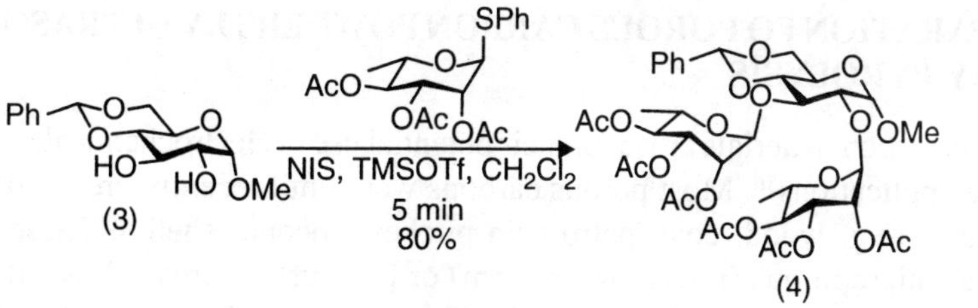

Scheme-40

The sonication procedure[50] can even work on a much less reactive glycosyl doner, glycosyl acetate (Scheme-41).

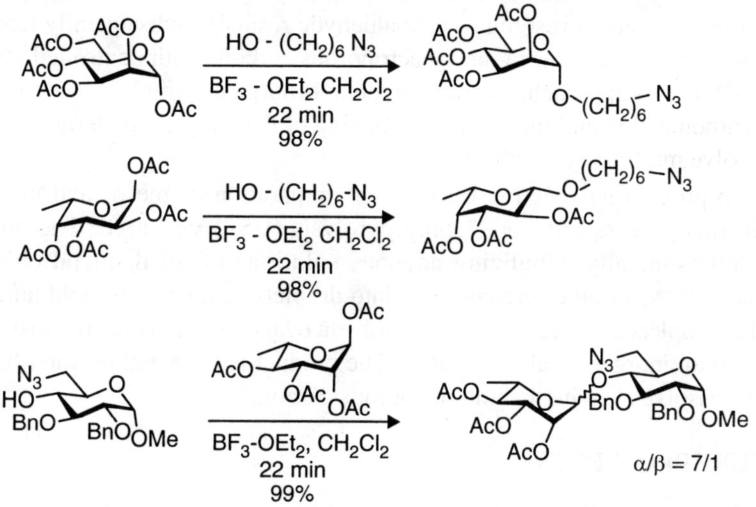

Scheme-41

Sonication[50], as expected is also applicable for more reactive, glycosyl trichloroacetimidate (1) and glycosyl bromide (2) (Scheme 42).

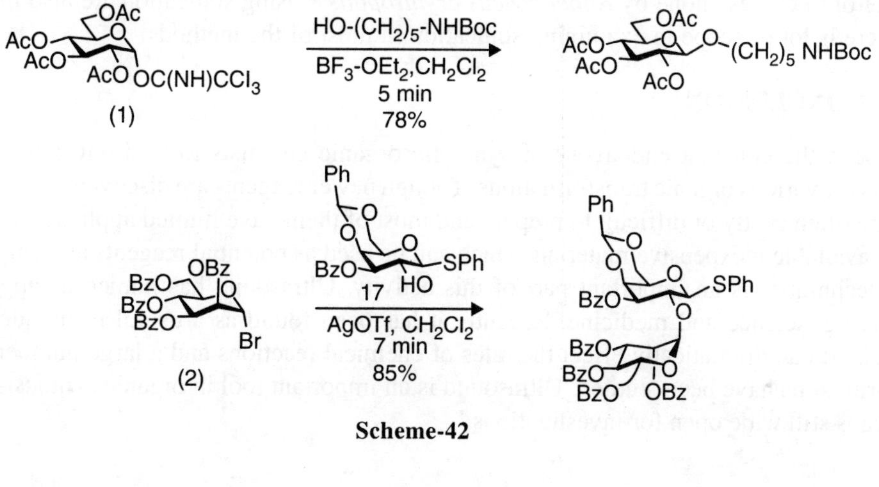

Scheme-42

10.18 PREPARATION FO POROUS CARBON POWDERS BY ULTRASONIC SPRAY PYROLYSIS

Interest in porous carbon materials arose for their potential use in electrochemical, catalytic, adsorbent and gas-storage applications[56]. Most porous carbons were until recently prepared by carbonization of raw materials[57] (e.g. wood, coal, petroleum pitches, coconut shells). These procedures gave carbons either as microporous (micro size < 2 nm) or low surface area solids with broad pore-size distribution. Meso- (pore size 2–20 nm) and macroporous (> 20 nm) carbons have come into prominance due to their applications involving large molecules (e.g., separations) and high diffusion rates (e.g., electrodes for double layer saparators, catalyst supports)[58]. Such porous materials were originally prepared by carbonization of either block Copolymers with thermally unstable parogens or organic aerogels based on resorcinol-formaldehyde resins[59]. Subsequently templates were used and their removal to produce meso- and macroporous carbons with controlled and, in same cases, periodic pores[58]. In this approach, a carbon precursor/inorganic template composite is first formed, followed by carbonization and the chemical leaching of the template material. Such procedures are tedius and involve multiple synthetic steps.

A one step process has been developed for preparation of meso- and macroporous carbon powders[60]. In this process ultrasonic spray pyrolysis (USP) was used. The porous carbons are prepared by ultrasonically nebulizing aqueous solutions of alkali metal chloroacetate (CA), dichloroacetate (DCA) or other organic salts into droplets using a household humidifier. As argon flow carries the droplets into a furnace, where solvent evaporates and the precursors are decomposed producing a carbon/inorganic salt composite. The product is collected in water bubblers, when the generated salt dissolves leaving behind the porous carbon.

10.19 FUTURE PROSPECTS

Application of sonochemical principles in polymer chemistry[61] and in coal liquification[62] are possible. Combination of sonochemistry with other specific methods such as photochemistry[63] and electrochemistry[64] are quite promising. The use of sonication to stimulate microbiological reactions are also possible. Thus, the cyclisation of squalene oxide with bakers yeast[65] and oxidation of cholesterol to cholestenone by *Rhodococcus erythropolis*[66] using sonication are also important. In fact, there is lot of scope in examining sonication in most of the methods.

10.20 CONCLUSION

It has been the constant endeavour of synthetic organic chemists to find out newer and better methods for various organic transformations. Though newer reagents are discovered very frequently, these are often costly or difficult to prepare and most of them have limited applications. Search for readily available inexpensive materials which can be used as potential reagents and employment of newer techniques is an important part of this activity. Ultrasound has numerous applications in engineering, science and medicine. Recently it has been found as a useful technique in organic synthesis. It can dramatically effect the rates of chemical reactions and a large number of organic transformations have been studied. Ultrasound is an important tool in organic synthesis. However, this area is still wide open for investigations.

NOTES AND REFERENCES

1. K. Shibata, H. Kondo, K. Urano and M. Matsumi, *Chemistry Express*, 1987, **2**, 117 and 169.
2. K.S. Suslick, P.F. Schubert and J.W. Grodale, *J. Am. Chem. Soc.*, 1981, **103**, 7324.
3. K.S. Suslick, J.W. Goodale, P.F. Schubert and H.H. Wang, *J. Am. Chem. Soc.*, 1983, **105**, 5783.
4. A.M. Horton, D.M. Hollinshead and S.V. Ley, *Tetrahedron*, 1984, **40**, 1737.
5. M.J. Begley, S.G. Puntambekar and A.H. Wrigit, *J. Chem. Soc. Chem. Commun.*, 1987, 1251.
6. S.Y. Ley, C.M.R. Low and A.D. White, *J. Organomet. Chem.*, 1986, 302, C13.
7. J. Einhorn, C. Einhorn and J.-L. Luche, *Synthesis*, 1989, 798.
8. K.S. Suslick and D.J. Casadonte, *J. Am. Chem. Soc.*, 1987, **109**, 3459.
9. J.L. Luche, C. Petrier and C. Dupuy, *Tetrahedron Lett.*, 1985, **26**, 753.
10. T. Chou and M. You, *Tetrahedron Lett.*, 1985, **26**, 4495.
11. T. Chou and M. You, *J. Org. Chem.*, 1987, 52, 2224.
12. C. Roger, P. Marseille, C. Salus, J.R. Hamon and C.J. Lapinte, *Organomet. Chem.*, 1987, 336, C13.
13. D.D. Sternbach, J.W. Hughes, D.F. Burdi and B.A. Banks, *J. Am. Chem. Soc.*, 1985, **107**, 2149.
14. A.V. Kuchin, R.A. Nurushev and G.A. Tolstikov, Zh. Obshch. Khim, 1983, **53**, 2519; C.A. 1984, **100**, 103426.
15. P. Yang, K. Liou and Y. Lin, *J. Organometal. Chem.*, 1986, **307**, 273.
16. P. Knochel and J.F. Normant, *Tetrahedron Lett.*, 1984, **25**, 1475.
17. Y. Inou, J. Yamashita and H. Hashimoto, *Synthesis*, 1984, 244.
18. C. Petrier, J.C. de Souza-Barboza, C. Dupuy and J.-L. Luche, *J. Org. Chem.*, 1985, 30, 5761.
19. J.-L. Luche, C. Petrier, J.P. Lansard and A.E. Greene, *J. Org. Chem.*, 1983, **48**, 3837.
20. C. Petrier, J.-L. Luche and C. Dupuy, *Tetrahedron Lett.*, 1984, **25**, 3463.
21. A.E. Greene, J.P. Lansard, J.-L. Luche and C. Petrier, *J. Org. Chem.*, 1984, **49**, 931.
22. A.B. Smith and T.L. Leenay, *Tetrahedron Lett.*, 1988, **29**, 2787.
23. H.C. Broun and U.S. Racheria, *Tetrahedron Lett.*, 1985, **26**, 4311.
24. K. Liou, P. Yang and Y. Lin., *J. Organometal. Chem.*, 1985, 294, 145.
25. C. Petrier, A.L. Gemal and J.-L. Luche, *Tetrahedron Lett.*, 1982, **23**, 3361.
26. O. Repic and S. Vogt, *Tetrahedron Lett.*, 1982, **23**, 2729.
27. O. Repic, P.G. Lee and N. Giger, *Org. Prep. Proc. Int.*, 1984, **16**, 25.
28. E.C. Friedrich, J.M. Domek and R.Y. Pong, *J. Org. Chem.*, 1985, **50**, 4640.
29. X. Xu, Z. Li, Y. Na and G. Liu, Yingyong Huaxue, 1987, **4**, 73; C.A. 1988, **109**, 54337.
30. J. Yamashita, Y. Inoue, T. Kondo and H. Hashimoto, *Bull. Chem. Soc.* Japan, 1984, **57**, 2335.
31. S.V. Ley and J.-L. Luchi, *Tetrahedron Lett.*, 1986, **27**, 501.
32. J. Einhorn and J.-L. Luchi, *Tetrahedron Lett.*, 1986, **27**, 5363.
33. T. Fujita, S. Watanabe, M. Sakamoto and H. Mashimoto, *Chem. Ind.* (London), 1986, 427.
34. M.A. Palominos, R. Rodriguez and J.C. Vega, *Chemistry Let.*, 1986, 1251.
35. B.H. Han and P. Boudjouk, *J. Org. Chem.*, 1982, **47**, 751.
36. R. Boudjouk, R. Sooriyakumaran and B.H. Han, *J. Org. Chem.*, 1986, **51**, 2818.
37. B.H. Han and P. Boudjouk, *J. Org. Chem.*, 1982, **47**, 751.
38. L. Xu, F. Tao and Y. Yu, *Tetrahedron Lett.*, 1985, **26**, 4231.

39. J. Brennan and F.H.S. Husain, *Synthesis*, 1985, **8**, 749.
40. T. Yamashita, Y. Inoue, T. Kondo and H. Hashimoto, *Bull. Chem. Soc.*, Japan, 1985, **58**, 2709.
41. B.H. Han and P. Boudjouk, *Tetrahedron Lett.*, 1982, 23, 1643.
42. S. Roucher and P. Klein, *J. Org. Chem.*, 1981, **46**, 3558.
43. H. Bonnermann, D.W. Bogdanovic, D.W. Hi. Brinkman and H. Spliethoff, *Angew. Chem. Int. Ed. Eng.*, 1983, **22**, 728.
44. G.C. Mailap, *Ph.D. Thesis*, Delhi University, 1991.
45. A.C. Weymouth-Wilson, *Nat. Prod. Rep.*, 1997, 99–100.
46. R.A. Devek, *Chem. Rev.*, 1996, **96**, 683–720.
47. C.R. Bertozzi and L.L. Kiessling, *Science*, 2001, **291**, 2357–2364.
48. S. Hanessan (Ed.), *Preparative Carbohydrate Chemistry*, Marcel Dekker, New York, 1997.
49. T. Hudlicky, D.A. Entwistle, K.K. Pitzer and A. Thorpe, *J. Chem. Rev.*, 1996, **96**, 1195–1220.
50. S. Deng, U. Gangadharmath and C. -W.T. Chang, *J. Org. Chem.*, 2006, **71**, 5170–5185.
51. S. Deng and C.-W.T. Chang, *Synlett*, 2006, 756–760.
52. J. Wang, C.-W.T. Chang in *Carbohydrate Drug Design.*, A.A. Witezak and D. Platt (Eds.), *ACS, Symp.* Ser. No. 932. American Chemical Society, Washington, DC, 2005.
53. J.P. Malkinson, R.A. Falconer and I. Toth, *J. Org. Chem.*, 2000, **65**, 5249–5252.
54. C. Shie, Z. Tzeng, S.S. Kulkarni, B. Uang and C. Hsu. (Eds.), *Angew. Chem., Int.* 2005, **44**, 1665–1668.
55. A.V. Demehenko, *Synlett.*, 2003, **9**, 1225–1240.
56. L.R. Radovic, (Ed.), *The Chemistry and Physics of Carbon*, Marcel Dekkar, New York, 2001; Vol. 27;
    F. Rodriguez-Reinoso and A. Sepulveda-Escribano, *Hand. Surf. Interfaces Matter*, 2001, **5**, 309–355.
57. F. Rodriguez-Reinoso, *Handb.* Porous Solids, 2002, **3**, 1766–1827.
58. B. Sankintuna and Y. Yürüm, *Ind. Eng. Chem. Res.*, 2005, **44**, 2893–2902; J. Lee, S. Han and T. Hyeon, *J. Mater. Chem.*, 2004, **14**, 478–486; R. Ryoo, S.H. Joo, M. Knik and M. Jaronice, *Adv. Mater.*, 2001, **13**, 677–681; S.H. Joo, S. Choi, I. Oh, J. Kawk, Z. Liu, O. Terasaki and R. Ryoo, *Nature*, 2001, 412, 169–172; H.C. Foley, *J. Microporous Mater.*, 1995, **4**, 407–433.
59. J. Fricks and R. Petricevic, *Handb. Porous Solids*, 2002, **3**, 2037–2062; R.W. Pekala, C.T. Alviso and X. Lu, *Non-Cryst. Solids*, 1995, **188**, 30–40; T. Kyotani, *Carbon*, 2000, **38**, 269–286.
60. S.E. Skrabalak and K.S. Suslick, *J. Am. Chem. Soc.*, 2006, **128**, 12642–12643.
61. P. Kruus, *Ultrasonics*, 1987, 25, 20.
62. M. Miyake, R. Uematsu and M. Nomura, *Chemistry Lett.*, 1984, 535.
63. A. Gaplowsky, J. Donovalova, S. Toma and P. Hrnciar, *Chem. Listy.*, 1986, **80**, 989.
64. B. Gautheron, G. Tainturier and C. Degrand, *J. Am. Chem. Soc.*, 1985, **107**, 5579.
65. J. Bujons, R. Guajardo and K.S. Kyler, *J. Am. Chem. Soc.*, 1988, **110**, 604.
66. R. Bar, Biotechnol. Bioeng, 1988, **32**, 655.

11
SONOCHEMICAL TREATMENT OF POLLUTED WATER

11.0 INTRODUCTION

Ultrasonic procedures are employed in a number of industrial processes, such as welding of thermoplastics and metals, homogenization of emulsions, dispersal of paints, cleaning and degreasing[1]. However, chemical reactions produced by sonication of liquids has not so far received much attention for large scale chemical processes. Chemical reactions can be induced in high-intensity acoustic irradiation of liquids at frequencies that produce cavitation in the range 20–1000 KHz. Sonochemical methods are potentially useful for a variety of industrial application. This is because chemical reactions can be produced by relatively simple piezoelectrical devices in a controlled regime, without addition of reagents. The fact that extremely high temperatures (of the order of thousands of K) are reached in localized hot-spots in a liquid (while the bulk of the fluid remains at low (i.e. room) temperature and pressure, makes sonochemistry a very versatile technique. Some applications of sonochemistry for large scale chemical synthesis have been reported[2–4].

11.1 TREATMENT OF POLLUTED WATER

In order to meet the more demanding water quality standards for ground water decontamination and industrial wastewater discharge, there is continuous upgradation of water treatment technologies particularly in the field of advanced oxidation processes (AOPs). Two approaches are followed for the treatment of polluted water. These are:
— Oxidation of contaminants like chlorinated and aromatic hydrocarbons, dyes, surfactants, pesticides and herbicides etc.
— Adsorption of trace metals like Cu(II), Mn(II), Hg(II) and Cr (VI) on charcoal.

11.1.1 Oxidation of Contaminants

A variety of water contaminants like chlorinated and aromatic hydrocarbons, dyes, surfactants, pesticides, herbicides and chlorinated volatile contaminants like dichloromethane (DCM) and

11.2 ■ Alternate Energy Processes in Chemical Synthesis

trichloroethylene (TCE) and the non volatile azo dye methyl orange (MO) have been removed by sonochemical degradation. Ultrasonic frequencies in the range 100–1000 KHz have been shown to be more practical than the lower frequencies around 20 KHz. In case the lower frequency (20 KHz) is used, the transductors have to be periodically replaced.

There are a number of reports relating to the sonolysis of organic pollutants in water[5-12]. Volatile chlorinated hydrocarbons are most suited to sonochemical degradation, since they are readily concentrated within the vapor phase of the cavitation bubbles, where they react readily via the thermolytical cleavage of C-Cl bonds at a relatively low temperatures during cavitational bubble collapse[5-12]. The reaction rate for the depletion of these molecules increases with the liquid temperatures in proportion to their intrinsic vapour pressures. This indicates that high-temperature gas-phase pyrolysis reactions take place. The chlorinated methanes are not susceptable to reaction with most oxidants in aqueous medium. However, sonolysis seems to be a convenient alternative to standard oxidation procedures.

Non-volatile substrates can also be subjected to sonochemical degradation. In this case, the degradation is mediated by active species generated (in the thermolysis of solvent molecules inside the cavitation bubbles) particularly $\cdot OH$ (aq), $\cdot HO_2$ (aq) and $\cdot O_2^-$ (aq) radicals and H_2O_2 (aq)[11, 13, 14]. These species take part in low temperature oxidation process involving the target molecules. The concentration of active species increases with decreasing temperature. Thus, the sonolysis rates of nonvolatile compounds decrease with increasing temperature; This is opposite effect than that is observed with volatile solutes.

An piolet plant has been fabricated for the sonochemical purification of polluted water. Following reactions takes place during the sonochemical reaction.

$$CH_2Cl_2 \text{ (g)} \xrightarrow{\Delta} \cdot CClH_2 + \cdot Cl \text{ (g)} \quad (1)$$

$$C_2HCl_3 \text{ (g)} \xrightarrow{\Delta} \cdot C_2HCl_2 + \cdot Cl \text{ (g)} \quad (2)$$

$$MO \text{ (aq)} + \cdot OH \text{ (aq)} \longrightarrow \text{Colourless product} \quad (3)$$
(MO is azo dye methyl orange)

Reactions 1 and 2 take place in the gas-phase inside the cavitation bubbles while reaction 3 occurs in solution.

11.1.2 Adsorption of Trace Metals on Charcoal

The release of metal ions into the environment by several industries has caused serious problems. The recovery of metal ions from waste water has been attempted by several investigators. The main techniques that have been used include precipitation by lime, ion exchange, membrane processing and electrolytic methods. However, adsorption seems to be an economical alternative method for removing trace metals from water. Activated carbon has been effectively used[16-21] to remove low concentrations of metal ions from water and industrial effluents. After exhausting activated carbon is regenerated or replaced by fresh carbon. The first option is preferred in view of the high cost of the adsorbent. It has, however, been found more advantageous to use ultrasound[23] for the desorption of metal ions from activated carbon. Ultrasound waves of 16 KHz to 500 MHz frequency can be transmitted in the form of mechanical energy through any medium having elastic properties such as water, gas-saturated water and aqueous particulate suspensions. It is believed that the effects induced by ultrasonic waves in aqueous solution are attributed to acoustic cavitation, which leads to not only mechanical action between solid and liquid interfaces, but also to chemical effects. During ultrasonic irradiation, bubbles are nucleated and knocked in the liquid phase. High temperatures of the order of several thousand Kelvin and few hundred bars are produced for a few microseconds in the liquid phase during the adiabatic collapse of cavitation bubbles. Chemical reactions like the cracking of water into hydrogen and hydroxyl radicals are promoted under the extreme conditions of temperature and pressure. The impacts created by shock waves from cavitation in liquid-solid slurries produce high velocity interparticle collusion, which are sufficient to melt most metals.

On the basis of investigations[23] it has been found that ultrasonic irradiation improves the desorbed amount and the rate of desorption. The amount of desorbed metal ions increases with increasing temperature and the intensity of ultrasonic irradiation. Enhancement of the desorption rates is caused by adding ethanol to the desorption system. Also, acidification of the regenerating solution to pH 1 leads to increase of the adsorption of Cu (II). Ultrasound enhances surface diffusivity.

To sum up the metal ions in ground water or waste water are first adsorbed on activated carbon and then the adsorbed metals are desorbed from activated carbon in the presence of ultrasound. Especial instruments have been fabricated[7] for this purpose.

NOTES AND REFERENCES

1. T.J. Mason and J.P. Lorimer, *Sonochemistry: Theory, Applications and Uses of Ultrasound in Chemistry*, Wiley, New York, 1988.
2. L. Thompson and L. Dovaiswamy, *Ind. Eng. Chem.Rev.*, 1999, **38**, 1215–1249.
3. P.D. Martin and L.D. Ward, *Trans. Inst. Chem. Eng.*, 1992, **70A**, 296–303.
4. Anonymous, *Chem. Eng.* (London), 1990, **480**, 15.
5. I. Hua and M. Hoffmann, *Environ. Sci. Technol.*, 1996, **30**, 864–871.
6. H. Hung and M. Hoffmann, *J. Phys. Chem.*, A1999, **103**, 2734–2739.
7. L. Weavers, F. Ling, M. Hoffmann, *Environ. Sci. Technol.*, 1998, **32**, 2727–2733.
8. L. Weavers, N. Malmstadt, M. Hoffmann, *Environ. Sci. Technol.*, 2000, **34**, 1280–1285.
9. H. Destaillats, H. Hung and M. Hoffmann, *Environ. Sci. Technol*, 2000, **34**, 311–317.
10. J. Joseph, H. Destallars, H. Hung and M. Hoffmann, *J. Phys. Chem.*, A2000, **104**, 301–307.
11. H. Destaillarts, A. Colussi, J.M. Joseph and M.R. Hoffmann, *J. Phys. Chem.*, A2000, **104**, 8930–8935.
12. H. Destaillats, T.W. Alderson II and M.R. Hoffmann, *Environ. Sci. Technol.*, 2001, **35**, 3019–3024.
13. A. Colussi, L. Weavers and M. Hoffmann, *J. Phy. Chem.*, A1998, 102, 6927–6934.
14. I. Hua and M. Hoffmann, *Environ. Sci. Technol.*, 1997, **31**, 2237–2243.
15. H. Destaillats, T.M. Lesko, M. Knowlton, H. Wallace and M.R. Hoffmann, *Amer. Chem. Soc.*, 2006 in press.
16. M.F. Yardim, T. Budinova, E. Ekinel, N. Petrov, M. Razvigorova and V. Minkova. Removal in mercury (II) from aqueous solution by activated carbon obtained from furfural, *Chemosphere*, 2003, **52**, 835.
17. N.R. Bishnoi, M. Bajaj, N. Sharma and A. Gupta, Adsorption of Cr (VI) on activated Rice Husk Carbon and Activated Alumina, *Bioresour. Technol.*, 2004, **91**, 309.
18. M. Pasavento, A. Profumo, G. Alberti and F. Conti, Adsorption of lead (II) and copper (II) on activated carbon by complexation with surface functional groups, *Anal. Chem. Acta.*, 2003, **480**, 171.
19. M. Kobya, Removal of Cr (VI) from aqueous solutions by adsorption onto hazelnut shell activated carbon, Kinetic and Equilibrium Studies, *Bioresour. Technol.*, 2004, **91**, 317.

20. J.P. Chen and L. Wang, Characterisation of metal adsorption kinetic properties in Batch and Fixed-Bed Reactors. *Chemosphere*, 2004, **54**, 397.
21. D. Savova, N. Petrov, M.F. Yardim, E. Ekinci, T. Bodinova, M. Razvigorova and V. Minkova. The influence of the texture and surface properties of carbon adsorbents obtained from Biomass products as the adsorption of manganese ions from aqueous solution, *Carbon*, 2003, 41, 1879.
22. O. Hamdaoui, R. Djeribi and E. Naffrechoux, *Ind. Eng. Chem. Res.*, 2005, **44**, 4737–4744 and the references cited there in.

Part-III

SOLID STATE SOLVENT FREE ORGANIC SYNTHESIS

Part-III

SOLID STATE SOLVENT FREE ORGANIC SYNTHESIS

12

SOLID STATE ORGANIC SYNTHESIS AT ROOM TEMPERATURE

12.0 INTRODUCTION

There is distinct advantage of conducting organic reactions under solvent free conditions over conventional methods in terms of time, yield and relatively benign conditions[1-3]. Also, there is ease of manupulation and formation of clean products. Even the use of volatile solvents are avoided[2, 4].

Solid state organic reactions can either be performed at room temperature by simple mixing (mechanochemical mixing) the reactants or on slight warming.

Following are given some of the reactions of this type.

12.1 SYNTHESIS OF β-KETO SULFONES FROM KETONES

In organic synthesis, β-keto sulfones are an important class of compounds[5]. A number of useful compounds have been prepared using β-keto sulfones as intermediates. Some of such compounds include olefins[6], disubstituted acetylenes[6], vinyl sulfones[7], allenes[8] and polyfunctional 4H-pyrans[9]. β-keto sulfones are also precursors for optically active β-hydroxy sulfones[10]. β-keto sulfones are found to possess fungicidal activity[11].

β-keto sulfones are normally synthesised by the oxidation of β-keto-sulfides[12], reaction of sulfonyl chlorides with silyl enol ethers[13], reaction of diazo sulfones with aldehydes[14], alkylation of arene sulfinate salts with α-haloketones[15], acylation of alkyl sulfones, reaction of alkyl sulfones with N-acylbenzotriazoles[16] and the reaction of sulfonyl chloride with arylacetylenes[17]. This procedures require toxic substrates, multi step synthesis and more reactions time and the yields are moderate.

A one pot synthesis[18] of β-keto sulfones involve the condensation of ketones with relatively benign [hydroxy (tosyloxy) Iodo] benzene and sodium arene sulfinates in the presence of a phase transfer catalyst (tetrabutyl ammonium bromide, T BAB) and sodium benzene sulfinate. Intimate grinding of the above reaction mixture gave the β-keto sulfones in high yields (Scheme-1).

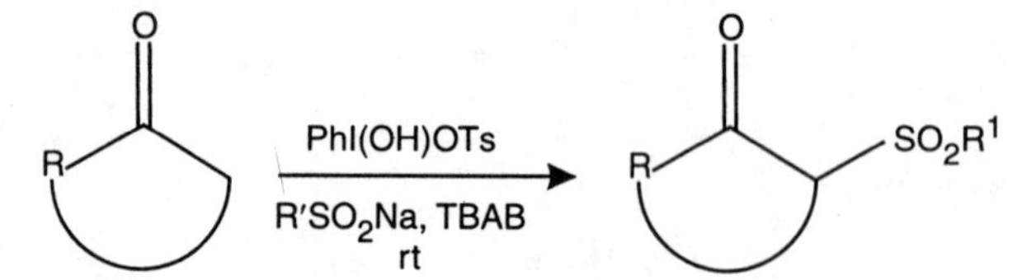

12.4 ■ Alternate Energy Processes in Chemical Synthesis

Scheme-1.

Using the above procedure, following β-keto sulfones were prepared (Table 1).

Table 1: Synthesis of β-keto sulfones from ketones using hydroxy (tosyloxy) iodobenzene[a]

Entry	Substrate	Product	Time (min)	Yield[b] (%)
1	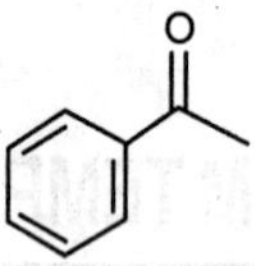	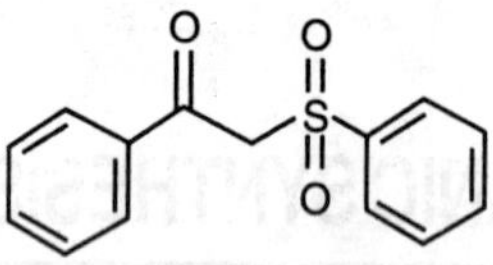	7	91
2	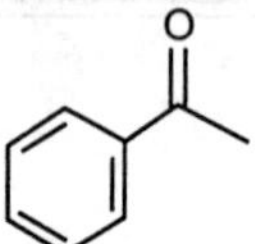	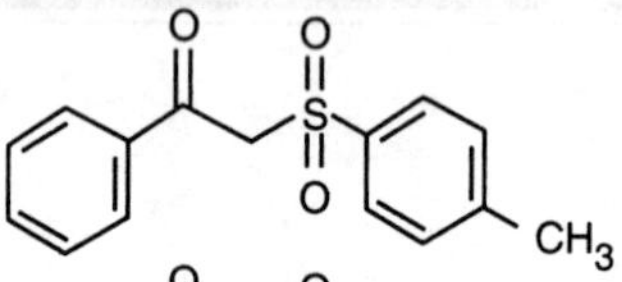	8	92
3	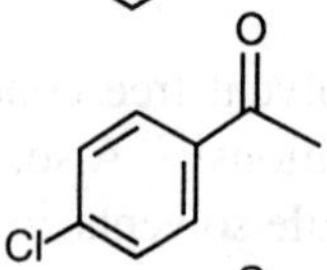	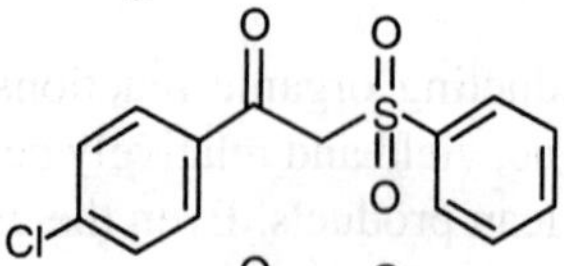	5	87
4	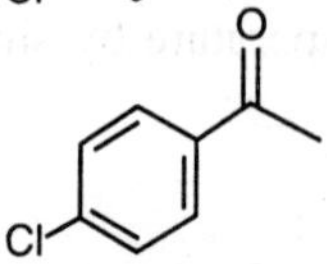	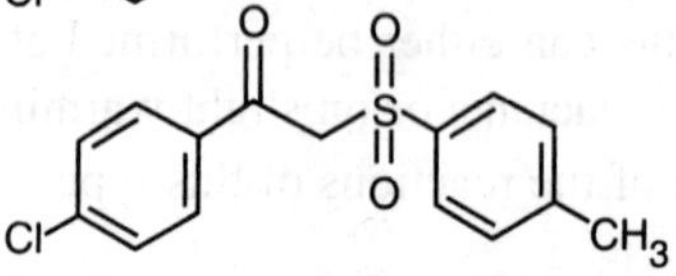	5	89
5	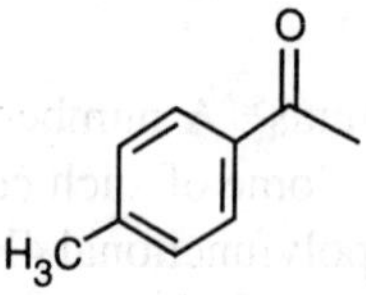	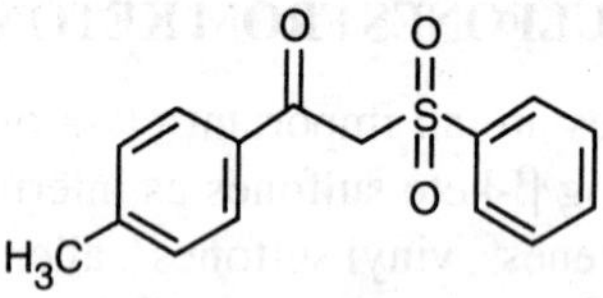	8	88
6	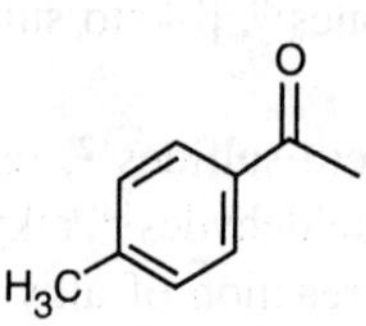	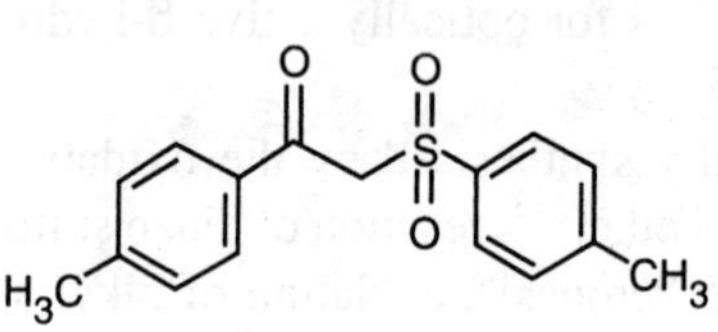	7	90
7	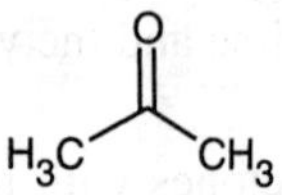	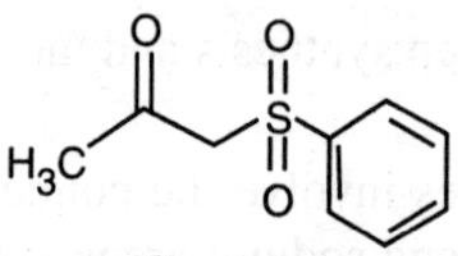	6	84
8	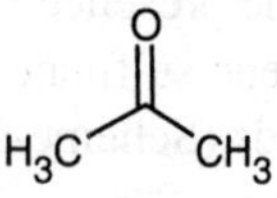	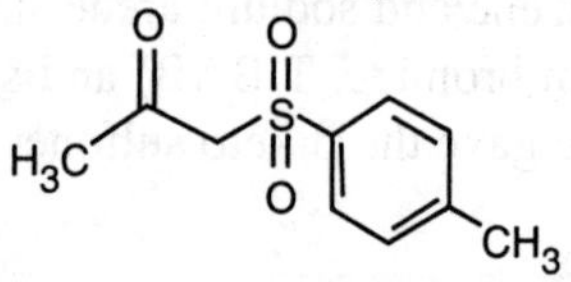	5	85

contd.

Entry	Substrate	Product	Time (min)	Yield[b] (%)
9	 cyclopentanone	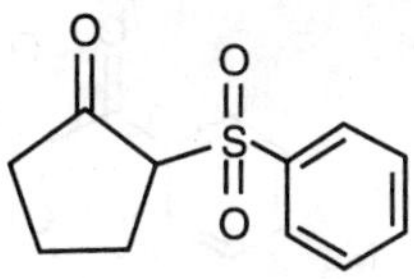 2-(phenylsulfonyl)cyclopentanone	6	82
10	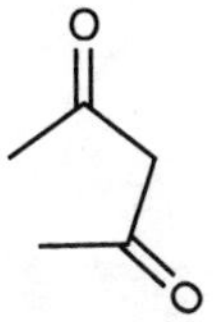 pentane-2,4-dione	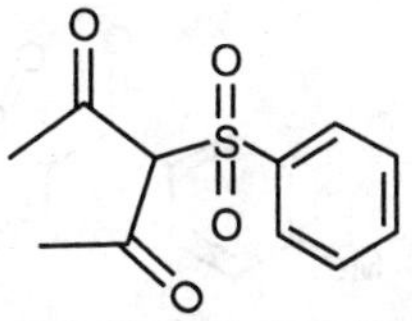 3-(phenylsulfonyl)pentane-2,4-dione	7	81

Note: Data taken from reference 18.

12.2 SYNTHESIS OF α-TOSYLOXY β-KETO SULFONES

α-Tosyloxy β-keto sulfones are potentially very useful precursors in organic synthesis, particularly for heterocyclic compounds[19]. Most of the reported procedures[20] for α-functionalization of β-keto sulfones required longer reaction times under refluxing conditions and give the required products in moderate yields. A simple procedure for the synthesis of α-tosyloxy β-keto sulfones consists in grinding together a neat mixture of α-benzenesulfonyl-acetophenone and [hydroxy (tosyloxy) iodo] benzene [H.T.I.B, commonly known as Koser's reagent] at room temperature using a pestle and mortar. Within, 5 min, the reaction gave the required α-tosyloxybenzenesulfonyl-acetophenone in 94% yield via the formation of a eutetic semi-liquid phase (Scheme-2).

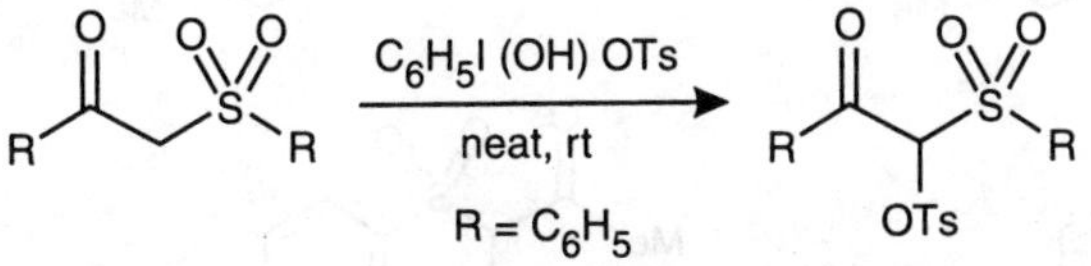

$$R\text{-CO-CH}_2\text{-SO}_2\text{-R} \xrightarrow[\text{neat, rt}]{C_6H_5I(OH)OTs} R\text{-CO-CH(OTs)-SO}_2\text{-R}$$

$R = C_6H_5$

Scheme-2.

Using this procedure, following α-tosyloxy β-keto sulfones were synthesised[20] (Table 2).

Table 2: Synthesis of α-tosyloxy β-keto sulfones using [hydroxy(tosyloxy)iodo] benzene

Entry	β-Keto sulfones	Product[a]	Time (min)	Yield[b] (%)
1	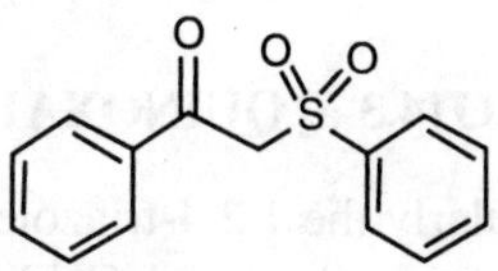 PhCOCH$_2$SO$_2$Ph	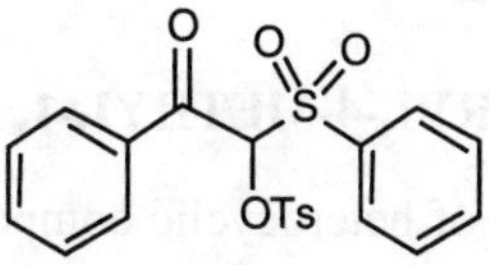 PhCOCH(OTs)SO$_2$Ph	7	94
2	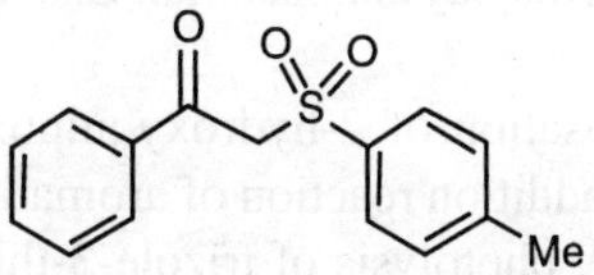 PhCOCH$_2$SO$_2$C$_6$H$_4$Me	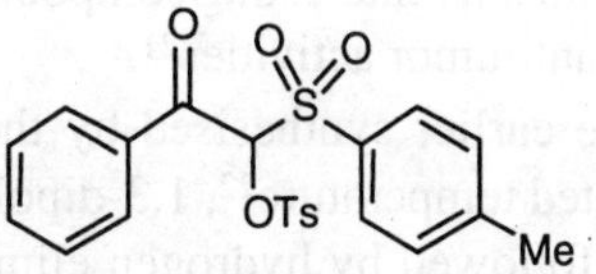 PhCOCH(OTs)SO$_2$C$_6$H$_4$Me	9	92

contd.

12.6 ■ Alternate Energy Processes in Chemical Synthesis

Entry	β-Keto sulfones	Product[a]	Time (min)	Yield[b] (%)
3			6	94
4			5	90
5			5	90
6			5	86
7			10	72
8			4	85
9			5	88

Note: Data taken from reference 20.

12.3 SYNTHESIS OF 1-ARYL-4-METHYL-1,2,4-TRIZOLO [4,3-a] QUINOXALINES

Triazoles, an important class of heterocyclic compounds, particularly the 1,2,4-triazole nucleus is an integral part of therapeutically interesting compounds which display antibacterial, CNS stimulative, sedative, antifungal and antitumor activities[21].

1,2,4-Triazoles were earlier synthesised by the condensation of 2-hydroxyquinoxaline with carboxylic acids at elevated temperature[22], 1,3-dipolar cycloaddition reaction of aromatic nitriles in presence of strong base followed by hydrogen elimination[23], photolysis of trizole-3-thiones[24] and oxidation of arylhydrazones[25]. Most of these methods involve multistep harsh reaction conditions, toxic reagents and require longer reaction times.

1-Aryl-4-methyl-1,2,4-trizolo [4,3-a] quinoxalines have been synthesised[26] by mixing (in a pestle mortar) of arenecarbaldehyde 3-methylquinoxalin-2-yl-hydrozones with iodobenzene diacetate (Scheme-3).

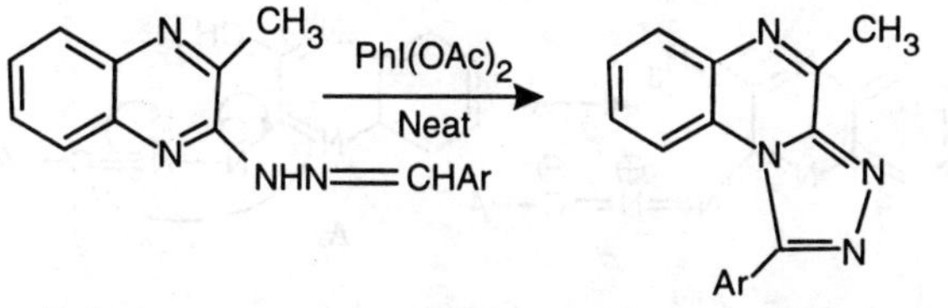

Ar = C_6H_5, p-$CH_3C_6H_4$, p-ClC_6H_4, p-$OCH_3C_6H_4$
m-$OCH_3C_6H_4$, p-$N(CH_3)_2C_6H_4$

Scheme-3

The above oxidative conversion involves a thorough mixing of the substrate with iodobenzene at room temperature (slightly warming in some cases) via an exothermic reaction. Hydrazone derivatives containing an electron-donoting group undergo ready conversion without warming. Following 1-aryl-4-methyl-1,2,4-triazolo[4,3-a]quinoxalines were obtained by this procedure (Table 3).

Table 3: Synthesis of some 1-aryl-4-methyl-1,2,4-triazolo[4,3-a]quinoxalines

Entry	Ar	mp/°C	Yield (%)[a]
2a	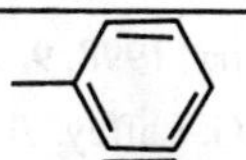 –C₆H₅	205–206 (203)	65
2b	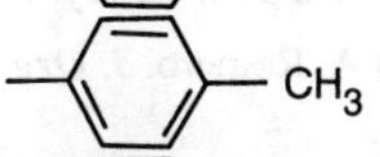 –C₆H₄–CH₃	212–213	76
2c	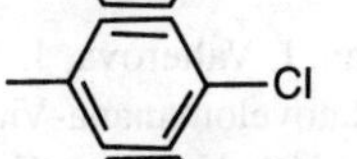 –C₆H₄–Cl	220–222	75
2d	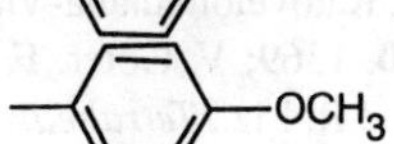 –C₆H₄–OCH₃	186–187	74
2e	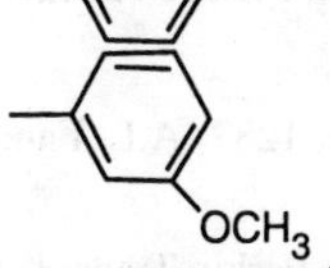 –C₆H₄(OCH₃)	168–187	72
2f	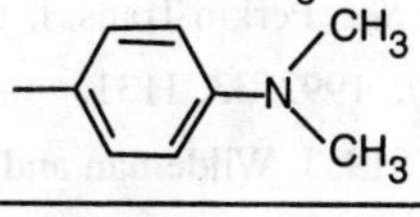 –C₆H₄–N(CH₃)₂	258–260	69

*Unoptimized yields are for the isolated products.
Data taken from Ref. 26.

A plausible pathway for this transformation is given in Scheme-4.

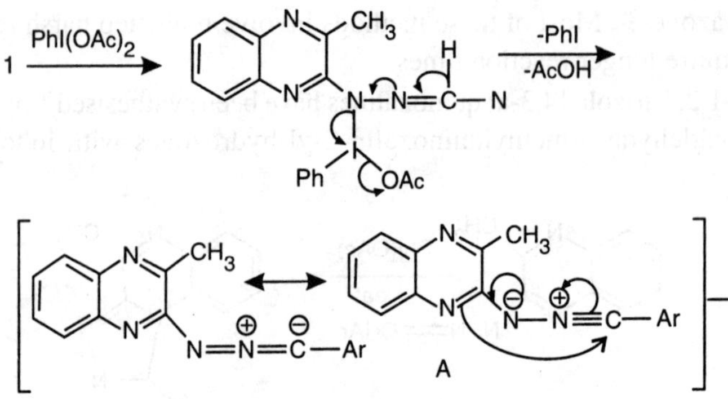

12.8 ■ Alternate Energy Processes in Chemical Synthesis

Scheme-4

NOTES AND REFERENCES

1. G. Rothenberg, A.P. Dournil, C.L. Raston and J.L. Scott, *J. Am. Chem. Soc.*, 2001, **123**, 8701.
2. F. Toda and K. Tanaka, *Chem. Rev.*, 2000, **100**, 102J.
3. K. Tanaka, Solvent Free Organic Synthesis, Wiley-VCH, Weinheim, 2003.
4. R.S. Varma, *Green Chem.*, 1993, 43;
   R.S. Varma, *Pure Appl. Chem.*, 2001, **73**, 193.
5. N.S. Simpkins, *In Sulfones in Organic Synthesis*, J.E. Baldwin, (Ed.), Pergamon Press; Oxford, 1993.
6. M. Ihara, S. Suzuki, T. Taniguchi, T.Yokunaga and K. Fukumoto, *Tetrahedron*, 1955, **51**, 9873; P.A. Bartlett, F.R. Green. III, and E.H. Rose, *J. Am. Chem. Soc.*, 1978, **100**, 4852; T. Mandai, T. Yanagi, K. Araki, Y. Morisaki, M. Kawada and J. Otera, *J. Am. Chem. Soc.*, 1984, **106**, 3670; B. Lythgoe and I. Waterhouse, *Tetrahedron Lett.*, 1978, **29**, 2625.
7. S. Sengupta, D.S. Sarma and S.Mondal, *Tetrahedron Asymmetry*, 1998, **9**, 2311.
8. J.E. Baldwin, R.M. Adlington, N.P. Crouch, R.L. Hill, and T.G. Laffey, *Tetrahedron Lett.*, 1995, **36**, 7925.
9. J.L. Marco, I. Fernandez, N. Khiar, P. Fernandez, and A. Romero, *J. Org. Chem.*, 1995, **60**, 6678; J.L. Marco, *J. Org. Chem.*, 1977, **62**, 6575.
10. A. Satos, Z. Hukova, V. Kren, M. Hoskovee, D. Syaman, I. Valterova, J. Verkoc, and B. Koutek, *Tetrahedron Asymmetry*, 1996, **7**, 1285; P. Berlus, P. Phansavath, V. Ratovelomanana-Vidal, J.P. Genet, A.R. Touati, T. Homri and B.B. Hassine, *Tetrahedron Asymmetry*, 1999, **10**, 1369; V. Goter, F. Rebolledo and R. Liz, *Tetrahedron Asymmetry*, 1999, **10**, 1369; V. Goter, F. Rebolledo and R. Liz., *Tetrahedron Asymmetry*, 2001, **12**, 513.
11. W.M. Wolf, *J. Mol. Struct.*, 1999, **474**, 113.
12. B.M. Trost and D.P. Curran, *Tetrahedron Lett.*, 1981, **22**, 1287; A.L. Fan, S. Cao and Z. Zhang, *J. Heterocycl Chem.*, 1997, **34**, 1657.
13. N. Kamigata, K. Udodaira and T. Shimizu, *J. Chem. Soc.* Perkin Trans. I, 1997, 783.
14. C.R. Holmquist and E.J. Roskamp, *Tetrahedron Lett.*, 1992, **33**, 1131.
15. G.E. Vennstra, and B. Zwaneburg, *Synthesis*, 1975, 519; J. Wildeman and A.M. Van Leusen, *Synthesis*, 1979, 733.
16. A.R. Katrizky, A.A.A. Abdel-Fattah and M. Wang, *J. Org. Chem.*, 2003, **68**, 1443.
17. C. Lai, C. Xi, Y. Jiang and R. Hua, *Tetrahedron Lett.*, 2005, **46**, 513.
18. Dalip Kumar, Swapna Sundaree, V.S. Rao and R.S. Varma, *Tetrahedron Lett.*, 2006, **47**, 4197.
19. K.F. Koser, A.G. Relenyl, A.N. Kalos, L. Rebrovic and R.H. Wettach, *J. Org. Chem.*, 1982, **47**, 2487; R.M. Moriarty, R.K. Vaid and G.F. Roser, *Synlett*, 1990, 365.

20. Dalip Kumar, M. Swapna Sundaree, G. Patel, V.S. Rao and R.S. Varma, *Tetrahedron Lett.*, 2006, **47**, 8239 and the references cited there in.

21. N.D. Heindel and J.R. Reid, *J. Heterocycl. Chem.*, 1980, **17**, 1087; B.S. Holla, B. Kalluraya, K.R. Sridhar, E. Drake, L.M. Thomas, K.K. Bhandry and M. Levmi, *Eur. J. Med. Chem.*, 1994, **29**, 301; M.A. Ghannoum, N.F. Ewiss, A.A. Bahajaj and M.A. Qureshi, *Microbios*, 1983, **37**, 151; S. Demirayak, K. Benkli and K. Guven, *Eur. J. Med. Chem.*, 2000, **35**, 1037.

22. D.I. Shiho and S. Tagami, *J. Am. Chem. Soc.*, 1960, **82**, 4044.

23. Z. Guolin and H. Yongzhou, *J. Chem. Res. (S)*, 2002, 560.

24. G. Jayanthi, S. Muthusamy, R. Paramasivam, V.T. Ramakrishnan, N.K. Ramasamy and P. Ramamurthy, *J. Org. Chem.*, 1997, **62**, 5766.

25. O.V. Singh and V. George, *Synth. Commun.*, 1944, **24**, 2627; D. Kumar, O. Prakash and S.P. Singh, *J. Chem. Res.(S)*, 1993, 244.

26. Dalip Kumar, V.G. Kondapalli, Chandra Sekhar, H. Dillon, V.S. Rao and R.S. Varma, *Green Chem.*, 2004, **6**, 156.

13

SOLID STATE ORGANIC SYNTHESIS BY SLIGHT WARMING OF THE REACTANTS

Following are given some of the reactions which can be conducted by slight warming of the reactants:

13.1 OXIDATION OF HYDROXYLATED ALDEHYDES AND KETONES TO HYDROXYLATED PHENOLS USING UREA-HYDROGEN PEROXIDE ADDUCT (UHP)

In organic synthesis, phenols and their derivatives are extensively used. The conversion of hydroxylated benzaldehydes to hydroxylated phenols has been achieved using alkaline hydrogen peroxide[1], although other oxidants such as peroxybenzoic acid[2] and peroxyacetic acid[3] have also been frequently used. However, solid-state oxidation of hydroxylated benzaldehydes with urea-hydrogenperoxide adduct is superior alternative in terms of shorter reaction time, cleaner product formation and ease of manupulation (Table 1).

Table 1. Solid-State Oxidation of Aldehydes and Ketones using UHP

Entry	Startin Material	Product	Reaction condition		Yield
			Temp. (°C)	Time	
1	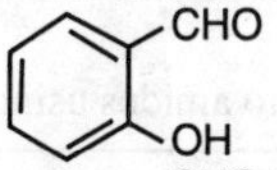 2-hydroxybenzaldehyde	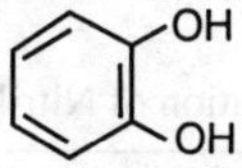 catechol	55	1.5h	85
2	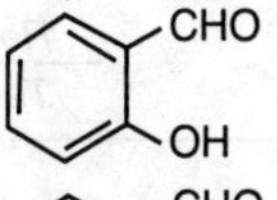 2-hydroxybenzaldehyde	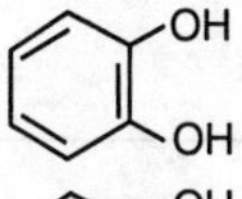 catechol	85	20 min	80
3	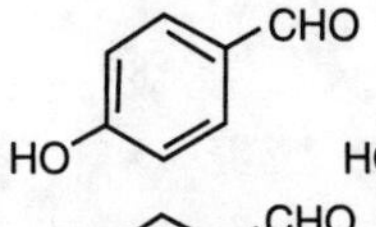 4-hydroxybenzaldehyde	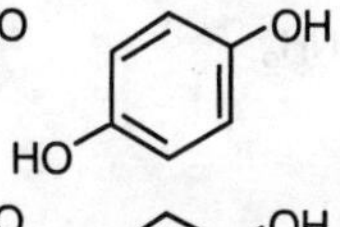 hydroquinone	55	3h	83
4	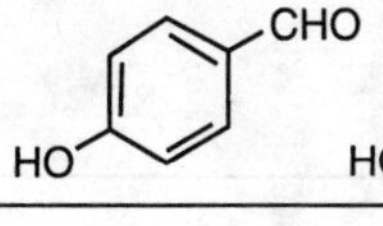 4-hydroxybenzaldehyde	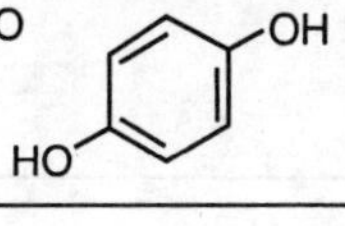 hydroquinone	85	75 min	82

contd.

13.2 ■ Alternate Energy Processes in Chemical Synthesis

Entry	Startin Material	Product	Reaction condition Temp. (°C)	Time	Yield
5	2-hydroxyacetophenone	catechol	85	1 h	86
6	4'-hydroxyacetophenone	hydroquinone	85	1 h	80
7	2-hydroxy-3-nitrobenzaldehyde	3-nitrocatechol	85	25 min	83
8	2-methoxybenzaldehyde	2-methoxyphenol	85	45 min	80
9	benzaldehyde	benzoic acid	85	20 min	95

Data taken from Ref. 4.

13.2 OXIDATION OF NITRILES TO AMIDES USING UPH

The hydrolysis of nitriles is generally used for the preparation of carboxylic acid amides. Usual methods for the hydration of nitriles involve the use of strong mineral acids[5], although some metals[6] and their oxides[7] and complexes[8] have also been used for the conversion of carbonitriles to amides. Some other reagents like titanium tetrachloride in acetic acid[9], potassium fluoride, on alumina[10], potassium hydroxide in tertiary butyl alcohol[11], sodium percarbonate in aqueous methanol[12] and some hydroxylamine derivatives[13] have also been used. The alkaline hydrolysis is not used, since it gives carboxylic acids. The base catalysed hydrolysis of nitriles was originally used by Rdziscewski, but it worked with aromatic nitriles, and the reaction gave only dismal yields in case of aliphatic nitriles[14]. Hydrogen peroxide in dimethyl sulfoxide[15] or under phase transfer conditions[16] has also been used. Many of these methods suffer from disadvantages of selectivity, longer reaction time and the use of hazardous reagents.

Eco-friendly hydrolysis of nitriles to amides has been achieved using UHP under solvent free conditions[4] (Table 2).

Table 2: Solid-State Oxidation of Nitriles to amides using UHP

Starting Material	Product	Reaction conditions Temp. (°C)	Time	Yield[a]
benzyl cyanide (PhCH$_2$CN)	phenylacetamide (PhCH$_2$CONH$_2$)	85	1 h	80
benzonitrile (PhCN)	benzamide (PhCONH$_2$)	85	1.5 h	85

Data taken from Ref. 4.

13.3 SELECTIVE OXIDATION OF SULFIDES TO SULFOXIDES OR SULFONES USING UHP

Selective oxidation of sulfides to sulfoxides or sulfones is a challenging task for synthetic organic chemist. Some of the oxidants that have been used for this conversion include hydrogen peroxide[17], chromic acid[18], nitric acid[19], manganese dioxide[18], ozone[20], peracids[21], selenium dioxide[22], sodium periodate[23], hypervalent iodine reagents[24], sodium perborate[25] and dinitrogen tetroxide[26]. Most of these procedures suffer from drawbacks like extended period of time, using corrosive acids, hazardous peracids and toxic metallic compounds.

A safer, viable protocol for the oxidation of sulphides to sulfoxides or sulfones is to use urea-hydrogen peroxide adduct (UHP) under solvent free conditions[4] (Table 3).

Table 3: Solid-State Oxidation of Sulfides using UHP

Entry	Startin Material	Product	Reaction condition Temp. (°C)	Time	Yield
1	Ph-SMe	Ph-SOMe	85	15 min	80
2	Ph-SMe	Ph-SOMe	85	1h	87
3	$[CH_3(CH_2)_3]_2S$	$[CH_3(CH_2)_3]_2SO$	85	10 min	85
4	$[CH_3(CH_2)_3]_2S$	$[CH_3(CH_2)_3]_2SO_2$	85	45 min	90
5	tetrahydrothiophene	tetrahydrothiophene S=O	85	7 min	87
6	tetrahydrothiophene	tetrahydrothiophene SO$_2$	85	1h	88
7	Ph-(CH$_2$)$_2$S	Ph-(CH$_2$)$_2$SO	85	30 min	87
8	Ph-(CH$_2$)$_2$S	Ph-(CH$_2$)$_2$SO$_2$	85	2h	90

Data taken from Ref. 4.

On the basis of the results obtained, it is seen that oxidation of sulfides to sulfoxides (entries 1,3,5 and 7) and sulphones (entries 2, 4, 6, 8) are summarised in Table 3 above.

13.4 OXIDATION OF NITROGEN HETEROCYCLES TO N-OXIDES USING UPH

Using UPH, nitrogen heterocyclic compounds could be oxidised to the corresponding N-oxides (Table 4).

13.4 ■ Alternate Energy Processes in Chemical Synthesis

Table 4: Solid-State Oxidation of Nitrogen heterocycles to N-Oxides using UHP

Entry	Starting Material	Product	Reaction conditions		Yield
			Temp. (°C)	Time	
1	pyridine	pyridine N-oxide	85	45 min	87
2	tetramethylpyrazine	tetramethylpyrazine N-oxide	85	45 min	92

Data taken from Ref. 4.

The reactions described have been conducted by slight warming of the reactants. It is, however, believed that all these reactions could be carried out using microwave irradiation for few seconds to few minutes.

NOTES AND REFERENCES

1. C.H. Hasall, *Org. React.*, 1957, **9**, 73.
2. Y. Ogata and Y. Sawala, *J. Org. Chem.*, 1969, **34**, 3985.
3. J. Boeseken, W.D. Cohen and C.J. Kip, *Recl. Trav. Chim. Pays-Bas*, 1936, **53**, 815.
4. R.S. Varma and K.P. Naicker, *Org. Letters*, 1999, **2**, 189.
5. R.E. Steiger, *Org. Synth.*, 1955, **3**, 66; J.H. Jones, *Comprehensive Organic Chemistry*, Pergaman Press, Oxford, U.K., 1979, Vol. 2, p. 964.
6. E.N. Zilberman, Usp. Khim, 1984, **53**, 1523; *Chem. Abstr.* 1984, **101**, 210082 **h**.
7. K.T. Liu, M.A. Shih, H.W. Huang and C.J. Hu, *Synthesis*, 1988, 715.
8. S.E. Diamond, B. Grant, G.M. Tom and H. Taube, *Tetrahedron Lett.*, 1974, 4025.
9. T. Mukaiyama, K. Kamio, S. Kobayashi, and H. Takei, *Chem. Lett.*, 1973, 357.
10. C.G. Rao, *Synth. Commun.*, 1982, **12**, 177.
11. J.H. Hall and M. Gisler, *J. Org. Chem.*, 1976, **41**, 3769.
12. G.W. Kabalka, S.M. Deshpande, P.P. Wadgaonkar and N. Chatta, *Synth. Commun.*, 1990, **20**, 1445.
13. T. Miyazawa, T. Endo and M. Okawarna, *Synthesis*, 1984, 1034.
14. B. Radziszewski, Ber. Dlysch. Chem. Ges., 1884, **17**, 1289.
15. V. Sawaki and Y. Ogata, *Bull. Chem. Soc. Jpn.*, 1981, **54**, 793; A.R. Katritzky, B. Pilarski and L. Urogdi, *Synthesis*, 1989, 949.
16. S. Cacchi and D. Misiti, *Synthesis*, 1980, 243.
17. M. Gazdar and S. Smiles, *J. Chem. Soc.*, 1908, **93**, 1833; D.A. Peak and T.I. Watkins, *J. Chem. Soc.*, 1950, 445.
18. D. Edwards and J.B. Stenlake, *J. Chem. Soc.*, 1954, 3272; R. Knoll, *Prakt. Chem.*, 1926, **113**, 40.

19. C. Marcker, Justus Liebigs *Ann. Chem.*, 1865, **136**, 75; F.G. Bordwell and P.J. Boutan, *J. Am. Chem. Soc.*, 1957, **79**, 717.
20. L. Horner, H. Schaefer and W. Ludwig, *Chem., Ber.*, 1958, **91**, 75; D. Baranard, *J. Chem. Soc.*, 1957, 4547.
21. C.G. Overberger and K.W. Cummins, *J. Am. Chem. Soc.*, 1953, **75**, 4250.
22. N.N. Melnikov, Usp. Khim, 1936, **5**, 443.
23. R.G. Hiskey and M.A. Harpold, *J. Org. Chem.*, 1967, 32, 3191.
24. D.H.R. Barton, G.R.A. Godfrey, J.W. Morzycki, W.B. Motherwell and A. Stobie, Tetrahedron Lett. 1982, **23**, 957.
25. A. McKillop and J.A. Tarbin, *Tetrahedron Lett.*, 1983, **24**, 1505.
26. R.W. Whitaker and H.H. Sisler, *J. Org. Chem.*, 1960, **25**, 1038.

14

MISCELLANEOUS REACTIONS

Following are given some miscellaneous reactions which have been conducted in solid state without the use of any solvent.

14.1 ALDOL CONDENSATION

The aldol condensation of lithium enolate of methyl 3,3-dimethylbutanoate with aromatic aldehyde gives[1] 8:92 mixture of syn and anti products in 70% yield (Scheme-1).

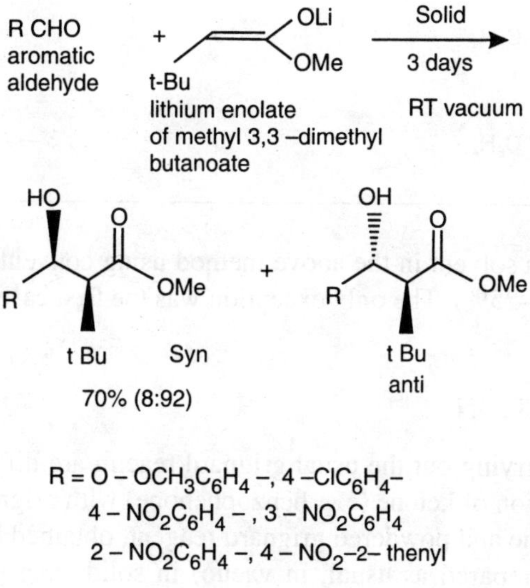

R CHO
aromatic aldehyde

+

t-Bu lithium enolate of methyl 3,3–dimethyl butanoate (OLi, OMe)

→ Solid, 3 days, RT vacuum

Syn product (HO, R, tBu, C(O)OMe) + anti product (OH, R, tBu, C(O)OMe)

70% (8:92)

$R = 4-OCH_3C_6H_4-, 4-ClC_6H_4-$
$4-NO_2C_6H_4-, 3-NO_2C_6H_4$
$2-NO_2C_6H_4-, 4-NO_2-2-\text{thenyl}$

Scheme-1

The above reaction (Scheme-1) is carried out by mixing freshly ground mixture of the starting materials in vacuum for 3 days at room temperature.

14.2 ■ Alternate Energy Processes in Chemical Synthesis

In the absence of any solvent, some aldol condensations proceed[2] more efficiently and steroselectively. In this procedure, appropriate aldehyde and ketone and NaOH is ground in a pestle and mortar at room temperature for 5 min.

The product obtained is the corresponding chalcone. In this method, the initially formed aldol dehydrates easily to the chalcone in the absence of a solvent (Scheme-2).

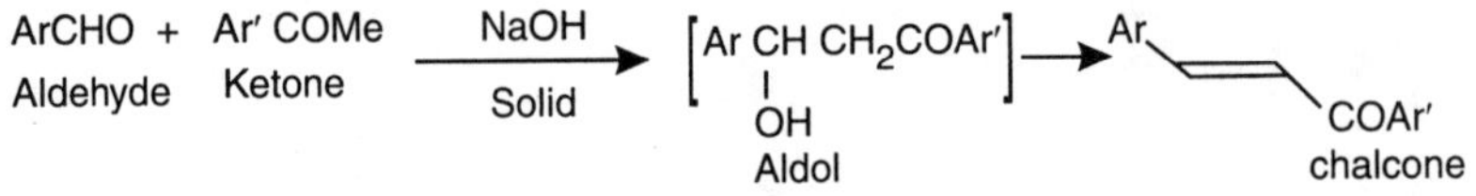

$$ArCHO + Ar'COMe \xrightarrow[\text{Solid}]{NaOH} [Ar\,CH(OH)\,CH_2COAr'] \longrightarrow Ar-CH=CH-COAr'$$
Aldehyde Ketone   Aldol   chalcone

Scheme-2

Using the above method, following chalcones were prepared:

Ar	Ar′	Reaction time (mm)	yield (aldol)	% yield of Chalcone
Ph	Ph	30	10	—
p-Me C$_6$H$_4$–	Ph	5	—	97
p-Me C$_6$H$_4$–	p-Me C$_6$H$_4$–	5	—	99
p-Cl C$_6$H$_4$–	Ph	5	—	98
p-Cl C$_6$H$_4$–	p-Me C$_6$H$_4$–	10	—	79
p-Cl C$_6$H$_4$–	p-Br C$_6$H$_4$–	10	—	81
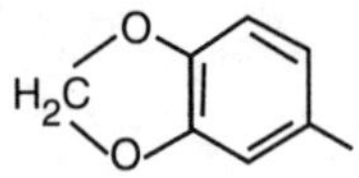 methylenedioxy-methylphenyl	p-Br C$_6$H$_4$–	10	—	91

The use of alcohol as a solvent in the above method using conventional procedure gives only the aldol in poor yields (10–25%). The only exception was the first case, where solid state reaction gave aldol in 10% yield.

14.2 GRIGNARD REACTION

The results obtained by carrying out the usual grignard reactin are different than that obtained in solid state[3]. Thus, the reaction of ketone (e.g. benzophenone) with grignard reagent (the reaction is carried out by mixing ketone and powdered grignard reagent, obtained by evaporating the solution of the grignard reagent, prepared as usual, in vacuo) in solid state gives more of the reduced product of the ketone than the alcohol (Scheme-3).

$$Ph_2O + R\,Mg\,x \xrightarrow[\text{Solid}]{0.5\,hr} \underset{\substack{\text{adduct}\\(A)}}{Ph_2RCOH} + \underset{\substack{\text{reduced product}\\(B)}}{Ph_2CHOH}$$

Scheme 3

Following results are obtained.

Grignard reagent RMgX		% products obtained in solid state	
R	X	(A)	(B)
Me	I	No reaction	
Et	Br	30	31
iPr	Br	2	20
Ph	Br	59	—

14.3 REFORMATSKY REACTION

Treatment of aromatic aldehydes with ethyl bromoacelate and Zn-NH$_4$Cl in the solid state gives[4] the corresponding Reformatsky reaction product (Scheme-4).

$$R\,CHO + Br\,CH_2CO_2Et \xrightarrow[\substack{\text{Solid State}\\3\,hr}]{Zn-NH_4Cl} \underset{80-90\%\text{ yield}}{RCH(OH)CH_2CO_2Et}$$

R = Ph, p-BrC$_6$H$_4$, 3,4 – methylenedioxyphenyl,

Ph – Ph,

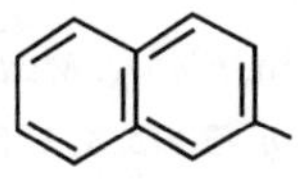

Scheme 4

14.4 POST SCRIPT

Quinoxaline and their derivatives are widely used in many fields, as curatorial intermediates, bacteriocides and insecticides[5]. Many synthetic methods for these heltrocycle compounds have been reported[5,6]. These procedures use catalysts and/or some special techniques[6]. Many of these methods are associated with a number of shortcomings such as long reaction times, expensive reagents, harsh conditions, low-product yields, formation of side products and difficulties in the recovery and reusability of the catalysts. An efficient synthesis of potential 1, 4-dihydro-quinoxaline-2, 3-diones has been reported[7] in a one-pot reaction at room temperature from substituted o-phenylene diamines and oxalic acid under solvent-free conditions by a simple grinding method at room temperature with unsupassed atom economy.

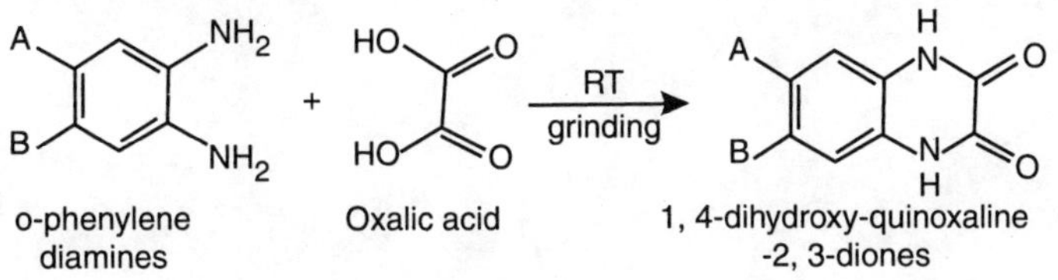

o-phenylene diamines + Oxalic acid $\xrightarrow[\text{grinding}]{RT}$ 1, 4-dihydroxy-quinoxaline -2, 3-diones

Scheme 5

14.4 ■ Alternate Energy Processes in Chemical Synthesis

Green Synthesis of quinoxaline derivatives

A	B	Time (b)	product obtained yield (%)
H	H	0.5	98
H	NO_2	3	82
H	Cl	2	87
H	Me	1	95
H	n-Pr	1	95
H	Ph	0.5	96
Cl	Cl	5	76

The yields refer to isolated products.
Data taken from reference 7.

NOTES AND REFERENCES

1. W. Wef, R. Bakthavatechalam, *Tetrahedron Lett.*, 1991, **32**, 1535.
2. F. Toda, K. Tanaka and K. Hamai, *J. Chem. Soc.* Perkin Trans. I, 1990, 3207.
3. F. Toda, H. Takumi and J.H. Yamaguchi, *Chem. Exp.*, 1980, **4**, 507.
4. H. Tanka, S. Kishigami and F. Toda, *J. Chem. Soc.*, 1991, **56**, 4333.
5. M.G. Moloney, *Nat. Prod. Rep.*, 2002, **19**, 597.
6. F.F. Jain and P.S. Zhao, 2004, *J. Mol. Struct.*, **705** 133;
   H.G. Cheon, C.M. Lee, B.T. Kimb and K.J. Hwangb, *Bioorg. Med. Chem. Lett.*, 2004, **14**, 2661;
   G. Kanpp and M.R. Naimi-Jamal, *Eur. J. Org.Chem.*, 2002, **8**, 1368;
   L. Wang, J. Liu and C. Qian, *Synth. Commun.*, 2004, **34**, 1349.
7. H. Thakuria and G. Das, *J. Chem. Sci.*, 2006, **118**, 425–428.

Part-IV

PHOTO INDUCED ORGANIC SYNTHESIS

15

PHOTO INDUCED ORGANIC SYNTHESIS

15.0 INTRODUCTION

We have known that for any chemical reaction or chemical transformation, certain amount of energy, known as activation energy has to be supplied to the molecules of the reactants. Besides direct heating, as already discussed, this energy can be supplied by microwaves or by ultrasound, commonly called sonication. However, it is some times possible to carry out a reaction by direct mixing of the reactants in the solid state or by slight warming.

Alternatively, the energy for a chemical reaction can be supplied photochemically. This involves absorption of electromagnetic radiation in the visible or ultraviolet region. Under these conditions, a molecule absorbs a quantum of light, the energy of which depends on the frequency of the radiation as expressed by the equation

$$E = h\nu = h\frac{c}{\lambda},$$

where h is the Plank's constant and c is the velocity of light. This energy is greater than the bond dissociation energy of a carbon-carbon σ bond (347 kJ mol^{-1}). Unlike thermal reaction, in photochemical reactions the absorption of light raises an individual molecule to an excited electronic state. In fact, the selective excitation of individual molecules is a special character of photochemical reactions. The chemistry of the excited molecules (in photochemical reactions) is quite different from the chemistry of the molecules in the ground state (in thermal reactions), even though the total energy supplied thermally is much more than what is introduced in photochemical process. Photochemical reactions, as we will see subsequently, have been used for synthesising highly strained, thermodynamically unstable compounds. Also the photochemical reactions are highly stereospecific. The products obtained by thermal and photochemical process normally differ in stereochemistry.

Photochemical reactions occur by absorption of electromagnetic radiation to produce electronically excited states. It is the electronically excited molecules which gives the product of

15.4 ■ Alternate Energy Processes in Chemical Synthesis

reactions. The process at raising molecules from ground state of minimum energy to an excited state of higher energy is called excitation. Excitation in the rotational, vibrational or electronic energy levels of molecules which result from absorption in microwave, infrared and ultraviolet region of the electromagnetic spectrum. The effects of various kinds of radiation on the molecules is given below:

	Type of radiation	Effect on absorbing molecules
increasing energy ↑	Ultraviolet Visible Infrared	Electronic excitation Vibrational excitation
	Microwaves and Radiowaves	Rotational

The ultraviolet radiation causes high energy electronic transitions in molecules. The resulting photo excited species can relax (react) via a variety of pathways. Traditionally, most photo reactions have been performed with UV lamp sources.

The electronic excitation is of great importance in organic photochemistry, although, it is also accompanied by increase in vibrational and rotational energies.

A chemical reaction produces an electronically excited species which emits a photon in order to reach to ground state, the phenomenon is known as **chemiluminescence**. When these type of reactions are encountered in biological system, the phenomenon is known as **bioluminescence**. The chemiluminescence is a rare phenomenon. An example of chemiluminescence is given below:

$$NO + O_3 \rightarrow NO_2^* + O_2$$
$$NO_2^* \rightarrow NO_2 + h\nu \ (\lambda = 600 - 2800 \text{ nm})$$

On the basis of what has been stated so far it can be said that thermal reactions occur through higher vibrational levels of the ground state, whereas, photochemical reactions take place through electronic excited states (singlet or triplet). Thermal reactions require less energy (63–209 kJ mol^{-1}). So light-induced or photochemical reactions are comparatively high energy processes. In photochemical reactions the light energy is absorbed by the reactants at room temperature (or even below room temperature). In case photochemical reactions lead to polymeric products due to high concentration of substrate, in such cases low concentration of substrate is used.

15.1 PHOTOCHEMICAL REACTIONS

The importance of photochemical reactions can hardly be overemphasised. these are an important tool in modern synthetic chemistry and lead to products virtually inaccessible by thermal reactions and proceed along the excited-state pathway. The earliest known photochemical natural reaction is the photosynthesis of sugars by plants using sun-light, CO_2 and H_2O in presence of chlorophyll. Some examples of common photochemical reactions are given below:

15.1.1 Photochemical Reactions of Some Carbonyl Compounds

Carbonyl compounds, particularly ketones undergo very interesting photochemical reactions. Some of these are discussed below:

15.1.1.1 *Photolysis of acetone (Propanone)*

The term photolysis is used when light absorption of molecules leads to cleavage of bond. Absorption of light by acetone leads to the formation of an excited state, which has sufficient energy to undergo cleavage of a C — C bond (the weakest bond in the molecule) and forms a methyl free radical and an acetyl free radical.

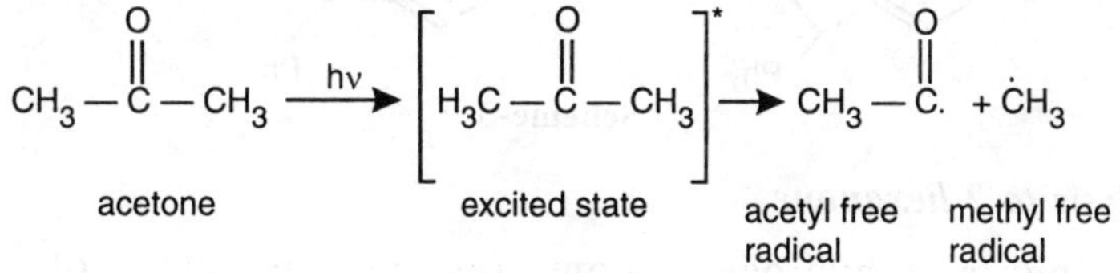

$$CH_3-\underset{\underset{\text{acetone}}{}}{\overset{O}{\overset{\|}{C}}}-CH_3 \xrightarrow{h\nu} \left[H_3C-\underset{\underset{\text{excited state}}{}}{\overset{O}{\overset{\|}{C}}}-CH_3 \right]^* \rightarrow \underset{\underset{\text{acetyl free radical}}{}}{CH_3-\overset{O}{\overset{\|}{C}}\cdot} + \underset{\underset{\text{methyl free radical}}{}}{\dot{C}H_3}$$

Scheme-1

At temperature much above room temperature the acetyl radical breaks down to give another methyl radical and carbon monoxide.

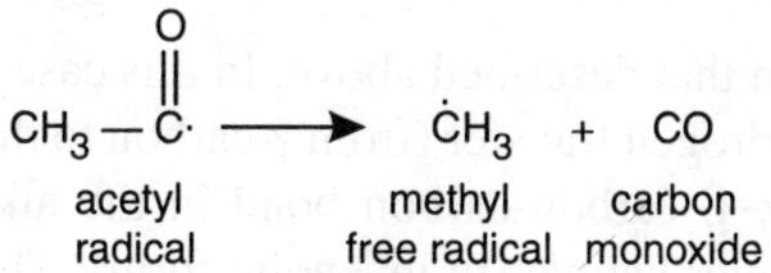

$$\underset{\underset{\text{acetyl radical}}{}}{CH_3-\overset{O}{\overset{\|}{C}}\cdot} \longrightarrow \underset{\underset{\text{methyl free radical}}{}}{\dot{C}H_3} + \underset{\underset{\text{carbon monoxide}}{}}{CO}$$

Scheme-2

The product obtained on completion of the reaction are ethane and carbon monoxide.

$$2\dot{C}H_3 \rightarrow CH_3-CH_3$$

In case of photolysis of 3-pentanone ($CH_3CH_2-\overset{O}{\overset{\|}{C}}-CH_2-CH_3$) the $CH_3\dot{C}H_2$ radical forms, which can also undergo disproporationation in addition dimerization.

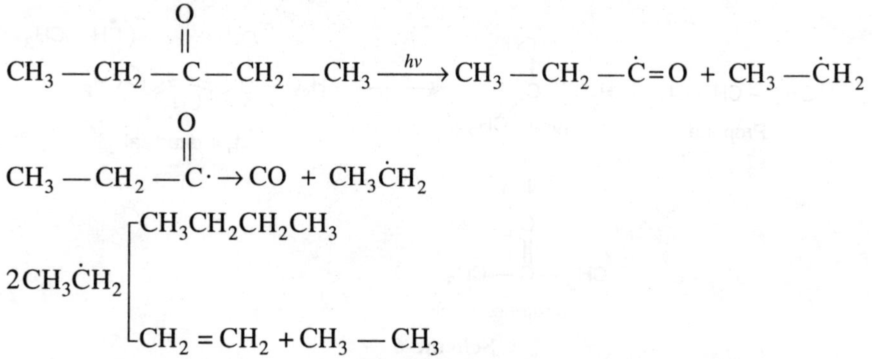

$$CH_3-CH_2-\overset{O}{\overset{\|}{C}}-CH_2-CH_3 \xrightarrow{h\nu} CH_3-CH_2-\dot{C}=O + CH_3-\dot{C}H_2$$

$$CH_3-CH_2-\overset{O}{\overset{\|}{C}}\cdot \rightarrow CO + CH_3\dot{C}H_2$$

$$2CH_3\dot{C}H_2 \begin{bmatrix} CH_3CH_2CH_2CH_3 \\ \\ CH_2=CH_2 + CH_3-CH_3 \end{bmatrix}$$

Scheme-3

In case of unsymmetrical ketones, splitting takes place in a way as to generate the more stable of the two possible radicals.

15.6 ■ Alternate Energy Processes in Chemical Synthesis

$$CH_3COCH_2CH_3 \xrightarrow{h\nu} CH_3\dot{C}O + \dot{C}H_2CH_3$$

$$CH_3COCH(CH_3)_2 \xrightarrow{h\nu} CH_3\dot{C}O + (CH_3)_2\dot{C}H$$

Scheme-4

Photolytic decarbonylation of cyclic ketones produces cyclic hydrocarbons.

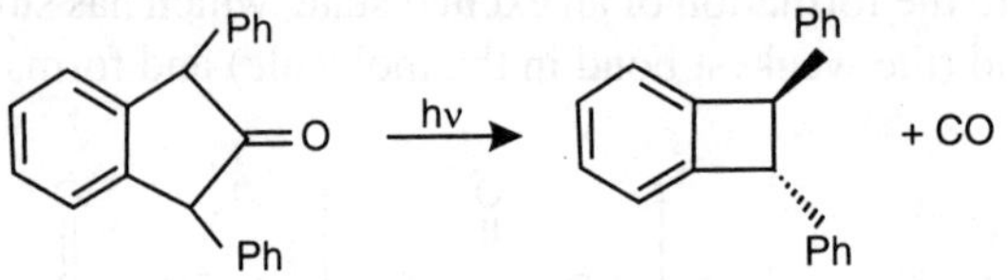

Scheme-5

15.1.1.2 Photolysis to 2-hexanone

Ketone possessing a γ-hydrogen atom (as in the case of a 2-hexanone,

$$CH_3 - \overset{O}{\underset{\|}{C}} - \overset{\alpha}{CH_2} - \overset{\beta}{CH_2} - \overset{\gamma}{CH_2} - CH_3)$$ undergoes an interesting photochemical reaction, the course of which is different than that described above. In this case, the excited state of 2-hexanone undergoes an intramolecular hydrogen transfer (from γ-carbon to the oxygen of the carbonyl group) with simultaneous fission of α–β carbon-carbon bond in the alkyl group. The products of this reaction is an alkene and a simpler ketone (in its enolic form). This process is known as **Norrish type II reaction**. This reaction proceeds by an initial γ-hydrogen abstraction by the oxygen atom yielding a 1,4-biradical, which subsequently gives an olefin and an enol.

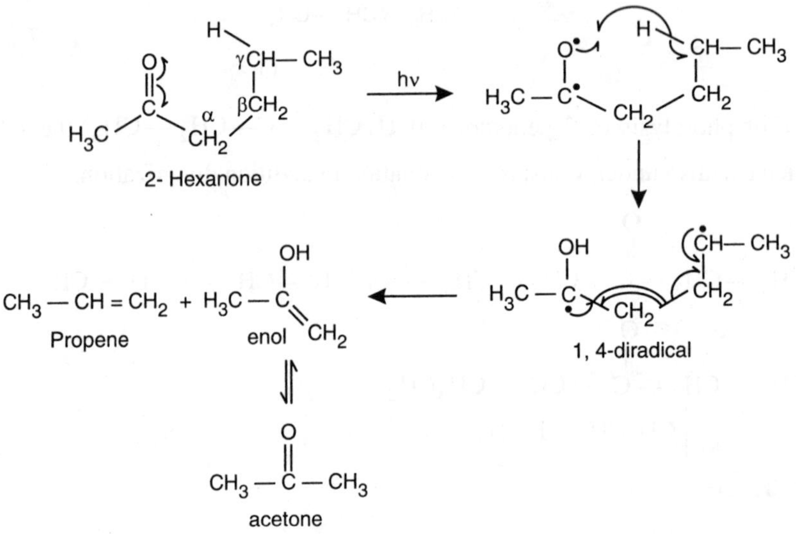

Scheme-6

The 1,4-diradical can undergoes ring closure to give cyclobutane derivative.

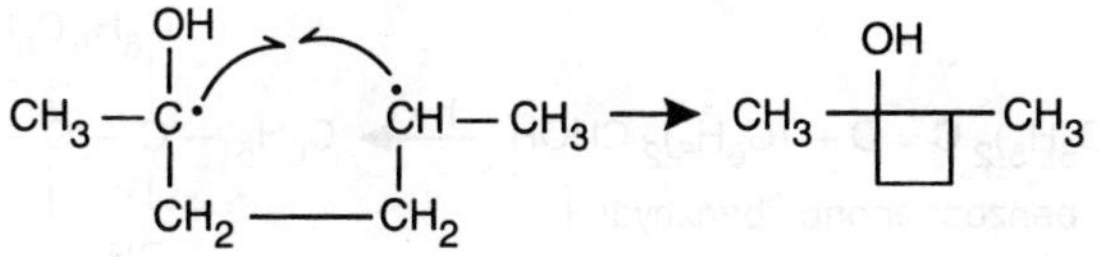

$$CH_3-\overset{OH}{\underset{\underset{CH_2 \longrightarrow CH_2}{|}}{\overset{|}{C}}}\overset{\curvearrowleft}{\cdot}\quad \overset{|}{\underset{}{C}H}-CH_3 \longrightarrow CH_3 -\hspace{-2mm}\underset{}{\boxed{}}\hspace{-2mm}- CH_3$$

Scheme-7

In the photolysis of 2-hexanone, Norrish type II cleavage is often accompanied by a Norrish type I process though to a lesser extent. Propanone, the product of photolysis of 2-hexanone can undergo Norrish type I reaction and finally a complex mixture of products consisting of propene, propanone, ethane and carbon monoxide are formed.

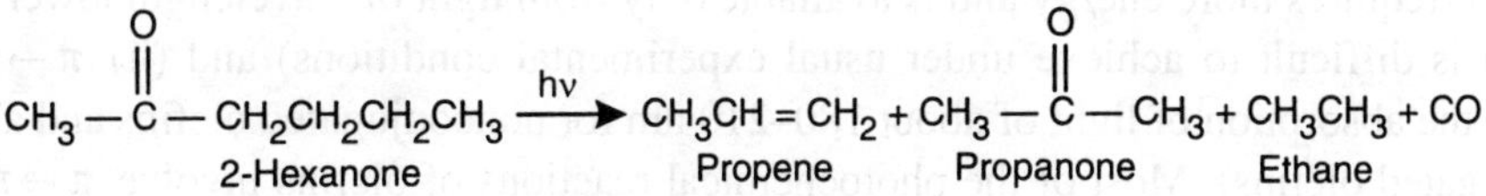

$$\underset{\text{2-Hexanone}}{CH_3-\overset{O}{\overset{\|}{C}}-CH_2CH_2CH_2CH_3} \xrightarrow{h\nu} \underset{\text{Propene}}{CH_3CH=CH_2} + \underset{\text{Propanone}}{CH_3-\overset{O}{\overset{\|}{C}}-CH_3} + \underset{\text{Ethane}}{CH_3CH_3} + CO$$

Scheme-8

Similarly, irradiation of 2-pentanone yields the following products:

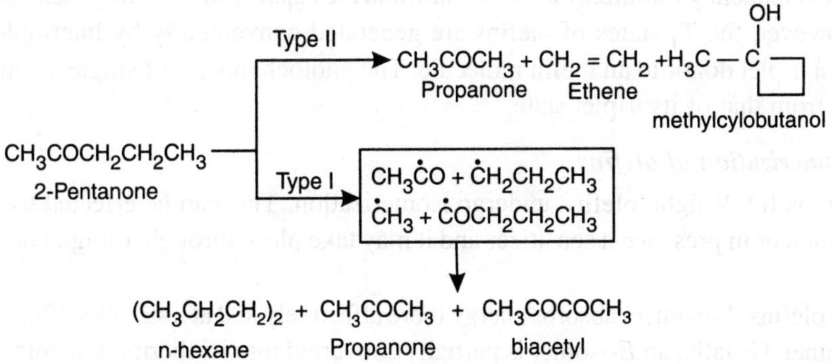

Type II → $CH_3COCH_3 + CH_2=CH_2 + H_3C-\overset{OH}{\underset{}{\overset{|}{C}}}\hspace{-2mm}\boxed{}$
Propanone Ethene
methylcylobutanol

$\underset{\text{2-Pentanone}}{CH_3COCH_2CH_2CH_3}$ — Type I → $\boxed{\begin{array}{c}CH_3\overset{\cdot}{C}O + \overset{\cdot}{C}H_2CH_2CH_3 \\ \overset{\cdot}{C}H_3 + \overset{\cdot}{C}OCH_2CH_2CH_3\end{array}}$

↓

$\underset{\text{n-hexane}}{(CH_3CH_2CH_2)_2} + \underset{\text{Propanone}}{CH_3COCH_3} + \underset{\text{biacetyl}}{CH_3COCOCH_3}$

Scheme-9

15.1.1.3 *Photolysis of benzophenone*

Photolysis of benzophenone (in sunlight) in presence of an alcoholic solvent (preferably isopropyl alcohol) gives benzopinacol in quantum yield unity.

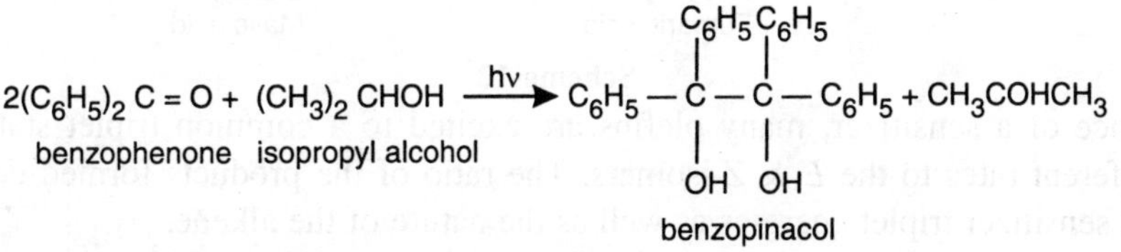

$$\underset{\text{benzophenone}}{2(C_6H_5)_2C=O} + \underset{\text{isopropyl alcohol}}{(CH_3)_2CHOH} \xrightarrow{h\nu} \underset{\text{benzopinacol}}{C_6H_5-\overset{C_6H_5}{\underset{OH}{\overset{|}{\underset{|}{C}}}}-\overset{C_6H_5}{\underset{OH}{\overset{|}{\underset{|}{C}}}}-C_6H_5} + CH_3COHCH_3$$

Scheme-10

It is well known that benzophenone by reduction with zinc and acetic acid gives benzopinacol. The photolysis reaction of benophenone is known as **photoreductive dimerization**.

The coupling reaction of benzophenone to benzopinacol is more efficient when the hydrogen donor is benzhydrol, the reduction product of the carbonyl compound (benzophenone).

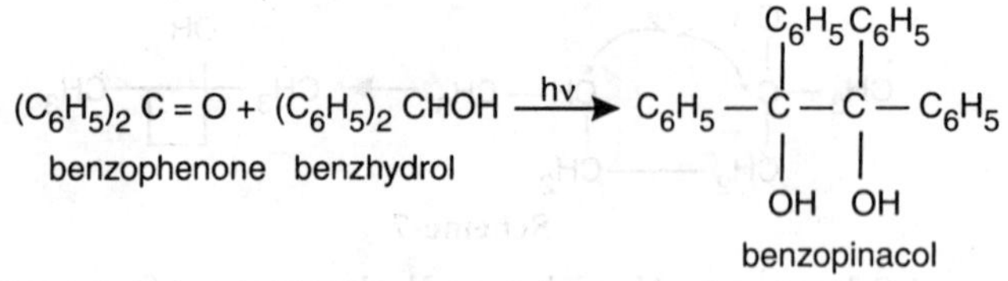

15.8 ■ Alternate Energy Processes in Chemical Synthesis

$$(C_6H_5)_2 C=O + (C_6H_5)_2 CHOH \xrightarrow{h\nu} \underset{\underset{\text{benzopinacol}}{\underset{OH\ \ OH}{|\ \ \ \ |}}}{\overset{\overset{C_6H_5\ C_6H_5}{|\ \ \ \ |}}{C_6H_5-C-C-C_6H_5}}$$

benzophenone benzhydrol

Scheme-11

15.1.2 Photochemical Reactions of Olefins

In case of olefins, photochemical reactions involves two types of electronic absorption (i) $\sigma-\pi^*$ excitation (requires more energy and is available only from light of wavelength lower than 150 nm and so it is difficult to achieve under usual experimental conditions) and (ii) $\pi \to \pi^*$ excitation (requires the absorption of light of about 180–210 nm for nonconjugated olefins and above 220 nm for conjugated olefins). Most of the photochemical reactions of olefins involve $\pi \to \pi^*$ excitation. The initial excitation generally occurs with no change in multiplicity and so a first singlet excited state is obtained. Also, this transition ($\pi \to \pi^*$) is **symmetry allowed**. The singlet excited states of olefins have less tendency for intersystem crossing and are capable of initiating many photochemical reactions. However, the T_1 states of olefins are generated conveniently by intermolecular energy transfer from a triplet donor to an olefin molecule. The photochemistry of singlet excited state of an olefin differs from that of its triplet state.

15.1.2.1 *Isomerization of olefins*

On irradiation with UV light, olefins undergo isomerization. This can be effected by irradiation of the olefin as such or in presence a sensitizer and it may take place through a singlet or triplet excited state.

In simple olefins, *E*-isomers absorbs energy more effectively and at a slightly different wavelength than the Z-isomer. Usually, an *E*-isomer is partially converted into its thermodynamically less stable Z-isomer. An example is the interconversion of fumaric and maleic acids.

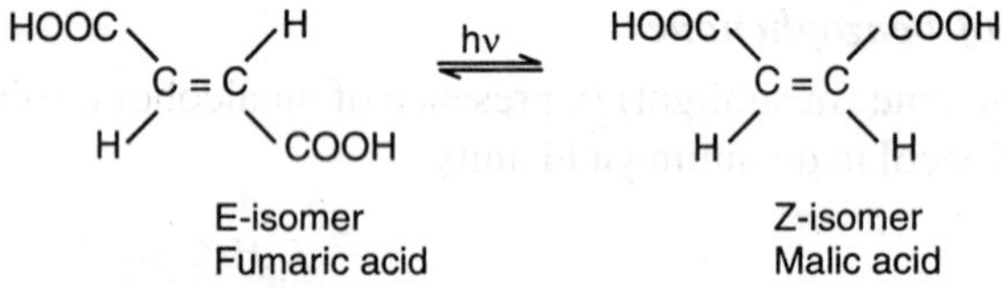

$$\underset{\underset{\text{Fumaric acid}}{\text{E-isomer}}}{\overset{HOOC}{H}}C=C\overset{H}{\underset{COOH}{}} \quad \underset{h\nu}{\rightleftharpoons} \quad \underset{\underset{\text{Malic acid}}{\text{Z-isomer}}}{\overset{HOOC}{H}}C=C\overset{COOH}{\underset{H}{}}$$

Scheme-12

In presence of a sensitizer, many olefins are excited to a common triplet state which then decays at different rates to the *E* & *Z*-isomers. The ratio of the products formed depends on the substrate and sensitizer triplet energies as well as the nature of the alkene.

A most commonly studied is the photoisomerization of stilbene (discussed in detail below).

15.1.2.2 *Photoisomerization of cis and trans-stilbene*

Olefines are known to exhibit geometrical isomerism. The photochemical *cis-trans* isomerization of stilbenes (1,2-diphenylethenes) provides the simplest case of light-induced geometrical isomerization. It is found that irradiation of trans stilbene in hexane in UV light results in the formation of the *cis*-isomer. After some time the *cis-trans* ratio becomes constant and does not change if irradiation is continued. This condition is called a **photostationary state** and is also reached on irradiation of the *cis*-isomer. The equilibrium favours the formation of thermodynamically less stable *cis* form. The *trans*-stilbene gives the *cis*- and *trans*-isomers in relative amounts of 10:1.

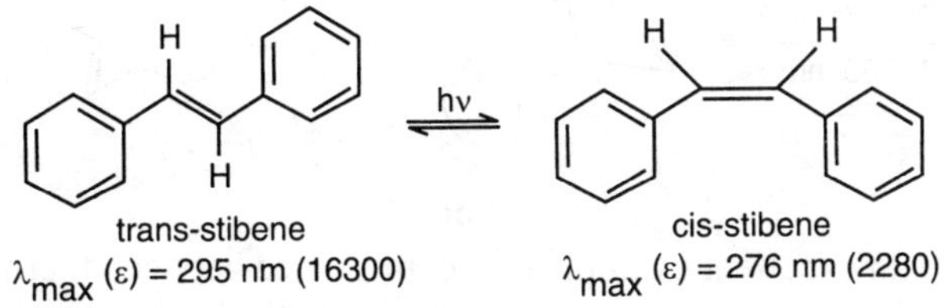

trans-stibene
λ_{max} (ε) = 295 nm (16300)

cis-stibene
λ_{max} (ε) = 276 nm (2280)

Scheme-13

15.1.3 Photochemical Cycloaddition Reactions

1,2- and 1,4-cycloadditions occur photochemically with or without sensitizers. Examples of both 1,2- and 1,4-cycloadditions are given below:

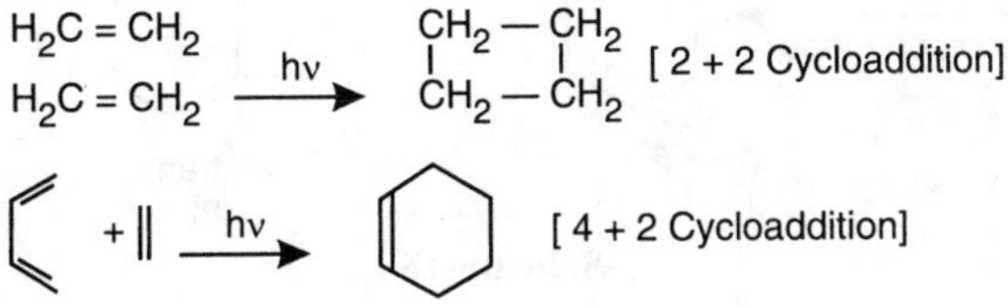

Scheme-14

Photochemical cycloaddition of olefins give four-membered ring is a synthetically useful process. One familiar example is the dimerization of cyclopentenone on irradiation with light in dichloromethane to give a mixture of 'head to head' and 'head to tail' dimers. These dimers may be formed via an excimer (excited dimer) derived from the ($\pi \to \pi^*$) cyclopentenone and a molecule of ground state cyclopentene.

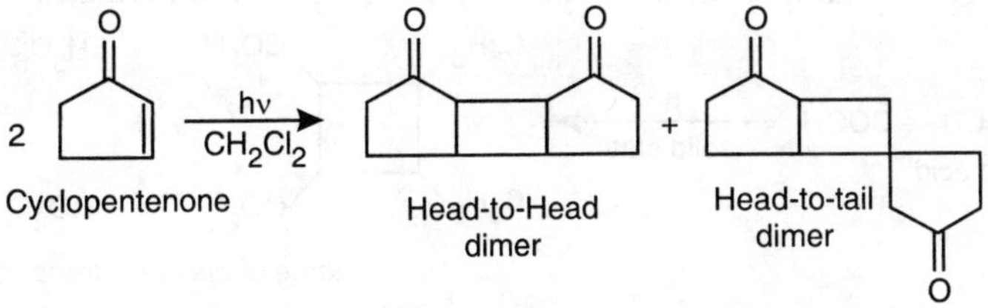

Scheme-15

The photocyclisation could also proceed in an intramolecular fashion as shown below.

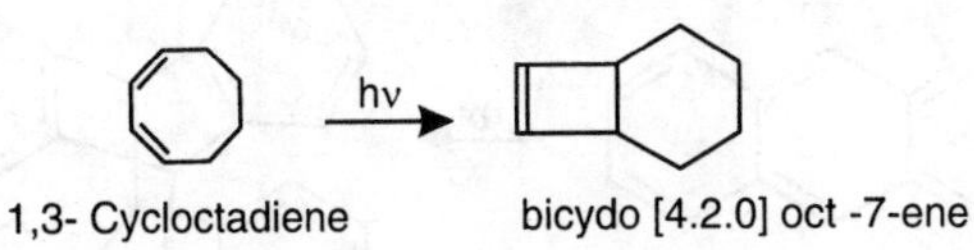

1,3- Cycloctadiene   bicydo [4.2.0] oct -7-ene

Scheme-16

15.10 ■ Alternate Energy Processes in Chemical Synthesis

Photodimerization of cyclopentadiene in the presence of benzophenone (sensitizer) gives mixture of products.

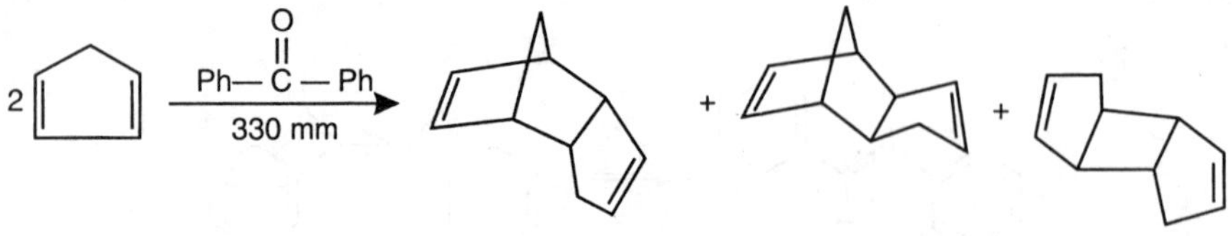

Scheme-17

Butadiene provides an excellent example of the difference between thermal and photochemical cycloadditions.

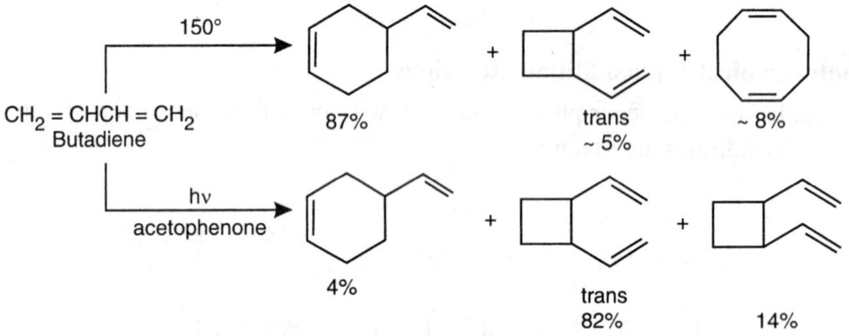

Scheme-18

It is found that the photochemical cycloaddition products of butadiene depend on the sensitizer used.

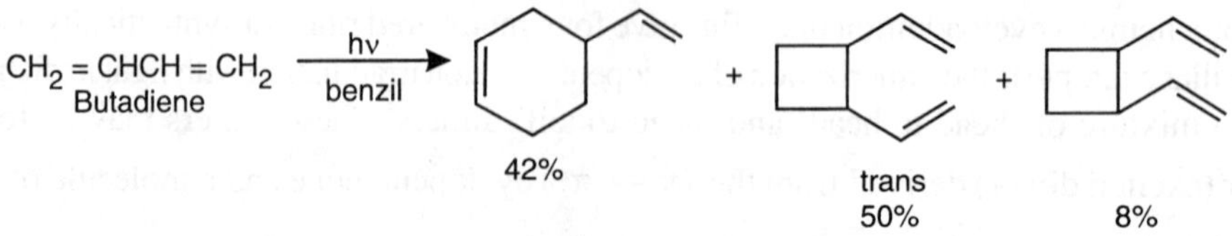

Scheme-19

Following are given examples of a number of photochemical dimerization reactions.

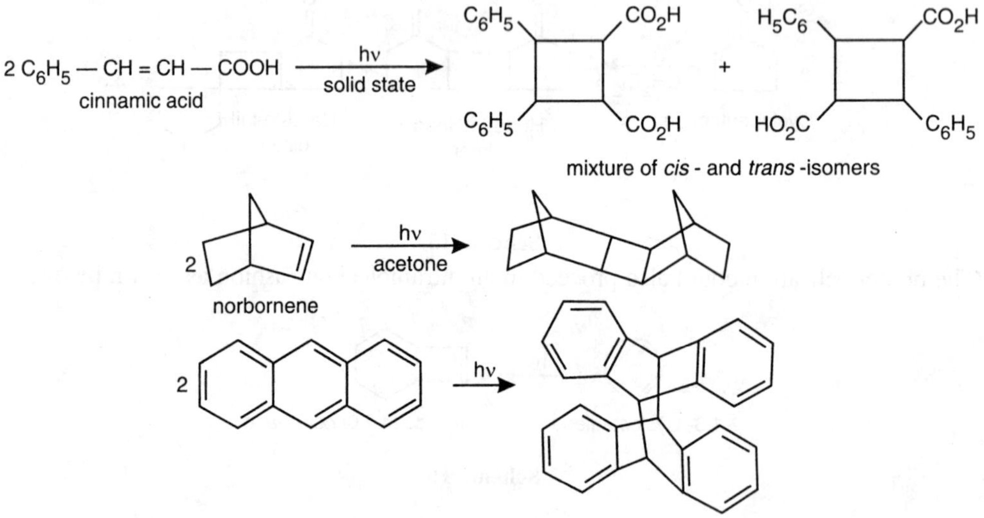

$2 C_6H_5-CH=CH-COOH$ (cinnamic acid) $\xrightarrow{h\nu, \text{ solid state}}$ mixture of cis- and trans-isomers

Scheme-20

Photocycloaddition may occur between an alkene or alkynes with carbonyl or aromatic compounds.

A very interesting photochemical cycloaddition reaction is the addition of carbonyl compounds to olefins to yield oxetanes (oxa-cyclobutanes).

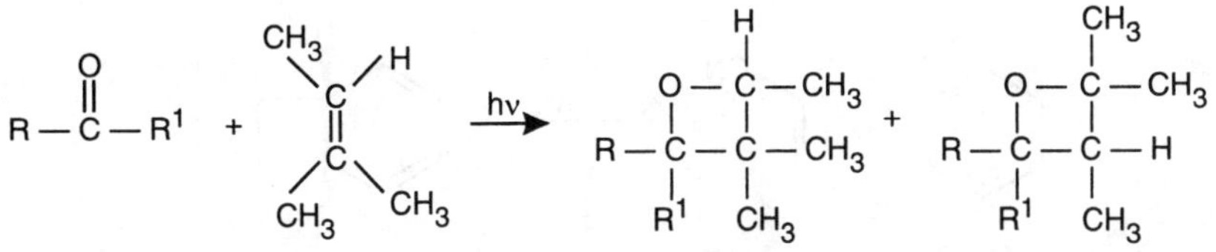

Scheme-21

This reaction is known as **Paterno-Büchi reactions**. For example, photocycloaddition of butyraldehyde to 2-methyl-2-butene yields mixture of 2,3,3-trimethyl-4-propyloxetane and 2, 2,3-trimethyl-4-propyloxetane.

$$CH_3-CH_2-CH_2-\overset{\overset{O}{\|}}{C}-H + (CH_3)_2C=CHCH_3$$

Butyraldehyde 2-Methyl-2-butene

$\downarrow h\nu$

2,3,3-Trimethyl-4-propyloxetane + 2,2,3-Trimethyl-4-propyloxetane

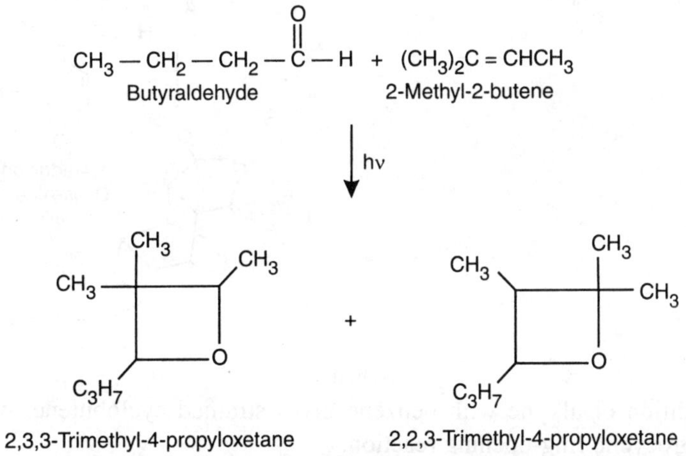

Scheme-22

The Paterno-Büchi reaction normally occurs by the cycloaddition of the triplet state of the carbonyl compound with the ground state of an alkene.

The photocycloaddition of benzophenone with *cis*- and *trans*-2-butene gives the same mixture of *cis*- and *trans*-oxetanes. This shows that the reaction is not stereospecific. The lack of stereochemical discrimination clearly shows that the reaction is not concerted and the ring is formed in two stages.

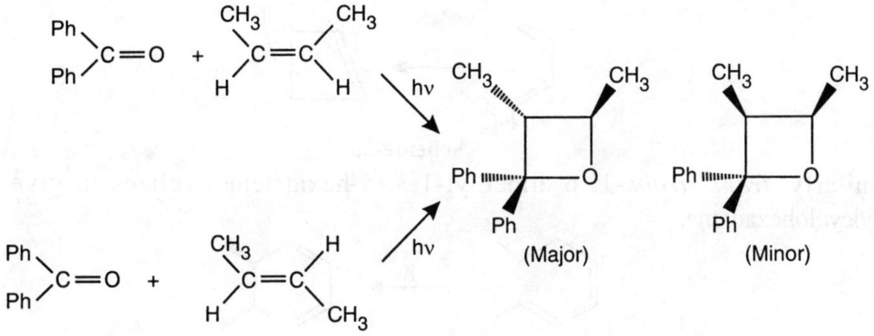

Scheme-23

15.12 ■ Alternate Energy Processes in Chemical Synthesis

The formation of mixture in both the cases (i.e. reaction of benzophenone with *cis-* and *trans-*2-butene) indicates that the time lag before the final spin-inversion is more than enough for rotation to occur about single bonds.

Examples of photochemical addition of alkene and alkynes with benzene are given below:

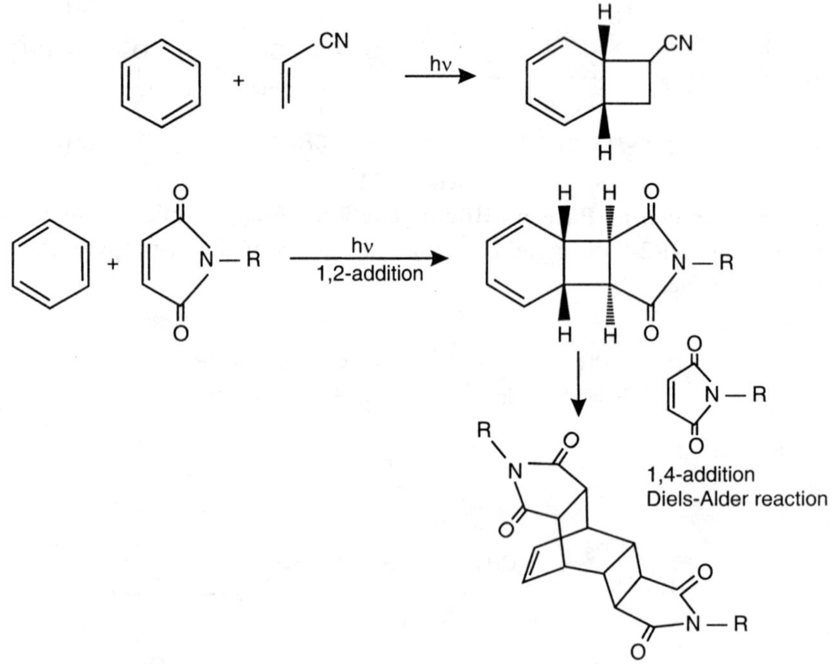

Scheme-24

1,2-Cycloaddition of alkyne with benzene gives strained cyclobutene, which undergoes a spontaneous electrocyclic ring-opening reaction.

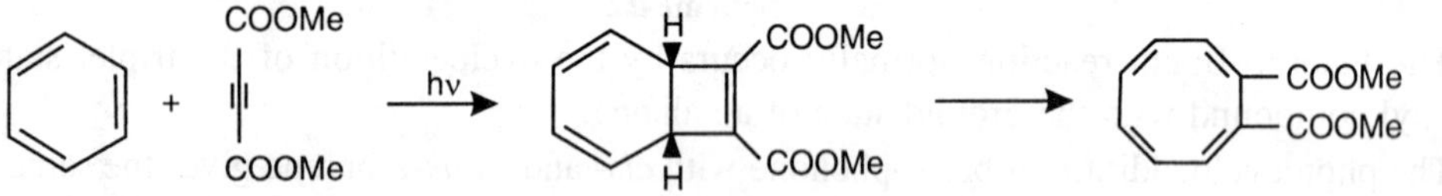

Scheme-25

The cycloaddition could also proceed in an intramolecular fashion. For example, irradiation of *trans, trans*-1, 4-dimethyl-1, 3-butadiene undergoes ring closure to give *cis*-3,4-dimethyl cyclobutene.

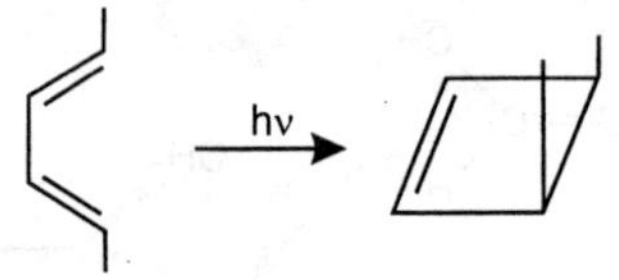

Scheme-26

Similarly, *trans, trans*-1, 6-dimethyl-1,3, 5-hexatriene cyclizes to give *trans*-5, 6-dimethylcyclohexadiene.

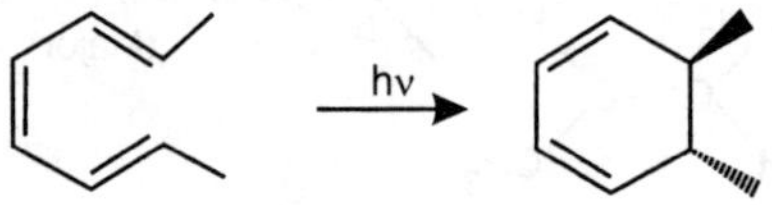

Scheme-27

Both the reactions are concerted and are electrocyclic type pericyclic reactions.

The intramolecular photocyclization reaction of 1,3-cyclooctadiene gives bicyclo [4 · 2 · 0] oct-7-ene.

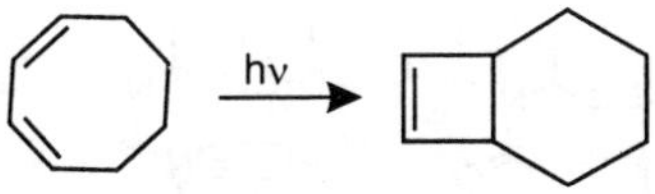

Scheme-28

However, when diene is not conjugated than a triplet sensitization is required as shown in the examples given below:

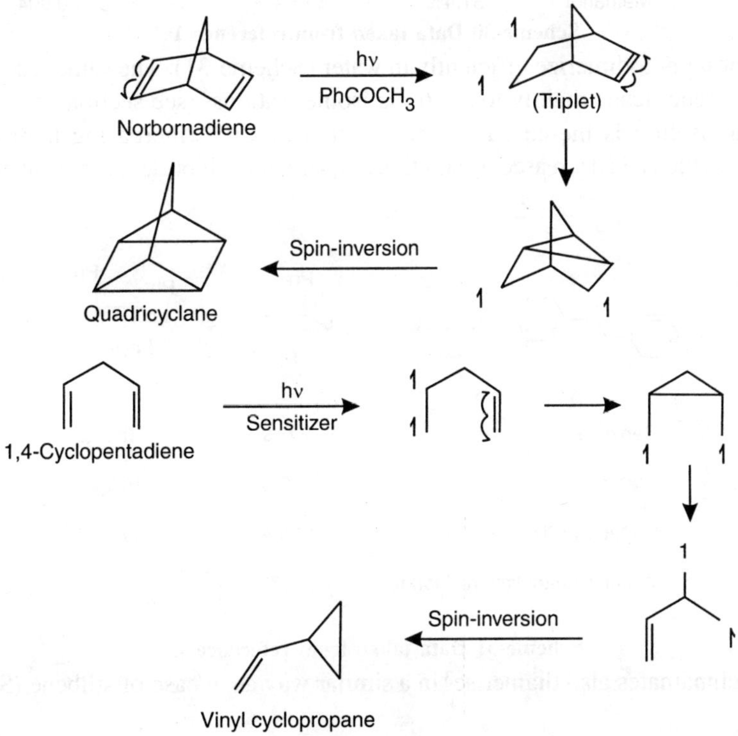

Scheme-29

There are numerous other examples. Most of the photochemical reactions are carried out in solvents like benzene. In view of the scare of the medium dependence of photochemical reactions, attempts were made for carrying out the reaction in water as a solvent and also in solid state.

15.1.4 Photochemical Cycloaddition Reactions in Water

A review of organic photochemistry in organized media, including aqueous, solvent has been reported by Ramamurthy[1]. Following are given some of the photochemical reactions performed in water.

(i) Photodimerisation of thymine, uracil and their derivatives increased considerably in water in comparison to other organic solvent. (Scheme-30). The increased quantum efficiency is due to the preassociation of reactants in the ground state.

15.14 ■ *Alternate Energy Processes in Chemical Synthesis*

Water	27.8%	63.1%	9.1%	∅ = 0.015
Acetontirile	24.9%	68.2%	6.7%	∅ = 0.0047
Methanol	31.4%	69.6%	—	∅ = 0.004

Scheme-30 Data taken from reference 1

(ii) Stilbenes, photodimerize efficiently in water (Scheme-31). The same reaction in solvents like benzene, leads mainly to *cis-trans* isomerisations[2] (see section 15.1.2.1). The yield of dimerisation is increased by the addition of LiCl (increasing hydrophobic effect. However, the yield decreased by addition of guanidium chloride (decreasing the hydrophobic effect.

Benzene	0%	0%
Water	12%	10%
Water + LiCl	25%	17%
Water + guanidinium chloride	8%	6%

Scheme-31 Data taken from reference 1

(iii) Alkyl cinnamates also diamerise[3] in a similar way as in case of stilbene (Scheme-32).

Scheme-32

(iv) Coumarin diamerised in water[4] more efficiently than in organic solvents (Scheme-33). The quantum yield of the dimerization in water is more than 100 times higher than in benzene and methanol.

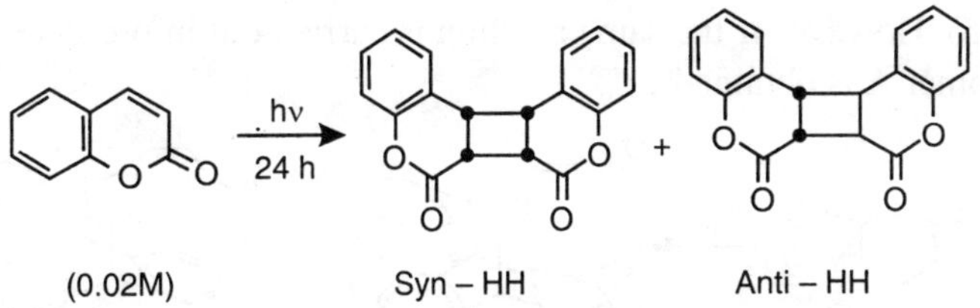

Solvent	Product	Quantum yield
Methanol	Anti – HH	< 10^{-5}
Benzone	Anti – HH	< 10^{-5}
Water	Anti – HH	2×10^{-3}

Scheme-33 Data taken from reference 4

(v) Photo oxidative dimerisation of capsaicin in aqueous ethanol gave[4a] 60% of the diamer within 20 min of irradiation (Scheme-34).

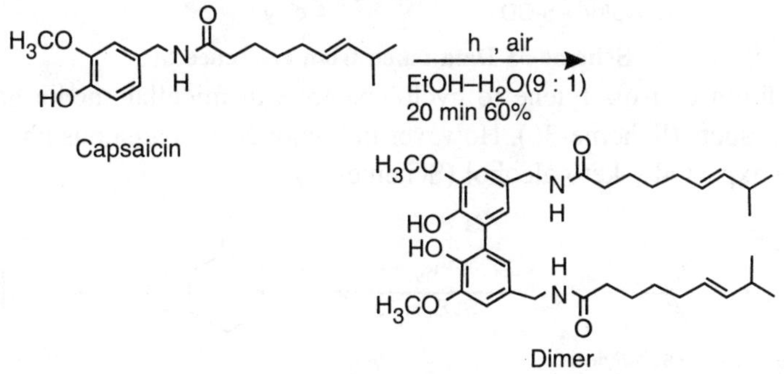

Scheme-34

(vi) Photooxidation of phenols is of interest in environmental chemistry. Phenols have been found to exist in all form of natural water. It is believed that singlet oxygen is involved in such oxidation process.

(vii) *Photo-fries rearrangement:*[4b] Phenolic esters in solution on photolysis give a mixture of *o*- and *p*-acylphenols. Photo Fries rearrangement unlike normal Fries rearrangement does not need a catalyst and is predominantly a intramolecular free radical process.

15.1.5 Photochemical Reactions in Micellar Media

In case a surfatant is added in water, it will aggregate in the formation of miscelles. It has been found[5] that the formation of such micelles has a significant effect on the regio- and stereoselectivity of photochemical reactions. In such a micellar case, the hydrophobic interior of miscelles provides a hydrophobic pocket within the bulk water solvent. A similar situation of hydrophobic pocket is to use cyclodextrin. Thus the selectivity in product formation can also be expected in this case also. Following are given some of the photochemical reactions in micellar media:

(i) *Photodimerization of anthracene-2-sulfonate*: The photodimerisation of anthracene-2-sulfonate in micellar media (water containing a surfactant) gives four products (A–D)

15.16 ■ *Alternate Energy Processes in Chemical Synthesis*

(Scheme-35). However if the same reaction is carried out in presence of β-cyclodextrin, only the isomer A is obtained[6].

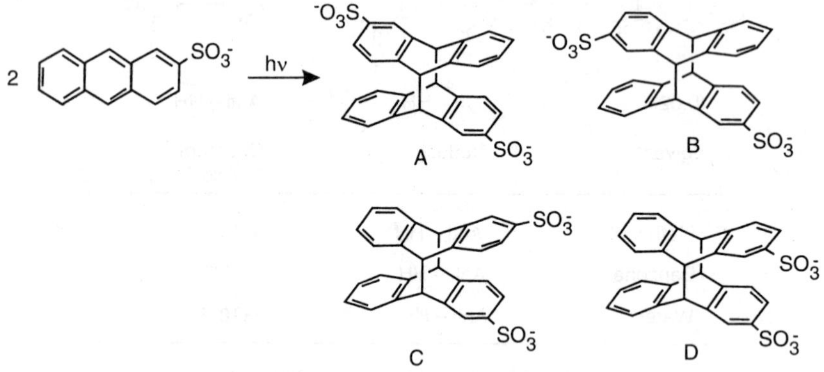

Medium	Product (A : B : C : D)
Water	1 : 0.8 : 0.4 : 0.05
Water + β-CD	A only

Scheme-35 Data taken from reference 6

(ii) Cycloaddition of isobutylene to cyclohexenone in micellar media gave[7] a mixture of bicycloproducts (Scheme-36). However in homogeneous aqueous phase, cyclohexenone gives the expected γ-keto alcohol (Scheme-36).

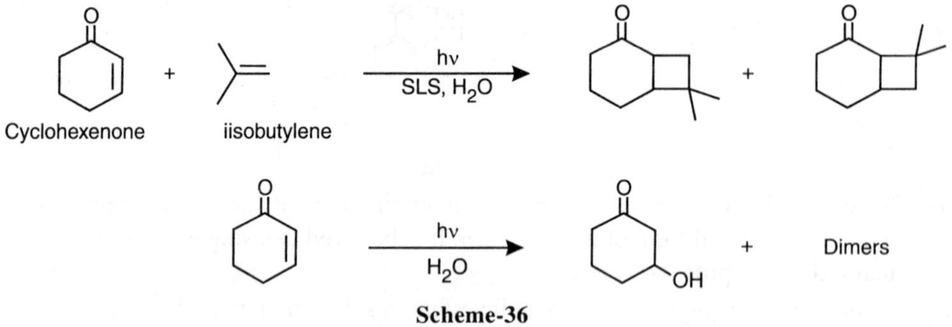

Scheme-36

(iii) The dimerization of isophorone has been found to be enhanced in micellar systems and microemulsions. The formation of the three products depends on the polarity of the solvent and the dominant phase respectively[5, 8] (Scheme-37).

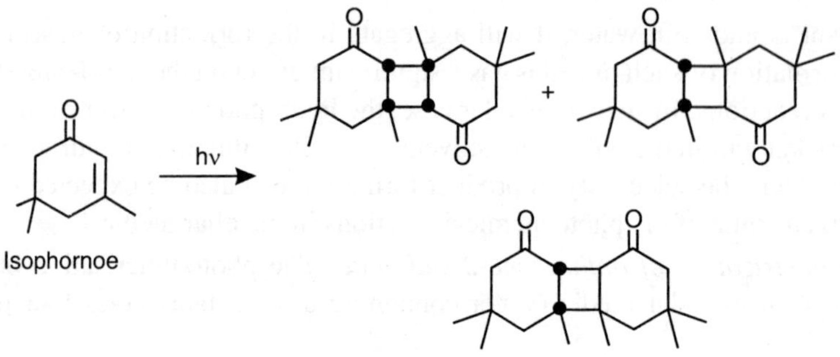

Scheme-37

(iv) 3-Alkylcylopentenone on photoexcition in solution of potassium dodecanoate (KDC), above the critical micelle concentration gave the dimerized product as in homogeneous solution such as benzene[9]. It is to be noted that the photodimerization in detergent solution occurs in a highly regiospecific manner; in the later case, the yield of the head to head (A) to head-to-tail (B) drimer is opposite to that observed in homogeneous solution (Scheme-38).

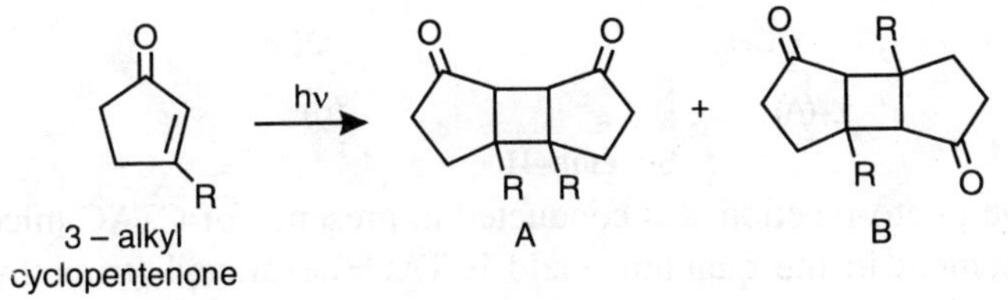

3 – alkyl cyclopentenone

R	solvent	A	:	B	yield
R = n– C_4H_9	KDC	98	:	2	100%
n– $C_{10}H_{21}$	C_6H_6	9	:	91	1–6%

Scheme-38

(v) Photodimerization of acenaphtylene in micelle containing solutions gives the dimeric product. Such products are not obtained in benzene[10] (Scheme-39).

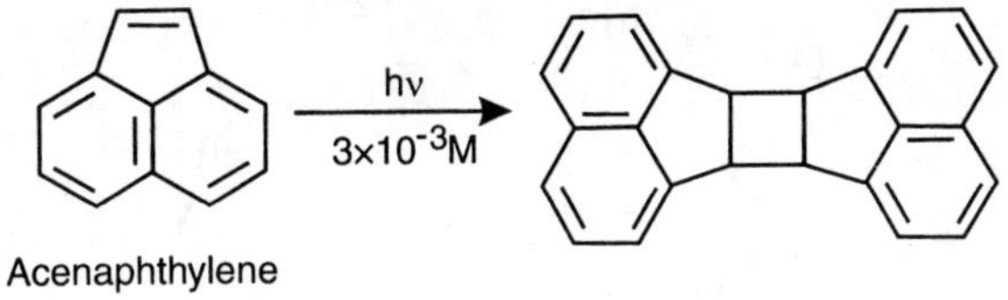

Acenaphthylene

hv
3×10^{-3}M

solvent	yield
SLS	96%
C_6H_6	0%

Scheme-39

(vi) In the photocrossed cycloaddition of acenaphthylene in acrylonitrile occurs in much higher yields in non-ionic detergent solution (e.g., in polyglycol ethers) (Scheme-40)[11].

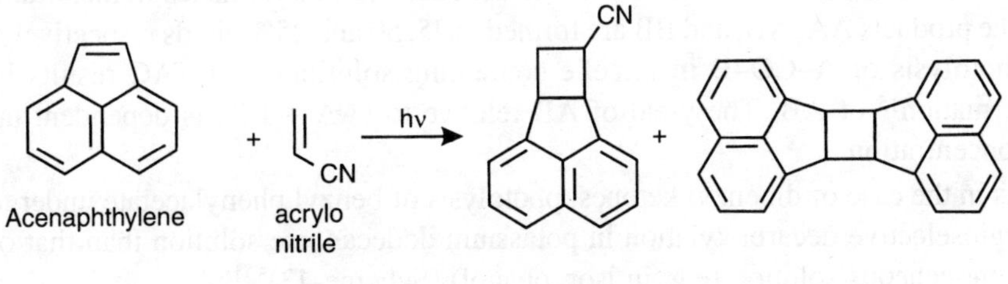

Acenaphthylene + acrylo nitrile

solvent	product 1	product 2
Nippol PBC – 34	30%	70%
C_6H_6	0%	30%

Scheme-40

(vii) Photoinduced substitution of aromatic compounds: Photoexcitation[12] of aqueous solution of 4-methoxy-1-nitronaphthalene (A) containing CN⁻ ions results in the formation of 4-methoxy-1-naphthalene carbonitrile (B) (Scheme-41).

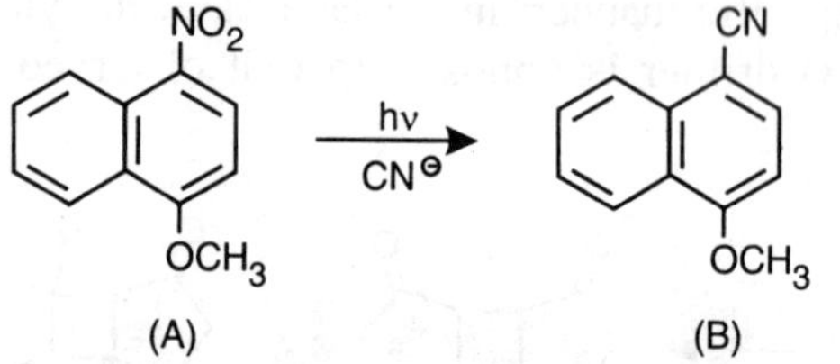

Scheme-41

When the above photo-reaction was conducted in presence of CTAC micelles there was a large enhancement in the quantum yield [CTAC=hexadecyl (trimethyl) ammonium bromide].

(viii) The photorearrangement of 4-nitrophenyl-nitromethane (A) into 4-nitrobenzaldehyde (B) was found to be much more efficient in a cationic detergent (CTAC) than in basic ethanol/water solution (Scheme-42)[13].

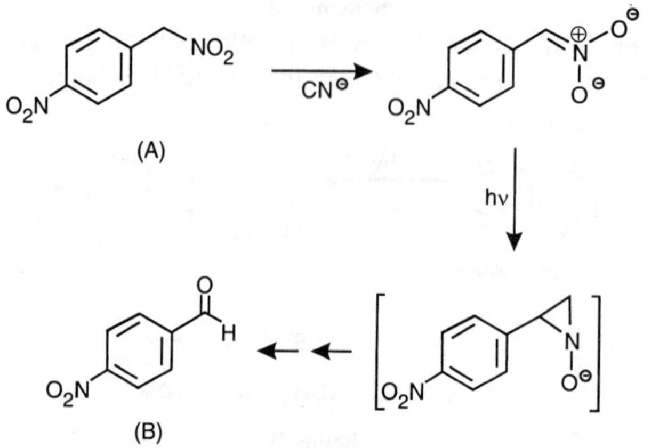

Scheme-42

(ix) Photodecarbonylation of unsymmetrical dibenzyl ketones A–CO–B in homogeneous solution occurs via a free radical mechanism to produce 1,2-diarylethanes in quantitative yield[14]. The products AA, AB, and BB are formed in 25, 50 and 25% yields respectively. However, photolysis of A–CO–B in micelle containing solutions of CTAC results in selective formation[15] of AB. The yield of AB relative to (AA + BB) is dependent upon CTAC concentration.

As in the case of dibenzyl ketones, photolysis of benzyl phenylacetate undergoes a more regioselective decarboxylation in potassium dodecanoate solution than that observed in homogeneous solution (e.g. in isopropanol) (Scheme-43)[5, 16].

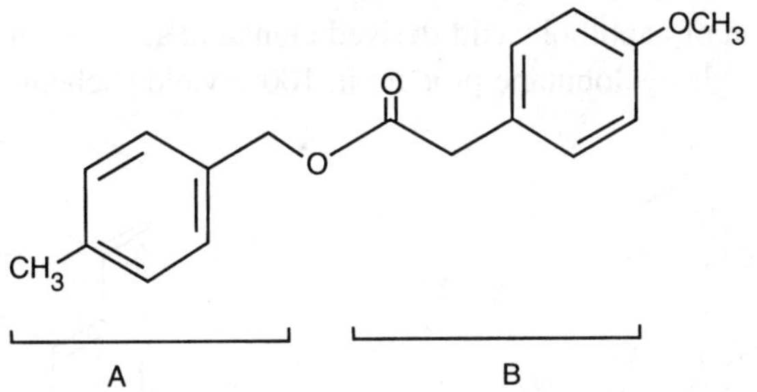

A – A	A – B	B – B	
~ 1	~ 50	~ 1	Detergent
~ 1	~ 5	~ 1	isopropanol

Scheme-43

15.1.6 Photochemical Reactions in Solid State

There are few reports of photochemical reactions in the solid state:
  (i) Cinnamic acid (single crystal) on photoirradiation gives[17] truxillic acid (Scheme-44).

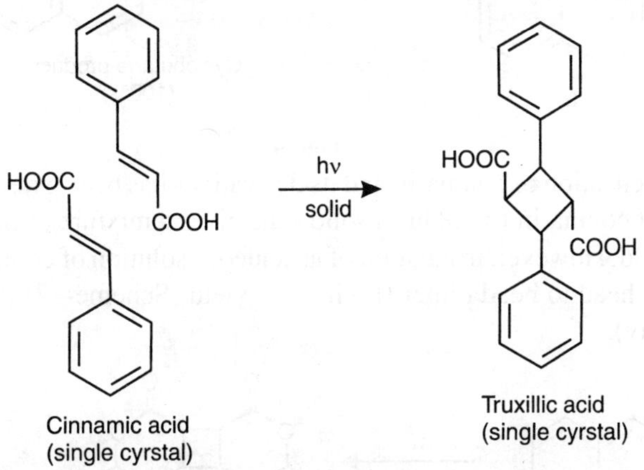

Scheme-44

The photodimerisation of cinnamic acid can be controlled by irradiation of its double salts with certain diamines in the solid states. Thus, the double salt crystal of cinnamic acid and o-diaminocyclohexane gave on irradiation in the solid state, β-truxinic acid as the major product[18] (Scheme-45).

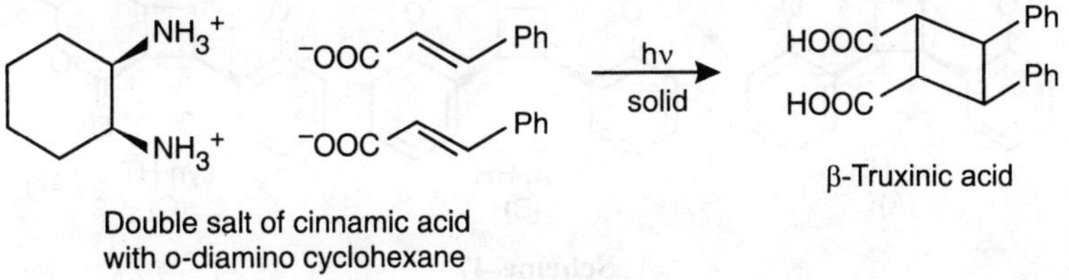

Scheme-45

15.20 ■ *Alternate Energy Processes in Chemical Synthesis*

(ii) Photoirradiation of naphthoic acid-derived cinnamic acid (A) in solid state for 20–50 hr. afforded[19] a single cyclobutane product in 100% yield (Scheme-46).

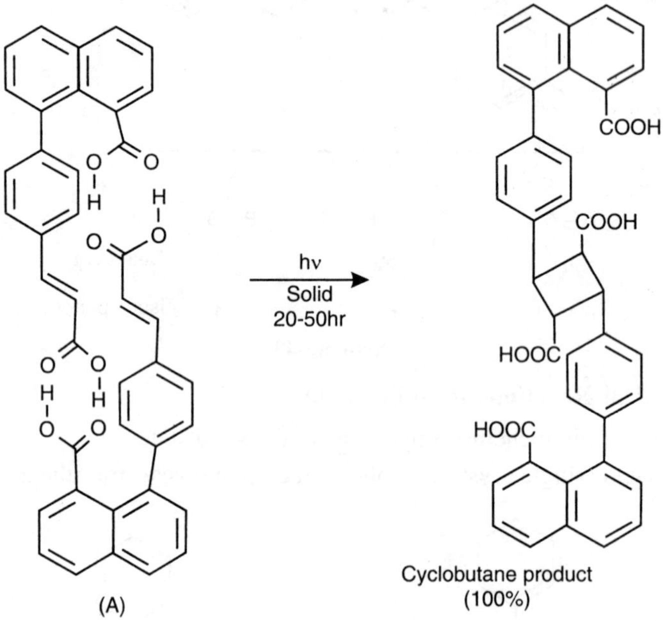

Scheme-46

(iii) The photocyclisation of coumarin and its derivatives has been extensively studied.[20] Thus, irradiation of coumarin for 48 hr in solid state gives a mixture of three products, A, B and C in 20% yield. However, irradiation of an aqueous solution of coumarin for 22 hr afforded only the syn-head to head dimer (D) in 20% yield (Scheme-47) (See also section 15.2.4 sub section iv).

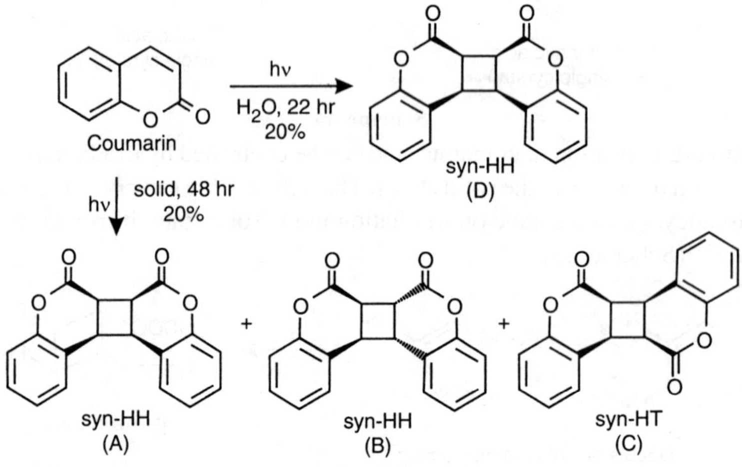

Scheme-47

(iv) Photoirradiation of 4,4′-dimethyl benzophenone in solid state gave a dimeric product[21]. It is, of course, well known that photoirradiation of 4,4′-dimethyl benzophenone in isopropyl alcohol give benzopinacol derivative in quantitative yield (Scheme-48).

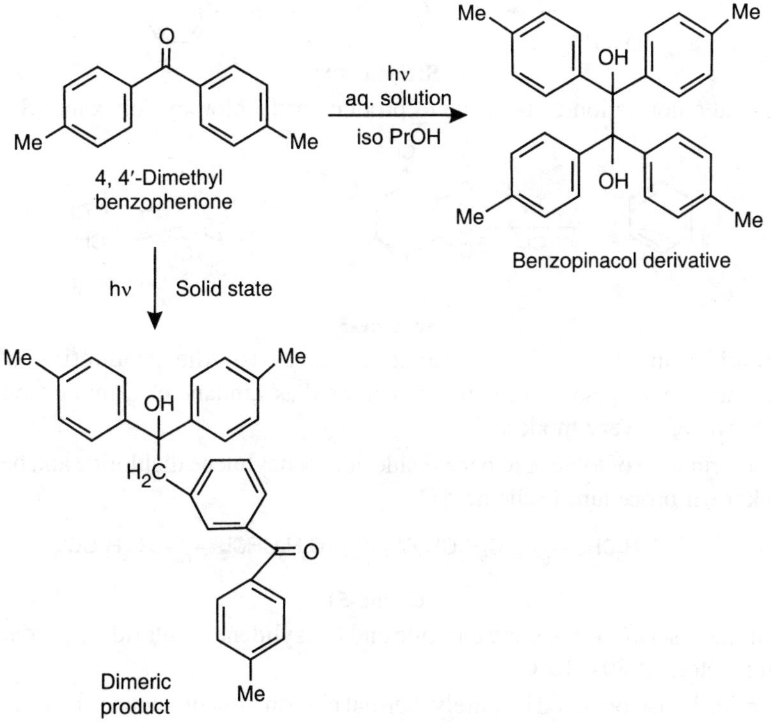

Scheme-48

Besides the reprentative examples of photochemical reactions in solid phase (given above), a large number of other illustrations are available[22].

15.2 PRINCIPAL INDUSTRIAL APPLICATIONS OF PHOTOCHEMISTRY

Most of the industrial applications of photochemistry are so far in the fields of free-radical chlorination, sulfochlorination, sulfoxidation and nitrosation. In addition, the photochemical reactions are being used on an increasing scale for the synthesis of vitamins, drugs and fragrances.

15.2.1 Free Radical Chlorination

The most important application of photochemical synthesis is in the area of radical chain reactions. The first industrial scale polychlorination[23] was recorded in 1940's. An example of industrial chlorination is externally irradiated glass tube was described by Philips corporation[24], viz., the production of monochloroalkanes from a C_{11-14} n-paraffin cut only 15 mole % of chlorine was introduced in the photochemical reactor tube. Formation of undesired dichloro compounds is avoided by stopping the reaction at an overall conversion to 30%. Optimum yield is obtained at temperature < 40°C.

$$RH + Cl_2 \xrightarrow{h\nu} RCl + HCl$$

RH = mixture of n-alkanes

15.22 ■ Alternate Energy Processes in Chemical Synthesis

The chloroalkanes find use for alkylatia of benzene to give alkylbenzenes, which are starting materials for alkylbenzene sulfonates[25], an important class of detergents (Scheme-40).

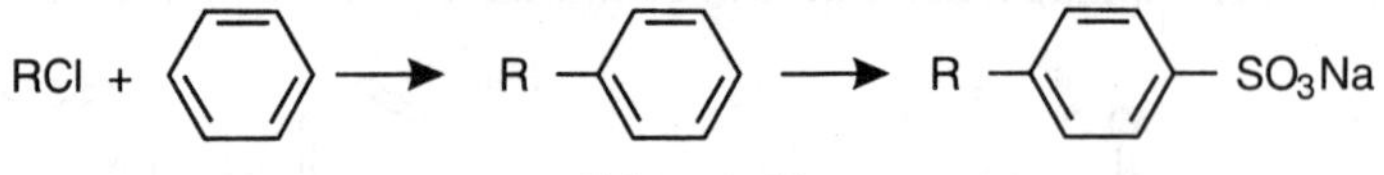

$$RCl + C_6H_6 \longrightarrow R{-}C_6H_5 \longrightarrow R{-}C_6H_4{-}SO_3Na$$

Scheme-49

Photochemical chlorination of benzene to produce hexachlorocyclohexane (Scheme-50).

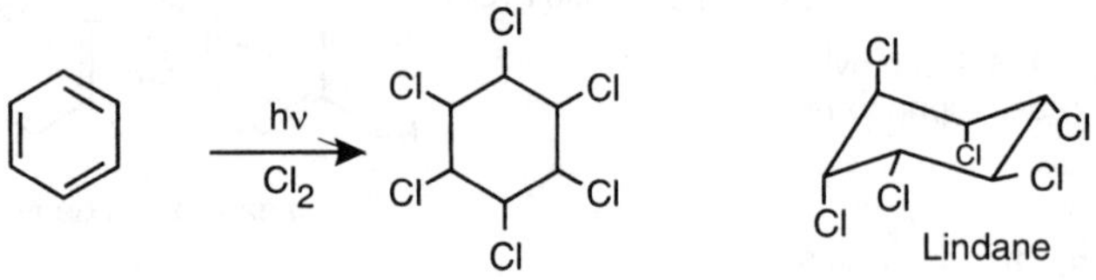

Scheme-50

The photoaddition of chlorine to benzene serves for the production of γ-isomer of hexachlorocyclohexane, a versatile insecticide marketed as Lindane or Gammexane (γ-BHC). The yield of 15%, however is very modest.

The photochlorination of toluene to benzyl chloride, benzylidene dichloride and benzotrichloride, is a very well known procedure (Scheme-51).

$$C_6H_5CH_3 \xrightarrow[Cl_2]{h\nu} C_6H_5CH_2Cl \xrightarrow[Cl_2]{h\nu} C_6H_5CHCl_2 \xrightarrow[Cl_2]{h\nu} C_6H_5CCl_3$$

Scheme-51

Mixtures of the first two, viz. benzyl chloride and benzylidene dichloride is produced in specially designed photoreactors at 80–110°C.

Benzyl chloride is also obtained by purely thermal chlorination of toluene and is mainly converted into benzyl alcohol, a well known fragrance. It is also used in the synthesis of drugs, disinfectants, and emulsifiers[28]. On hydrolysis or preferably on reaction with benzoic acid, benzylidene dichloride yields benzaldehyde; in the later case a valuable intermediate, benzoyl chloride is also obtained (Scheme-52).

$$C_6H_5CHCl_2 + C_6H_5COOH \longrightarrow C_6H_5CHO + C_6H_5COCl + HCl$$

Scheme-52

Benzotrichloride is a valuable intermediate for dyes of the triphenylmethane, xanthene, and anthraquinone series and also for plant protection agents.

15.2.2 Free Radical Sulfochlorination

Photochemical sulfochlorination of paraffins is of great industrial importance[25–29]. In sulfochlorination the function of light is for the formation of chlorine atoms from chlorine. The sulfonyl chloride group is distributed almost randomly over all the C atoms of hydrocarbon chain[30] (Scheme-53).

$$RH + SO_2 + Cl_2 \xrightarrow{h\nu} RSO_2Cl + HCl$$
RH = Mixture of n-alkanes
Scheme-53

In sulfochlorination process, di- and poly-sulfonyl chloride (having undesirable properties) are also formed. So the reaction must be stopped at conversion of 30 to 50%, yields of 80 to 90% are then obtained. Industrial sulfochlorination has been reviewed by Lindner[31]. The alkylsulfonyl chlorides, thus produced are hydrolysed by caustic soda to give water-soluble alkanesulfonates, which are mainly used as emulsifiers for polymerisations[25] (Scheme-54).

$$RSO_2Cl + 2NaOH \longrightarrow RSO_3Na + NaCl + H_2O$$

Scheme-54

Due to contamination of di- and poly-sulfonyl chlorides, the above process (Scheme-54) is not suitable for the manufacture of RSO_2Na, since it cannot be obtained in a pure state on industrial scale. So these type of alkenesulfonides are now not used in detergent manufacture.[32, 33]

The reaction of alkanesulfonyl chlorides with ammonia gives sulfonamides (Scheme-55), which are used as textile auxilliaries[32]. The sulfonamides further react with chloacetic acid to give a mixture of sulfonylaminoacetic acids (Scheme-55). These type of compounds serve as emulsifiers and as anticorrosion agents for mineral oils[32, 34].

$$RSO_2Cl \xrightarrow{NH_3} RSO_2NH_2 \xrightarrow[NaOH]{Cl\ CH_2\ CO_2H} RSO_2NHCH_2CO_2H$$

Scheme-55

15.2.3 Photochemical Sulfoxidation

A convenient synthesis of alkanesulfonates is by sulfoxidation process. In this procedure, oxygen serves as an oxidizing agent instead of chlorine used earlier (Scheme-56).

$$RH + SO_2 + \tfrac{1}{2}O_2 \xrightarrow{h\nu} RSO_3H$$

$$RSO_3H + NaOH \longrightarrow RSO_3Na + H_2O$$

$$RH = C_{14-18}\text{-}n\text{-alkane}$$

Scheme-56

The above procedure is used for Industrial preparation alkanesulfonates[25, 35, 36]. In the above procedure (Scheme-56), the primary sulfoxidation product is peroxysulfonic acid, which is trapped by water before it can undergo radical decomposition (Scheme-57).

$$RSO_2OOH + H_2O + SO_2 \longrightarrow RSO_3H + H_2SO_4$$

Scheme-57

15.2.4 Photonitrosation

Light induced reaction of nitrosyl chloride with cyclohexane gives cyclohexane oxime, which is a starting material for the synthesis of caprolactam, the monomer of nylon 6 (Scheme-58).

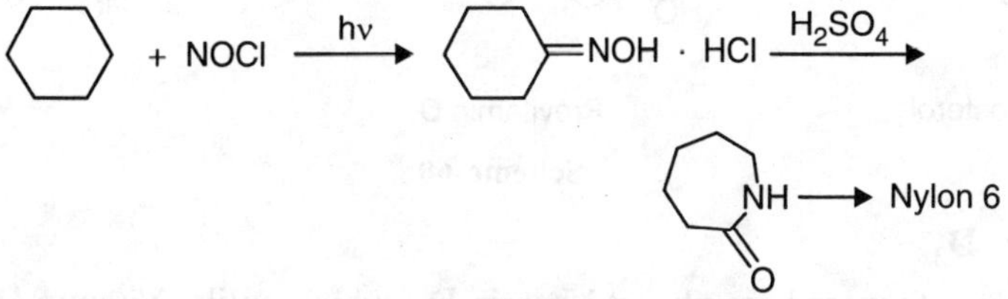

Scheme-58

A Japanese chemical company Toray finally decided to conduct the photochemical synthesis of caprolactam in a large production plant[37].

15.24 ■ *Alternate Energy Processes in Chemical Synthesis*

Another industrial application of photonitrosation is the manufacture of lauryllactam, which is a starting material for the production of Nylon 12. In this procedure, cyclododecane, [which is obtained from butadiene in two steps (Scheme-59)] is converted into the oxime by nitrosyl chloride in high yield. The oxime on reaction with acid gives lauryllactam[38], the monomer of nyclon-12 (Scheme-59).

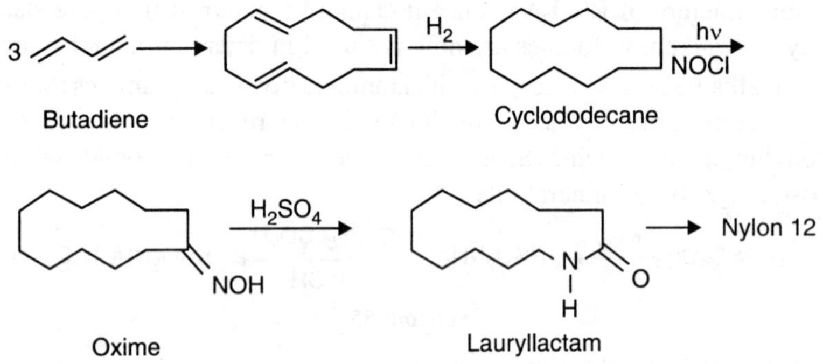

Scheme-59

Lauryllactam, due to its low density and low absorption of water, is used for a number of special purposes like production of dimensionally stable plastic components (automobile construction) and for the plastic coating of metals[39].

The photochemical lauryllactam was developed by ATO in france, who fabricated a 8000 t/a capacity plant[40].

15.2.5 Photochemical Synthesis of Vitamin D and Related Compounds

15.2.5.1 *Vitamin D_2*

One of the earliest non-radical industrial photoreactions is the synthesis of Vitamin D (commonly known as Vitamin D_2) from ergosterol (Scheme-60).

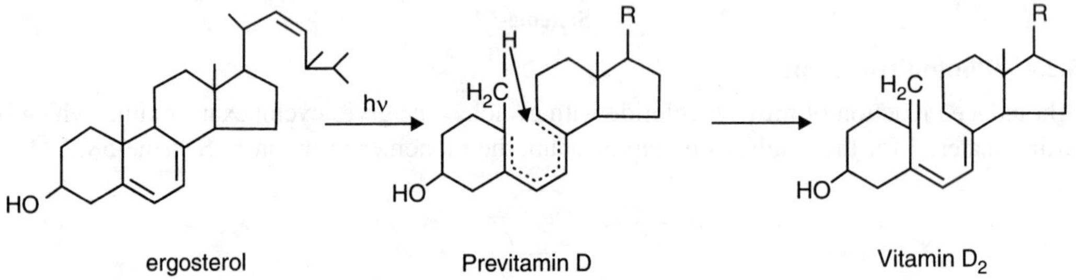

Scheme-60

15.2.5.2 *Vitamin D_3*

Interest is now a days focussed mainly on Vitamin D_3, which unlike Vitamin D_2, is also active in poultry. Most of the Vitamin D_3 now produced is not used to prevent or cure rickets (as Vitamin D_2) in children but instead it is used as an additive in animal nutrition.

Vitamin D_3 is produced by photoirradiation of 7-dehydrocholesterol, which is obtained from cholesterol in a 4 step synthesis[41]. On irradiation, 7-dehydrocholesterol undergoes ring opening leading to the formation of previtamin D_3, which on heating to 50–80°C gives thermodynamically more stable vitamin D_3 (Scheme-61). The irradiation of 7-dehydrocholesterol is carried out in dilute solutions (about 1%). Ethanol or ether serves as solvent[41, 42].

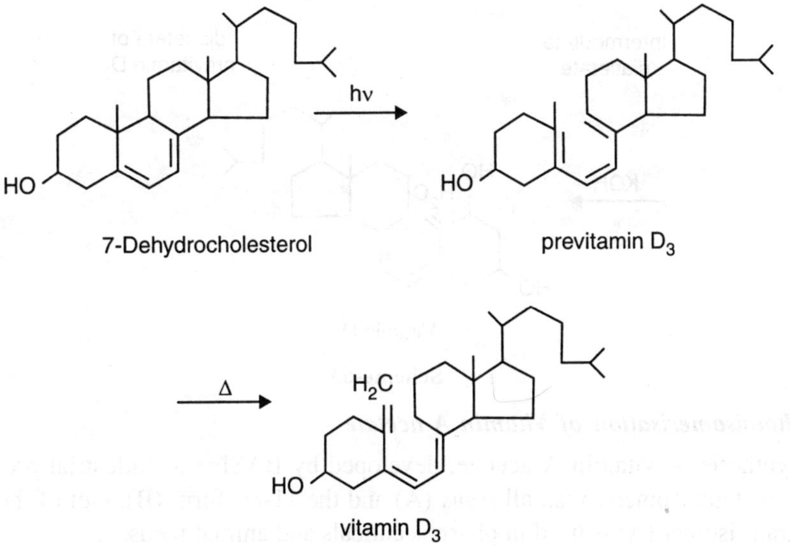

7-Dehydrocholesterol previtamin D_3

vitamin D_3

Scheme-61

15.2.5.3 Hydroxy derivatives of Vitamin D_3

It has been established[43] that it is not Vitamin D_3 which is responsible for regulating calcium metabolism. A metabolite formed in liver and kidney, 1α, 25-dihydroxyvitamin D_3 is in fact responsible for regulating calcium metabolism.

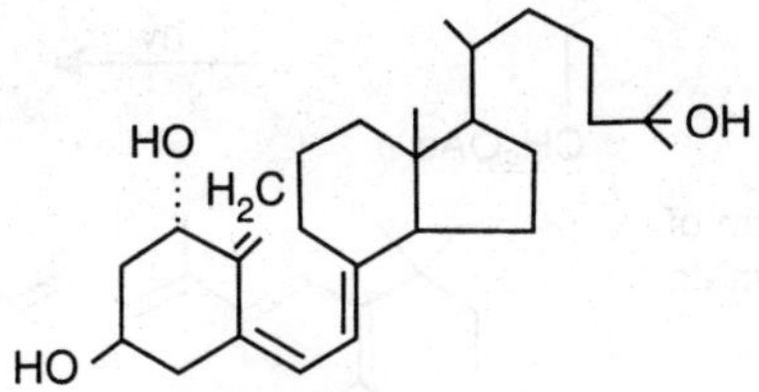

1α-25 dihydroxyvitamin D_3

Scheme-62

1α-Hydroxy Vitamin D_3, which is as active as 1α, 25-dihydroxy Vitamin D_3 is easier to produce. The intermediate (diacetate) required for the synthesis of 1α-hydroxy Vitamin D_3 is obtained[44, 45] in six steps from cholesterol. Photochemical ring opening of the starting diacetate gives previtamin D_3 derivative, which on thermal heating gave[44,45] the diacetoxy vitamin D_3. Final alkaline hydrolysis gave 1α-hydroxy Vitamin D_3[46] (Scheme-63).

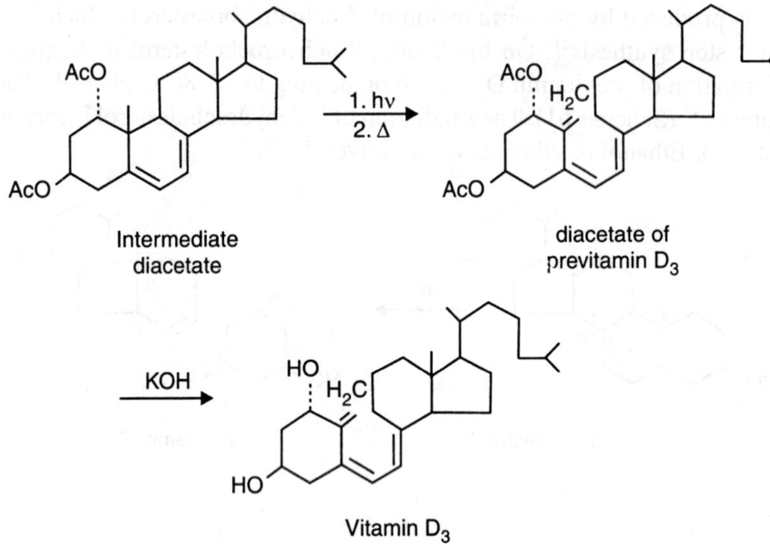

15.26 ■ *Alternate Energy Processes in Chemical Synthesis*

Scheme-63

15.2.5.4 *Photoisomerisation of Vitamin A acetate*

The wittig synthesis of Vitamin A acetate, developed by BASF[47] as industrial procedure gave a mixture of two stereoisomers, viz., all *trans* (A) and the 11-*cis* form (B). Out of the two isomers, only the all trans isomer (A) is used in pharmaceuticals and animal feeds.

A very convenient photochemical method has been developed by BASF[48] for converting 11-*cis*-(B) into all-*trans*-vitamin A acetate (A). The procedure consist in irradiating the stereoisomeric mixture with visible light in presence of a sensitizer such as chlorophyll or tetraphenylporphinatozinc. (Scheme-64).

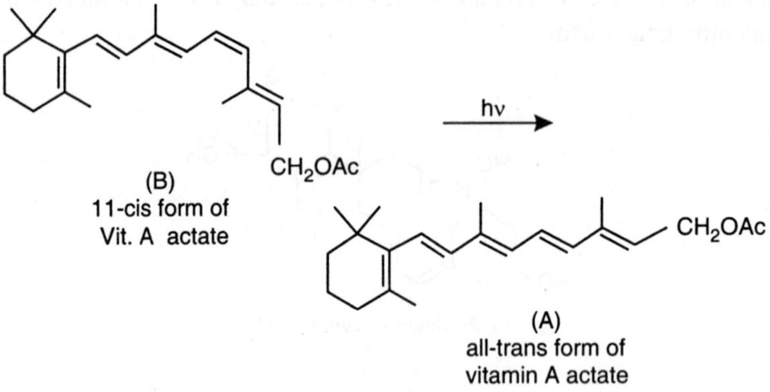

Scheme-64

15.2.6 Photo-oxygenation

During photo-oxygenation it is the singlet oxygen which is the reactive species. The most important method for the generation of singlet oxygen is the photoexcitation of the ground state of molecular oxygen in presence of a sensitiser (e.g. rose bengal). The formed hydroperoxide can be reduced to the alcohol.

15.2.6.1 *Rose Oxide*

A most promising field for industrial photochemistry is the synthesis of fragrances. Thus, photo-oxygenation of citronellol (1) with rose bengal as sensitizer gives a mixture of two isomeric hydropexoxides[49], which on reduction with sulphite give the corresponding alcohols (2) and (3). The major product (3) undergoes allylic rearrangement to (4) in acid solution. Final cyclisation of (4) gives rose oxide (5). Small scale production of rose oxide is being conducted at the Firmenich and Dragoeo companies (Scheme-65).

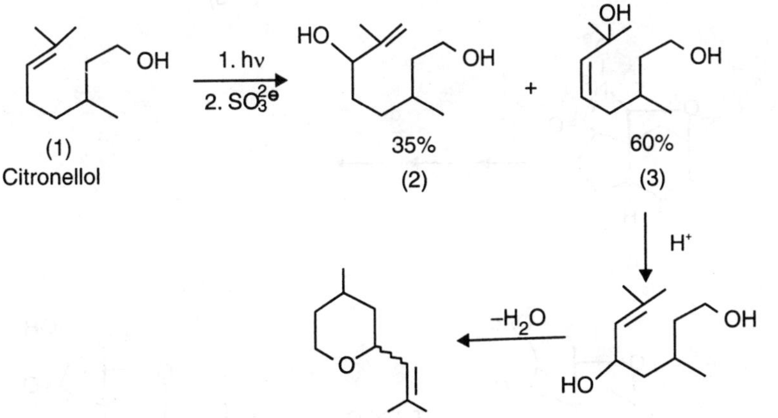

(1) Citronellol (2) (3) (4) (5) Rose oxide

Scheme-65

15.2.6.2 *Ascaridole*

Another example of photo-oxidation is the reaction of α-terpinene with singlet oxygen in presence of a sensitizer to give ascaridole. Though it was used earlier[50] as an fragrancy, but it is no longer used due to toxicity of this anthelmintic (Scheme-66).

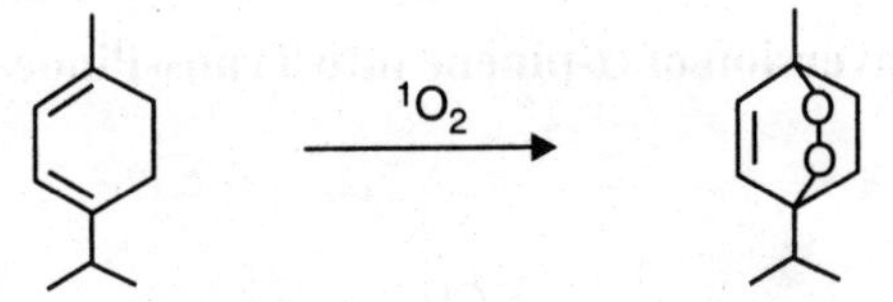

α-terpinene Ascaridole

Scheme-66

15.2.7 The Barton Reaction

The reaction, discovered by D.H.R. Barton[51] is used for the production of tritium-labelled aldosterone (1), which is used as a medical diagnostic aid. The reaction involves photochemical reaction of nitrite of pregnane (2) to give the C-18-oxime (3), which on heating gives the nitrone[52] (4). The nitrone can be converted in a series of conventional steps into 1,2-didehydroaldosterone acetate (5). This yields the radioactively labelled aldosterone (1) by catalytic tritiation and subsequent hydrolysis[53]. Various steps involved are shown in Scheme-67. The American company New England Nuclear produces (1) in a small scale for use as a medical diagnostic aid.

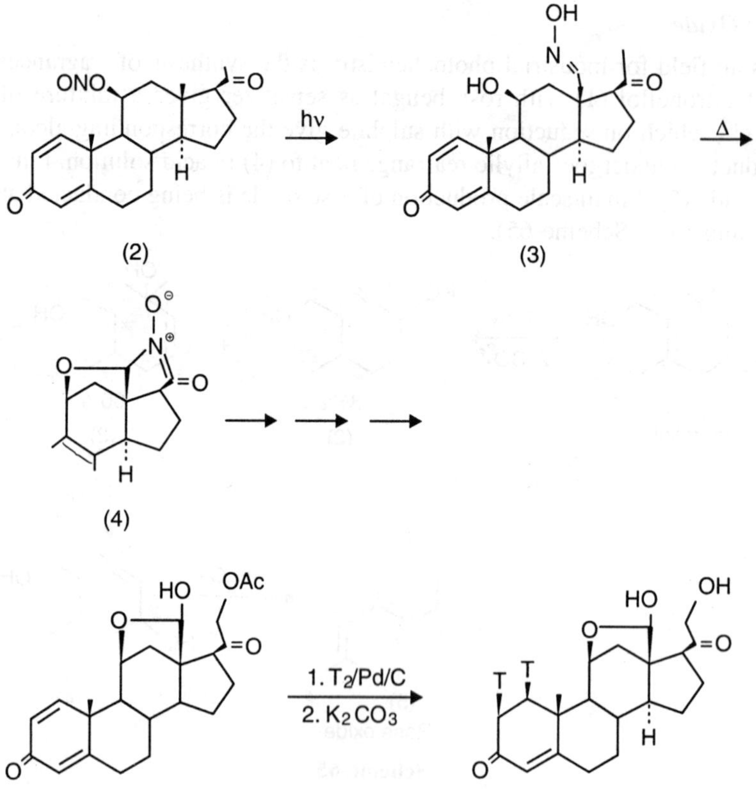

15.28 ■ *Alternate Energy Processes in Chemical Synthesis*

Scheme-67

15.3 MISCELLANEOUS PHOTOCHEMICAL REACTIONS

15.3.1 Photochemical Conversion of α-pinene into Trans-Pinocarveol using Singlet Oxygen (Scheme-68)

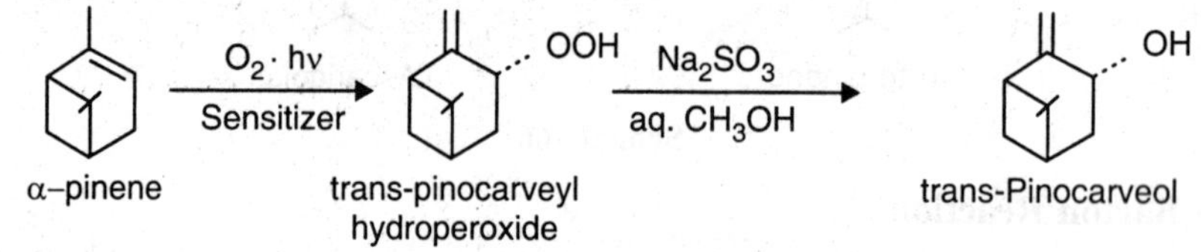

Scheme-68

16.3.2 Photoirradiation of Dibenzoyldiazomethane in Presence of Amino Acid[54] (Scheme-69)

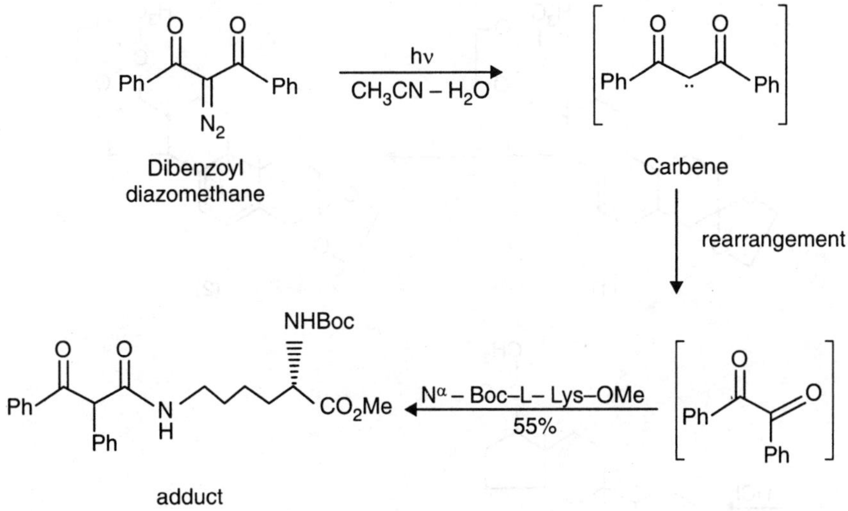

Scheme-69

15.3.3 Photochemical Aromatic Substitution

Irradiation of *o*-fluoroanisole in presence of aqueous potassium cyanide solution gives catechol monomethylether as the major product. However, irradiation of *p*-fluoroanisole in presence of aqueous potassium cyanide solution gives *p*-cyanoanisole as the major product[55] (Scheme-70).

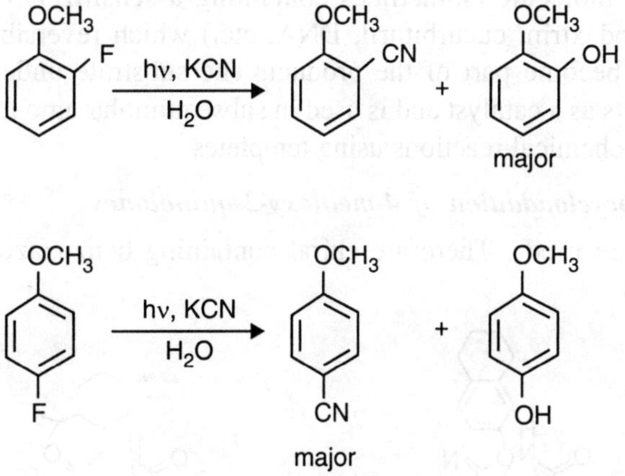

Scheme-70

In the case of o-fluoroanisol, the hydrogen bonding between water and the methoxy group was attributed to the hydroxylation reaction. The effect of such a hydrogen bonding on the product distribution is much less in the latter case.

15.3.4 Synthesis of Dydrogesterone

Irradiation of pregnadiene derivative (1) gives the retrosteroid (2), which is cleaved by alcoholic HCl to form dydrogesterone (3)[56] (a sexual hormone) (Scheme-71).

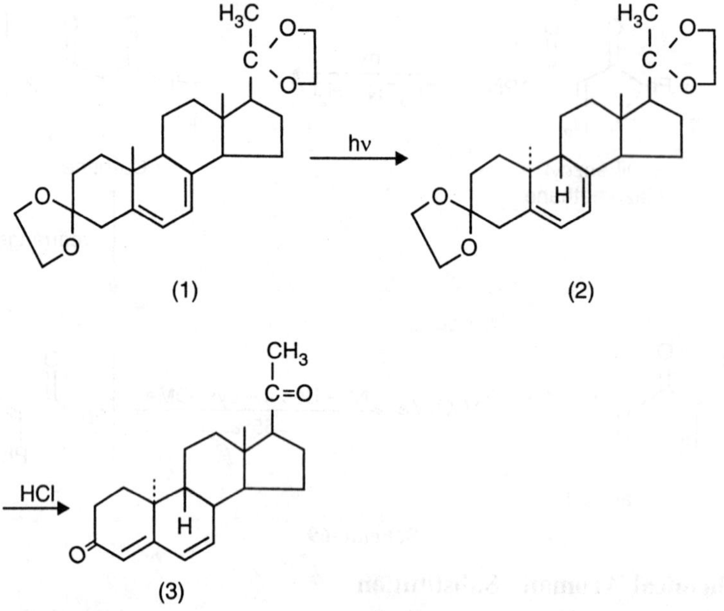

15.30 ■ *Alternate Energy Processes in Chemical Synthesis*

Scheme-71

15.3.5 Templeted Reactions in a Homogeneous Solution which are Mediated by Light

Template means a host molecule (sometimes containing a sensitizer) or various construction (a rational design, cyclodextrin, cucurbituril, DNA, etc.) which reversibly binds to a substrate molecule and does not become part of the product. The substrate undergoes a photochemical process. The template acts as a catalyst and is used in subsequimolar amounts. Following are given some examples of photochemical reactions using templetes.

15.3.5.1 *[2 + 2] Photocycloaddition of 4-methoxy-2-quinolones*

Templates (1) and (2) are used . There are chiral containing benzoxazole (host 1) or menthol (host 2) as steric shields.

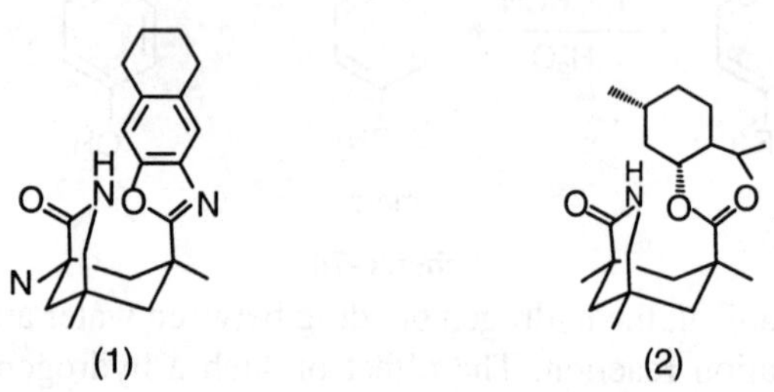

Scheme-72

Photocycloaddition of 4-methoxy-2-quinolone (3) with symmetrical olefin (4a) in presence of Chiral host (1) yielded[57] a single product (5) which was formed with 61% yield and 92% ee (Scheme-73).

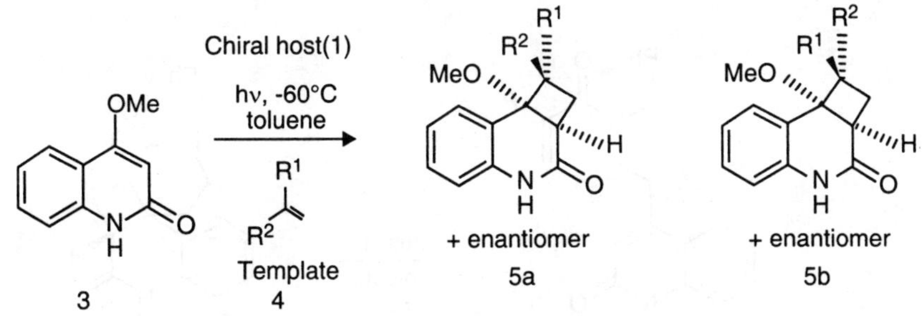

Scheme-73

	R	R'	5 (products)
(a)	Et	Et	a
(b)	CH$_3$CH$_2$CH$_2$OH	H	b
(c)	CH$_2$OAc	H	c
(d)	OAc	H	d
(e)	Ph	H	e
(f)	COOMe	H	f

Non symmetrical olefins 4b-f gave a mixture of two diasteromers 5. Olefins 4b-e, f showed diasteroselectivity in favour of isomer 5a (ratios 5a : 5b is 90 : 10, yields 80–84%) with ee values from 81% to 92%.

Styrene (5e) favoured formation of the corresponding endo-isomer (ratio 5a to 5b is < 5 : 95, yield 29% with ee 83%, Olefin 5d yielded a 63 : 27 mixture of the products 5a and 5b (overall yield 89%) (both product 5a and 5b formed with ee 93% and 98% respectively).

15.3.5.2 [2 +2] Photocyclisation of protected 4-(2'-aminoethyl)-quinolones with acrylates

Photocyclisation of 4-(2'-aminoethyl)-quinolones (6) with acrylates (7) could be performed by the template (1) (Scheme-74). In presence of I, the exo products were isolated with 44–86% yield and 70–81% ee[58].

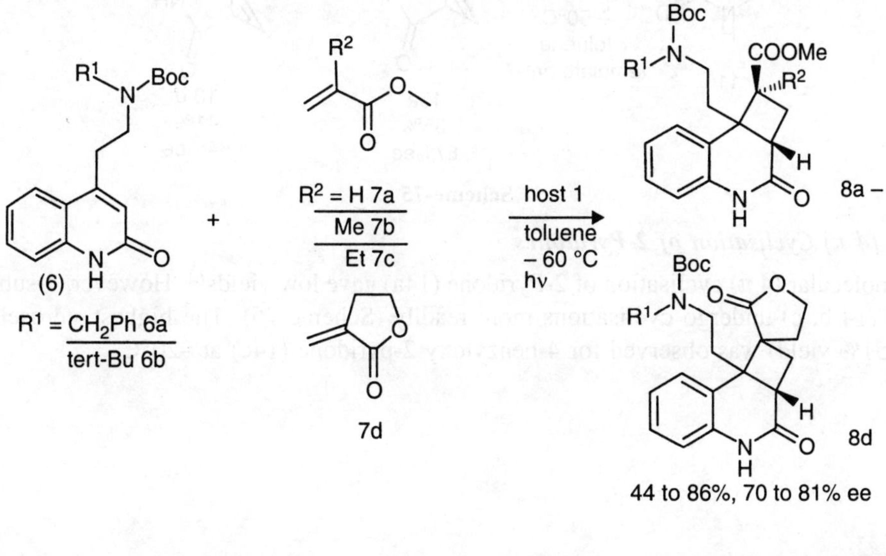

44 to 86%, 70 to 81% ee

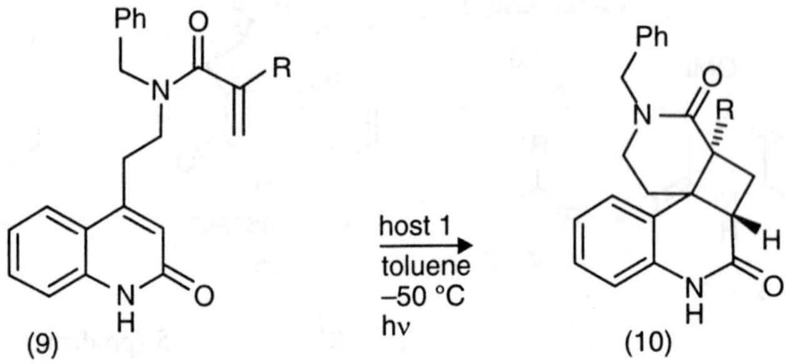

15.32 ■ *Alternate Energy Processes in Chemical Synthesis*

R = H	9a	10a	46%,	82% ee
Me	9b	10b	53%,	76% ee
Et	9c	10c	55%,	74% ee

Scheme-74

The quinolones (9) cyclised in presence of chiral host (1) to give the tetracyclic products (10) in an intramolecular fashion (Scheme-74).

The template (1) was also used for the steroselective Diels-Alder reaction of photochemically generated (E)-O-quinodimethane with alkanes[59] and also for an enantioselective radical cyclisation reaction of 4-[4'-iodobutyl]-quinolones[60].

15.3.5.3 *[4 + 4] Photocyclisation of 2-Pyridone to cyclopentadiene*

Photocyclisation of 2-Pyridone (11) with cyclopentadiene (12) in presence of chiral host (1) gave[61] a mixture of diastereomeric products (13), which were formed with significant ee value (Scheme-75).

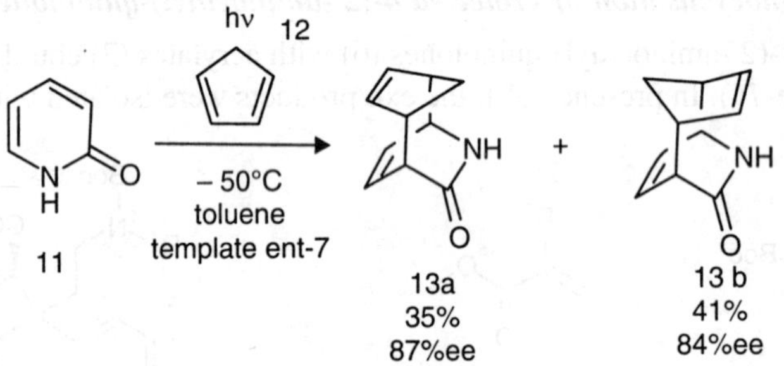

Scheme-75

15.3.5.4 *[4 π] Cyclisation of 2-Pyridones*

The intramolecular (4 π) cyclisation of 2-Pyridone (14a) gave low yields[61]. However, 4-substituted pyridones (14 b, c) undergo cyclisations more readily (Scheme-76). The highest steroselectivity (23% ee, 51% yield) was observed for 4-benzyloxy-2-pyridone (14c) at –20°C.

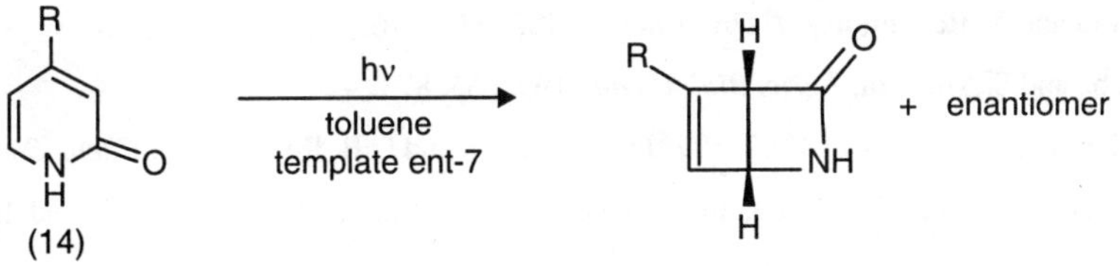

R = H	14a
OMe	14b
OCH$_2$Ph	14c

Scheme-76

15.3.5.5 *Steroselective Paterno-Büchi Reaction*

We have so far seen some of the reactions (discussed above) in which the template acted as a catalyst. There are however, some instances in which the template did not act as a catalyst. Thus, in the Paterno-Büchi reaction of 3,4-dihydro-1H-pyridin-2 one (15) with a covalently host-bound benzaldehyde[62, 63] (16), the substrate (15) was bound to chiral host (15) by two directional hydrogen bonds, and its exantiotopic faces were thus differentiated (Scheme-77). In this case the diasteroselectivity depended on solvent polarity. Polar solvents such as acetonitrile disturbed formation of the complex and lowered the diastereomeric ratio, while nonpolar solvents such as benzene favoured formation of the complex. Diasteromeric ratios upto 95 : 5 and upto 50% yields were observed. However, the N-methylated host (17), which does not allow face discrimination and showed no stercoselectivity (Scheme-77).

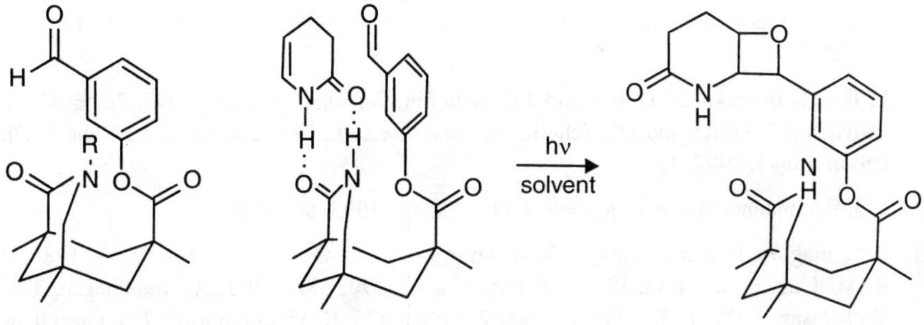

R = H 15
Me 17

16 – 15

Scheme-77

Besides what has been stated above a large number of templetal reactions in homogeneous solution mediated by light have been described[64].

NOTES AND REFERENCES

1. R. Ramamurthy, *Tetrahedron*, 1986, **42**, 5753.
2. M.S. Sayamala and V. Ramamurthy, *J. Org. Chem.*, 1986, **51**, 3712.
3. Y. Ito, K. Kajita, K. Kunimoto and T. Matsuura, *J. Org. Chem.*, 1989, **54**, 587.

4. K. Mathuramu and V. Ramamurty, *J. Org. Chem.*, 1982, **47**, 3976.

   (a) H. Tateba and S. Michara, *Agric. Biol. Chem.*, 1991, **55**, 873.

   (b) D. Bellus et al., *Chem. Rev.*, 1967, 599; J.C. Anderson and C.B. Reese, *Proc. Chem. Soc.*, 1960, 217.

5. N.J. Turro, M. Grätzel and A.M. Braun, *Angew. Chem.*, Int. Ed. Engl., 1980, **19**, 675-696 and the references cited there in.

6. T. Kamaki, *Chem. Lett.*, 1984, 53; T. Tamaki and T. Kokubu, *J. Inclusion Phenom.*, 1984, **2**, 815.

7. A.M. Braun and K. Ramsteiner, unpublished work.

8. I. Rico, M.T. Maurette, E. Oliveros, M. Riviere, and A. Lattes, *Tetrahedron Lett.*, 1978, 4795; R. Fargues, M.T. Maurette, M. Oliveros and A. Lattes, *Nouv. J. Chem.* 1979, **3**, 487.

9. K.H. Lee and P. de Mavo, *Chem. Soc. Chem. Commun.*, 1979, 493.

10. Y. Nakamura, Y. Imakura, K. Koto and Y. Morita, *Chem. Soc. Chem. Commun.*, 1977, 887.

11. Y. Nakamura, Y. Imakura and Y. Morita, *Chem. Lett.*, 1978, 965.

12. R.R. Hautala and R.L. Letsinger, *J. Org., Chem.*, 1991, **36**, 3762.

13. K. Yamada, K. Shigehiro, T. Kujozuka and H. Lida, *Bull. Chem., Soc.* Jpn., 1978, **51**, 2447.

14. P.S. Engel, *J. Am. Chem. Soc.*, 1970, **92**, 6074; W.K. Robins, R.H. Eastman, *J. Am. Chem. Soc.*, 1970, **92**, 6077; G. Quinkert, K. Opitz, W.W. Wiresdorf and J. Weinlich, *Tetrahedron Lett.*, 1963, 1863.

15. N.J. Turro and W.R. Cherry, *J. Am. Chem. Soc.*, 1978, **100**, 7432.

16. P. de Mago, unpublished results.

17. V. Ramamurthy, Ed., *Photochemistry of Organized and Constraint Media*, VCH, Weinheimn, Germany, 1991; CRC Handbook of *Organic photochemistry and Photobiology*, W.H. Horspool (Ed.), CRC Press, Boca Raton, Fl., 1995.

18. Y. Ilo, B. Borecka, J. Trotter and J.R. Scheffer, *Tetrahedron Lett.*, 1995, **36**, 6083; Y. ITo, B. Borecka, G. Olovasson, J. Trotter and J.R. Scheffer, *Tetrahedron Lett.*, 1995, **36**, 6087; Y. Ito and B. Olovasson, *J. Chem. Soc.* Perkin trans I, 1997, 127.

19. K.S. Feldmann and R.F. Campbell, *J. Org. Chem.*, 1995, **60**, 1924.

20. R. Granaguru, K. Ramasubba, K. Venkatesan and V. Ramamurthy, *J. Org. Chem.*, 1985, **50**, 2337; J.N. Moorthy, K. Venkatesan and R.G. Weiss, *J. Org. Chem.*, 1992, **57**, 3292; K. Venkatesan, T.N. Guru Row and K.J. Venkatasan, *J. Chem. Soc.* Perkin Trans 2, 1996, 1475; K. Vishnumurthy, T.N. Guru Row and K. Venkatisan, *J. Chem. Soc.* Perkin Trans 2, 1997, 615.

21. Y. Ito, T. Matsuura, K. Tabata and M. Ji-Ben, *Tetrahedron Lett.*, 1987, **43**, 1307.

22. K. Tanaka and F. Toda, *Chem. Rev.*, 2000, **100**, 1044-1066.

23. W.T. Anderson Jr., *Ind. Eng. Chem.*, 1947, **39**, 844; W.H. Shearon Jr; H.E. Hall and J.E. Steven Jr., *Ind. Eng. Chem.*, 1949, **41**, 1812.

24. T. Hutson, Jr., and R.S. Logan, *Chem. Eng. Prog.*, 1972, **68(5)**, 76.

25. F. Broich, Fette and Seifen, *Anstrichm*, 1970, **17, 22**.

26. L.J. Governate and J.T. Clarke, *Chem. Eng. Prog.*, 1956, **52(7)**, 281.

27. H.G. Haring and H.W. Knol, *Chem. Process Eng.* 1964, **45**, 560, 619, 690.

28. K.A. Lipper in: *Ullmamns Encykiopödie der technischen Chemie*, 4th Edn., Verlage Chemie, Weinheim, 1975, Vol. 9, 525.

29. H.I. Jaschek, *Chem.-Zig.*, 1969, **93**, 655.

30. F. Asinger, *Ber. Dtsch. Chem. Ges.*, 1944, **77**, 191.

31. K. Lindner, Tenside, Textihilfsmittel, Waschrohstoffe. *Wissenschafil, Verlagsges*, Stuttgart, 1964, Vol. 1, p. 705.

32. See ref. 31. p. 717.

33. H. Schuller in: *Ullmanns Encklopadie der Technischen. Chemi.*, 3rd Edn., *Urban and Schwarzenberg*, Muchehon, Vol. 16, p. 724.

34. F. Asinger, *Die Petro chemische Industrie. A. Kademie-Verlag*, Berlin, 1971, p. 731.

35. L. Orthner, *Angew. Chem.*, 1950, **62**, 302; R. Graf. Justus Liebigs *Ann. Chem.*, 1952, **50**, 578.

36. C. Beermann, *Eur. Chem. News*, Normal Paraffins Supplement, Dec. 2, 1966, p. 36; H. Hartig, *Chem-Ztg*, 1975, **99**, 179.

37. P. Turner, *Inf. Chins*, 1970, **9**,(**5, 6**) 51; P. Hurine, P.E. Turner, *Chem. Process Eng.*, 1967 (11), 96; Y. Ito, Y. Hara, DAS I 1962, **468**, 737, Toray.

38. *Inf. Chim*, 1970, 8, (3, 4), 47.

39. *Eur. Chem.* News, March 29, 1974, p. 18.

40. *Chem. Ind* (Düsseldorf), 1975, **27**, 735.

41. J. Fragner: Vitamine, Gustav Fisher Verlag, Jena, 1964, Vol. I, p. 613.

42. S.B. Greenbawm in Kirk-Othmer: *Encyclopedia of Chemical Technology*, 2nd Edn. Interscience, New York, 1970, Vol. 21, p. 549.

43. M.F. Hollick, E.J. Semmier, H.K. Schnoes and H.F. DeLuca, *Science*, 1973, **180**, 190.

44. D.H.R. Barton, R.H. Hesse, M.M. Pechet and E. Rizzardo, *J. Am. Chem. Soc.*, 1973, **95**, 2748; N. Müller, DOS 2 400189, BASF Ag.

45. A. Fürst, L. Labler, W. Muer and K.-A. Pfoertner, *Helv. Chim. Acta*, 1973, **56**, 1708.

46. *Eur. chem. News*, August 22/29, 1975.

47. W. Reif, H. Grassner, *Chem.-Ing. Tech.*, 1973, **45**, 646; H. Pommer, *Angew. Chem.*, 1977; **89**, 437; *Angew. Chem. Int.* Ed. Engl., 1977, **16**, 423.

48. M. Fischer and W.W. Weirsdorff, A. Nürrenbach, D. Horn, F. Feichimayr, DBP 2210800 BASF AG.

49. G. Ohloff, E. Klein and G.O. Schenck, *Angew. Chem.*, 1961, **73**, 578.

50. G.O. Schenck, *Dechema-Monographien*, 1955, **24**, 105.

51. M. Akthar, *Adv. Photochem.*, 1964, **2**, 263.

52. D.H.R. Barton, N.K. Basu, M.J. Day, R.H. Hesse, N.M. Pechet and A.N. Starrat, *J. Chem., Soc.* Perkin Trans. I, 1975, 2243.

53. K.R. Laumas, M. Gut, *J. Org. Chem.*, 1962, **22**, 314; D.H. R. Barton and J.M. Beaton, *J. Am. Chem. Soc.*, 1961, **83**, 4083.

54. K. Nakatani, J. Shirai, R. Tamaki, and I. Saito, *Tetrahedron Lett.*, 1995, **36**, 5363.

55. J.H. Liu and R.G. Weiss, *J. Org. Chem.*, 1985, **50**, 3655.

56. O.A. de Bruin, H.F.L. Schöler, J.N. Walop, Philips Tech. Rundsch, 1967, **28**(3,4), 72; M.P. Rappoldt and T.R. Rix, *Trav. Chim.* Pays-Bas, 1971, **90**, 27.

57. K. Bach, H. Bergmann, B. Grosch and H. Harms, *J. Am. Chem. Soc.*, 2002, **124**, 7982; T. Bach and H. Bergmann, *J. Am. Chem. Soc.*, 2000, **122**, 11525.

58. P. Selig and T. Bach, *J. Org. Chem.*, 2006, **71**, 5662.

59. B. Grosch, C.N. Orlebar, E. Herdtweck, M. Kaneda, T. Wadia, Y. Inoue and T. Bach, *Chem. Eur. J.*, 2004, **10**, 2179; B. Grosch, C.N. Orlebar, E. Herdtweck, W. Massa and T. Bach, *Angew. Chem. Int. Ed.*, 2003, **42**, 3693.

60. M. Dressel anu T. Bach, *Org. Lett.*, 2006, **8**, 3145.

61. T. Bach, H. Bergmann and K. Harms, *Org. Lett.*, 2001, **3**, 601.

62. T. Bach, H. Bergmann and K. Harms, *J. Am. Chem. Soc.*, 1999, **121**, 10650.

63. T. Bach, H. Bergmann; H. Brummerhop, W. Lewis, and K. Harms, *Chem. Eur. J.* 2001, **7**, 4512.

64. J. Svoboda and B. König, *Chem. Rev.*, 2006, **106**, 5412–5430 and the references cited there in.

INDEX

A

2-Aminothiophenol 3.12
2-Aryl-1, 2, 3, 4-tetrahydro-4-quinolones 4.27
4-Aminoquinazolines 4.28
4-Aryl-3, 4-dihydropyrimidine 2(1H) ones 3.5
Acetals and Diaxolanes 4.1
Activated MnO_2-Silica 4.10
Activation of Nickel Powder 10.3
Addition reactions 7.5
Additions 8.5
Adsorption of trace metals on charcoal 11.2
Aldol condensation 14.1
Alkenes 3.17
Alkylations 5.10
Alkylations 9.1
Alternative source of energy 1. 3
Aluminium Alkoxides 4.15
Anhydrides 5.1
Annulation 7.2
Anthraquinone 5.13
Arenes 4.13
Aromatic substitution 15.29
Aromatic substitution 5.8
Aromatic substitution reactions 3.7
Aromatisation 4.13
Arylamides 10.9
Ascaridole 15.27

Aspirin 5.16
Azidoglycosides 10.14
Aziridines 4.22
Azoles 4.25

B

Barbier Reaction of Carbonyl Compounds 9.12
Barton Reaction 15.27
Baylis-Hillman reaction 3.4
Beckmann Rearrangement 4.17
Benzamide 2.2
Benzil-Benzilic Acid Rearrangement 4.18
Benzimidazoles 4.22
Benzimidazoles 4.23
Benzodiazepin-2-ones 3.6
Benzyl chloride 2.1
Biginelli condensation 5.18
Bouveault Reaction 9.11
Bridgehead nitrogen heterocyclic compounds 5.3
Bromomethylation of Aromatics 10.12

C

[2 + 2] Cycloaddition reactions 7.7
[4 p] Cyclisation 15.32
3-Carbomoyl cephalosporin derivatives 4.31
C-alkylation 3.9
C-Alkylation 9.2

Cannizzaro Reaction 9.11
Carbohydrate synthesis 10.12
Carbohydrates 3.18
Carbohydrates 9.14
Carbon powders 10.18
Catalysis 10.12
Catalytic hydrogenation 3. 7
Cephalosporins 4.30
Chalcones 3.8
Chlorobenzene 1.4
Chromium Trioxide Supported on Wet Alumina 4.10
Cinnamic acid 15.19
Claisen rearrangement 3.3, 3.14
Clayfen 4.9
Combinatorial synthesis 5.16
Condensation reactions 4.19
Condensations 5.11
Continuous microwave reactor 3.12
Coumarin 15.14, 15.20
Coupling of amines with halides 2.3
Coupling reactions 9.8
Curtius rearrangement 7.1
Cyclisation reactions 3.16
Cycloaddition Reactions 7.8
Cycloaddition reactions 3. 6

D

1, 2-Dichlorobenzene 1.4
1, 2-Dichloroethane 1.4
1, 2-Diketones 4.11
1, 2-Dimethyl-3-hydroxy-pyrid-4-one 3.11
1, 3-Dipolar Cycloaddition 7.6
1, 3-Dipolar cydoaddition 10.15
1, 4-Dihydro-quinoxaline-2, 3-diones 14.3
Deacetylation 4.4
Debenzylation 4.4
Decarboxylations 3.8
Dehalogenation 10.10
Deoximation Reactions 4.6
Desilylation reactions 4.5
Dethioacetalization Reactions 4.6
Dethiocarbonylation 4.8
Dichlorocarbene 9.10
Dieckmann Cyclisation 9.13
Diels alder reaction 3.2

Diels-Alder Reaction 7.5
Diglyme 1.4
Dioxane 1.4
Dipole moment 1.4
Dydrogesterone 15.29

E

Electrophilic Addition 7.7
Enamines 4.13, 4.20
Enaminoketones 5.12
Enzyme catalysed reactions 5.5
Esterification 3.1
Esterification 8.1
Ethanol 1.4
Ethers 3.17
Ethylene glycol 1.4
Eugenol Isoeugenol Isomerisation 4.18

F

Ferrier rearrangement 3.10
Ferrocenyl oxime 3.17
Finkelstein exchange 8.4
Fishcher Indole Synthesis 5.9
Fisher Cyclisation 3.15
Flavones 4.32
Formamide 1.4
Free radical chlorination 15.21
Free radical sulfochlorination 15.22
Fries rearrangement 3.2
Furans 4.31

G

Glycosylation 10.16
Grignard Reaction 14.2
Grignard Reagents 7.3

H

Heterogeneous liquid-liquid reactions 8.1
Heterogeneous solid-liquid reactions 9.1
Hetro-Diels-Alder reaction 3.15
Hofmann elimination 2.1
Homogeneous sonochemical reactions 7.1
Hydrazone derivatives 4.21
Hydroboration 9.7
Hydrolysis of peptides 5.14

Hydrolysis/Solvolysis 8.3
Hydrosilation and Hydroalkylation 9.8
Hydroxy derivatives of Vitamin D_3 15.25

I

Imines 4.20
Iodobenzene Diacetate (IBD) 4.11
Ionic liquids 5.19
Isoflavan-3-enes 4.32
Isomerisation of Alkenes 7.2
Isomerisation of maleic acid to fumaric acid 7.2
Isomerisation reactions 4.18
Isomerization of olefins 15.8
Isopropylideneglycerol 3.11
Isoxazoles 4.26

J

Jusminaldehyde 3.10

K

β-keto sulfones 12.3
β-Keto-Thinoesters 10.10
Knoevenagel condensation 3.4
Knoevenagel Condensation 4.19

L

β-lactams 3.5
β-Lactams 4.30
Longifolene 3.15

M

Methanol 1.4
Methanolysis 3.9
Methyl benzoate 2.2
Methylenation 5.2
Michael addition 5.3
Microwaves 1.3
MW-assisted preparation of ionic liquids 5.20

N

β-nitrostyrene 5.2
N, N-dimethyl formamide 1.4
N-Alkylation 9.1
N-Alkylation Reactions 4.2
N-heterocyclisation 2.3
Norrish type II reaction 15.6
N-Phenyl benzamide 2.2
N-Sulfonylimines 4.21

O

O-Alkylation 9.2
Octylthiocyanate-Octylisothiocyanate Isomerisation 4.18
Organo zinc and palladium compounds 10.6
Organoaluminium Compounds 10.5
Organolithium Compounds 10.4
Organometallic compounds 10.4
Organometallic reactions 5.7
Organometallic reactions 7.2
Ortho-claisen rearrangement 3.3
Orthoester claIsen rearrangement 3.4
Ortho-ester claisen rearrangement 5.9
Oxadiazines 4.29
Oxazoborolidines 3.16
Oxidation 3.17
Oxidation of Contaminants 11.1
Oxidations 4.9
Oxidations 7.4
Oxidations 9.3
Oxidative coupling 5.2

P

[2 +2] Photocycloaddition 15.30
[2 +2] Photocyclisation 15.31
[4 +4] Photocyclisation 15.32
Paterno-Büchi reactions 15.11
Peptides 5.14
Pericyclic reactions 3.14
Pericyclic reactions 5.9
Phenylhydrazones 4.7
Phenylthioglycosides 10.16
Photo induced organic synthesis 15.3
Photo oxidative dimerisation 15.15
Photochemical cycloaddition reactions 15.9
Photochemical cycloaddition reactions in water 15.13
Photochemical Reactions 15.4
Photochemical reactions in micellar media 15.15
Photochemical reactions in solid state 15.19

Photochemical reactions of olefins 15.8
Photochemical sulfoxidation 15.23
Photochemical synthesis of Vitamin D 15.24
Photochemical synthesis of Vitamin D 15.24
Photodimerization of anthracene-2-sulfonate 15.15
Photoisomerisation of Vitamin A acetate 15.26
Photoisomerization of cis and trans-stilbene 15.8
photolysis of 2-pentanone 15.5
Photolysis of acetone (Propanone) 15.5
Photolysis of benzophenone 15.7
Photolysis to 2-hexanone 15.6
Photonitrosation 15.23
Photo-oxygenation 15.26
Photoreductive dimerization 15.7
Pinacol-pinacolone Rearrangement 4.17
Polluted Water 11.1
Potassium Superoxide 10.1
Protection and deprotection 5.14
Protection and deprotection reactions 4.1
Pyrazoles 4.23
Pyrazolo [3, 4-b] quinolines and pyrazolo [3, 4-c] 5.4
Pyridines 4.26
Pyrimidines 4.28
Pyrimidino [1, 6-a] benzimidazoles 4.28
Pyrimido [1, 6-a] benzimidazoles 5.12
Pyrroles 4.24

Q

Quinolines 4.27
Quinoxaline 14.3
Quinoxalines 4.27

R

Radical reactions 3.18
Radiolabelled compounds 5.6
Reductions 9.5
Reductive Amination 4.15
Reformatsky reaction 14.3
Reformatsky Reaction 9.11
Regioselective reductive cleavage 10.16
Ring Expansion 4.17
Rose Oxide 15.27

S

S-Alkylation 9.3
Saponification 4.9
Saponification 8.2
Semicarbazones 4.7
Silicon containing reagents 5.15
Singlet oxygen 15.28
Sodium borohydride (NaBH4) 4.14
Sodiumphenylselenide 10.9
Solid State Crossed Cannizzaro Reaction 4.15
Solid state organic synthesis at room temperature 12.3
Solid state organic synthesis by slight warming of reactants 13.1
Solid support synthesis 4.1
Solid-phase organic synthesis (spos) 5.5
Solvolysis and hydrolysis 7.4
Sonochemical Methylenation of alkenes 10.7
Sonochemical synthesis of ionic liquids 5.21
Sonochemical treatment of polluted water 11.1
Sonolysis of $Fe(CO)_5$ 10.1
Spiroketones 10.9
Sterosclective Paterno-Büchi Reaction 15.33
Strecker Reaction 7.6
Strecker Synthesis 9.11
Substitutions 8.4
Sulfides 4.12
Sulfones 4.12
Sulfoxides 4.12
Sulfur extrusion 7.1
Sulphones 5.11

T

1, 2, 4-Trichlorobenzene 1.4
α-Tosyloxy β-keto sulfones 12.5
Tetrapyrrole 5.13
Thiadiazepines 4.29
Thiazoles 4.26
Thioamides 10.12
Thioamides 4.8
Thioesters 4.8
Thioflavonoids 4.8
Thioglycosides 10.14
Thioketones 4.8

Thionotion Reactions 4.8
Tipson-Cohen reaction 3.18
Toluene 2.2
Trialkylboranes 10.7
Triglyme 1.4

U

α, β-Unsaturated Nitroalkenes 4.21
Ugi reaction 5.4
Ugi-type reaction 5.18
Ultrasonically dispersed potassium 10.3
Ultrasound 6.1
Unsaturated pyranosides 3.9
Ureahydrogen peroxide adduct (UHP) 13.3

V

(–)-Vincadifformine 3.10
Vitamin D_2 15.24
Vitamin D_3 15.24

W

Wittig Olefination Reactions 4.19
Wolff-Kichner reduction 4.21